Foundations of the
Economic Approach to Law

Interdisciplinary Readers in Law
ROBERTA ROMANO, *General Editor*

Foundations of Contract Law
RICHARD CRASWELL and **ALAN SCHWARTZ,** *Editors*

Foundations of Employment Discrimination Law
JOHN J. DONOHUE III, *Editor*

Foundations of the Economic Approach to Law
AVERY WIENER KATZ, *Editor*

Foundations of Tort Law
SAUL LEVMORE, *Editor*

Foundations of Environmental Law and Policy
RICHARD L. REVESZ, *Editor*

Foundations of Corporate Law
ROBERTA ROMANO, *Editor*

Foundations of Administrative Law
PETER SCHUCK, *Editor*

Foundations of the Economic Approach to Law

Edited by
AVERY WIENER KATZ

New York   Oxford
Oxford University Press
1998

Oxford University Press

Oxford   New York
Athens   Auckland   Bangkok   Bogota   Bombay
Buenos Aires   Calcutta   Cape Town   Dar es Salaam
Delhi   Florence   Hong Kong   Istanbul   Karachi
Kuala Lumpur   Madras   Madrid   Melbourne
Mexico City   Nairobi   Paris   Singapore
Taipei   Tokyo   Toronto   Warsaw

and associated companies in
Berlin   Ibadan

Published by Oxford University Press, Inc.
198 Madison Avenue, New York, NY 10016

Oxford is a registered trademark of Oxford University Press

Library of Congress Cataloging-in-Publication Data
Foundations of the economic approach to law / [edited by] Avery Wiener Katz.
        p.   cm.   —(Interdisciplinary readers in law)
    ISBN 0-19-509773-4; 0-19-509774-2 (pbk.)
    1. Law and economics.   I. Katz, Avery Wiener.   II. Series.
  K487.E3F66   1997
  330.1—dc21   97-4220

9 8 7 6 5 4 3 2

Printed in the United States of America
on acid-free paper

Preface

In putting together this reader, I have followed a somewhat different approach than most previous editors in the field, in that I have not seen my primary task as trying to survey economic analyses of the major fields of law. Instead, this book is organized around basic methodological concepts. It focuses on what is distinctive about economics as a way of thinking and on how economic analysis compares and contrasts with more traditional methods of legal reasoning. In my view, most lawyers exposed to law and economics do not acquire a solid grounding in these basics. Most major casebooks and textbooks in the field are concerned primarily with applications and deal with economic methodology only briefly or tangentially. As a result, both students and nonspecialists often lack a clear idea of what normative, descriptive, and interpretative positions the use of economics really commits them to; and while they may learn to use economics adequately or even expertly, they are less well equipped either to critique or to defend its use.

For this reason, as its title suggests, this book stresses the foundations. It begins with an introductory chapter on the building blocks of economic analysis: the model of rational choice, which provides economics with its descriptive theory of human behavior; the concept of efficiency, which provides its primary normative criterion; and the idea of positivist social science, which provides the underlying pragmatic justification for distinguishing between fact and value and for constructing simplified models of the world. The succeeding chapter introduces and compares the two main "schools" of law and economics—often caricatured as "Chicago" and "non-Chicago"—which I present as alternative economic models of legal relations, based on differing assumptions about the relative institutional efficacy of private exchange and governmental regulation.

With these fundamentals established, I turn to applications. The applications are organized not by field of law, however, but by analytical concepts of economics that cut across traditional doctrinal divisions. Still, those interested in doctrinal topics will find much here to read and discuss. The chapter entitled "A Survey of Basic Applications," for instance, focuses on standard problems of incentives—externalities, deterrence, collective action, and the like—in the context of the traditional first-year curriculum: tort, contract, property, criminal law, and procedure. The motive behind this chapter is that students are likely to be familiar with these problems from previous courses and that such a presentation will help them see the functional connections among incentive problems across different fields of law.

The next four chapters refine this basic approach by introducing a series of more advanced concepts that are central to understanding many legal institutions. Chapter 4 discusses the problem of strategic behavior; chapter 5 introduces the economics of risk, uncertainty, and insurance; chapter 6, the issue of incomplete or imperfect information; and chapter 7, the difficulties stemming from imperfectly rational behavior. The point of this organization is to demonstrate to a legal audience how studying economics can help them progress from simpler and more tractable representations of the world to more complex and realistic ones. The purpose is partly pedagogical (since the simpler models are a prerequisite to understanding more complex ones), but I also mean to help lawyers appreciate the methodology of using social science models generally—and to help them see that important lessons can be learned at all levels of complexity. As before, the presentation cuts across different fields of law to highlight the connections among functionally similar economic problems. In these later chapters I also draw examples from upper-level courses of study such as environmental law, bankruptcy, and products liability.

Chapter 8 surveys a variety of critiques of the economic approach. I have tried in this chapter to classify the critiques according to the perspectives they reflect, but I should make clear what I have in mind by this classification, since the perspectives overlap to some extent. By the *liberal* critique, I mean the objection that economics, by viewing the social interest as an aggregate, fails to do justice to individual persons or to respect their rights. What I call the *paternalist* critique calls into question the assumption that individuals are the best judges of their own interests; and the *sociological* critique questions whether individuals are motivated primarily by self-interest, narrowly defined, or by social relations more generally. The *radical* critique argues that law and economics is a deceptive apology for the status quo, in that it purports to offer a neutral mediating institution to resolve conflict among competing social interests where no such institution is possible. The *communitarian* critique argues that economics, by taking individual wants and preferences as givens, ignores the extent to which they are socially determined, and so disparages the possibility that legal institutions can help to work toward a better society. Finally, the *Legal Realist* critique argues that social reality is too complex to be adequately captured by formalistic approaches such as economics and that attempting to force legal analysis into any single framework will distort and impair its conclusions.

The final chapter presents three readings on the application of economics to family law, including Landes and Posner's controversial article analyzing adoption as a market for children. The critiques presented in chapter 8 can seem abstract, especially for those not well versed in social theory, and the adoption issue—along with the larger questions addressed by the economics of the family—raises them in a vivid and concrete fashion. Because the earlier chapters are organized around economic concepts, furthermore, this is the reader's first opportunity to look at a legal field on its own terms. It provides a good opportunity for the reader to synthesize the economic concepts that have been learned in previous chapters and exposes students to at least some applications that are on the frontier of the discipline. After all, it was not so long ago that law and economics itself was on the frontier.

A Note for Teachers

This reader is designed to serve as a primary text for a one-semester law school course in the economic analysis of law, and I have used it for that purpose myself. Alternatively, it could serve as a supplement to a casebook or textbook in such a course. Because of its focus on jurisprudential issues, it is less well suited for an audience of economics undergraduates, though economics graduate students curious about the methodological foundations of their discipline will find much here of interest. There is a lot of material here, however, so individual teachers using it may wish to pick and choose according to their interests; those using this book as a course supplement will need to be especially selective. Accordingly, each of the chapters is relatively self-contained and can be omitted without loss of continuity, with the exception of chapters 2 and 3. Additionally, some of the notes and questions to chapter 9 presuppose familiarity with the material covered in chapter 8.

I recognize that many teachers of law and economics are accustomed to organizing their courses around substantive law topics such as tort, contract, and property and that the structure of this book may make it more complicated to present sustained analyses of individual fields. Chapter 9 on family law is intended in part to address this concern. In my view, however, the traditional doctrinal organization obscures the economic approach's main contribution, which is to offer a new set of categories that cut across traditionally defined fields, allowing lawyers to recognize—and use— functional analogies and distinctions they had not previously appreciated. When an externality argument is used to analyze one legal doctrine, a risk-allocation argument another, a Coasian argument still another, and it is never made explicit why one argument is used here and another there, this larger lesson is lost. Students taught in this manner too often come to regard economics as an ad hoc tool, useful for reaching whatever results might seem appropriate on ulterior grounds, rather than as a systematic approach for explaining human behavior and clarifying normative choices. I hope, therefore, that even teachers who continue to organize their course along doctrinal lines will find some benefits in this reader's conceptual presentation.

Some teachers may prefer to defer the material in chapter 1, which deals with the underpinnings of the economic approach, until later in the course. There is room for a difference of opinion here. In my view, it is both better pedagogy and more intellectually honest to present economics' main positive and normative assumptions up front. I have found that when teaching good students who have no prior background in the subject, a policy of full disclosure helps to avoid the repeated detours and tangents that inevitably arise as the students discover and raise such questions on their own. Some students, however, may find this relatively theoretical material more accessible after having studied a number of more concrete applications in the fields of contract, tort, and property. Teachers who follow the latter approach may wish to place chapter 1's material either before or after chapter 7.

Additionally, when teaching the material in this reader, I have found it useful to provide students with some major legal cases that illustrate or provide opportunities to apply the ideas put forth by the individual readings. Different teachers may wish to use different cases as illustrations, depending on their backgrounds and interests, but I have found the following to offer good springboards for discussion:

Chapter 2: *Sturges v. Bridgman,* 11 Ch. 852 (1879); *Orchard View Farms v. Martin Marietta Aluminum,* 500 F.Supp. 984 (D. Or. 1980).

Chapter 3: *Boomer v. Atlantic Cement Co.,* 26 N.Y.2d 219, 309 N.Y.S.2d 312, 257 N.E.2d 870 (1970); *Spur Industries v. Del E. Webb Development Co.,* 494 P.2d 701 (Ariz., 1972); *U.S. v. Carroll Towing Co.,* 159 F.2d 169 (2d Cir. 1947); *The T. J. Hooper,* 6 F.2d 737 (2d Cir. 1932); *Peevyhouse v. Garland Coal Co.,* 382 P.2d 109 (Okla., 1962).

Chapter 4: *Austin Instrument v. Loral Corp.,* 29 N.Y.2d 124, 272 N.E.2d 533 (1971); *Batsakis v. Demotsis,* 226 S.W.2d 673 (Tex. Ct. Civ. App., 1949); *Williams v. Walker-Thomas Furniture Co.,* 350 F.2d 445 (D.C. Cir., 1965).

Chapter 5: *Taylor v. Caldwell,* 3 Best & S. 826 (Q.B., 1863); *Escola v. Coca-Cola Bottling Co.* 24 Cal. 2d 453, 150 P.2d 436 (1944); *Almota Farmers Elevator and Warehouse Co. v. United States,* 409 U.S. 470 (1973); *Alcoa v. Essex,* 499 F. Supp. 53 (W.D. Pa. 1980).

Chapter 6: *Laidlaw v. Organ,* 15 U.S. (2 Wheat.) 178 (1817); *Hadley v. Baxendale,* 9 Exch. 341 (1854); *EVRA Corp. v. Swiss Bank Corp.,* 673 F.2d 951 (7th Cir. 1982).

Chapter 7: *Henningsen v. Bloomfield Motors,* 32 N.J. 358, 161 A.2d 69 (1960); *Helling v. Carey,* 519 P.2d 981 (Wash. 1974).

Chapter 9: *In the matter of Baby M,* 109 N.J. 396, 537 A.2d 1227 (N.J., 1988).

Bibliographical Note

The citations in the notes at the end of each chapter are intended to be representative rather than exhaustive. Readers wishing a more systematic bibliography may wish to consult Boudewijn Bouckaert and Gerrit De Geest, *Bibliography of Law and Economics* (Dordrecht: Kluwer Academic Publishers, 1992). Two separate multivolume encyclopedias of law and economics, both comprising individual articles contributed by leading researchers in the field, are currently in preparation: the *Encyclopedia of Law and Economics,* edited by Professors Bouckaert and De Geest and scheduled to be published over the Internet in late 1997 as well as in bound form by Edward Elgar Press; and the *New Palgrave Dictionary of Economics and the Law* (London: Macmillan Press, forthcoming 1998), edited by Peter Newman. Those seeking a one-volume survey of legal applications of economics can do no better than to consult Richard Posner's classic text *Economic Analysis of Law* (Boston: Little, Brown, 1992), now in its fourth edition. Robert Cooter and Thomas Ulen's *Law and Economics* (Glenview, Ill.: Scott, Foresman, 1988) offers a survey of technical models suitable for advanced undergraduates in economics, and A. Mitchell Polinsky's *An Introduction to Law and Economics* (Boston: Little, Brown, 2d. ed. 1989) provides a brief and exceptionally clear introduction to the issues raised in this reader. Readers in search of a basic introduction to the tools and principles of microeconomics should consult either Steven Crafton and Margaret Brinig, *Quantitative Methods for Lawyers* (Durham, N.C.: Carolina Academic Press, 1994), or for a more comprehensive treatment, Robert Pindyck and Daniel Rubinfeld, *Microeconomics* (Englewood Cliffs, N.J.: Prentice Hall, 1995).

Acknowledgments

Many people assisted me in the planning and preparation of this volume. I am grateful to Roberta Romano and to Helen McInnis of Oxford University Press for encouraging me to undertake the project and for much helpful critical and editorial advice along the way; to many colleagues, students, and teachers for helpful discussions of the ideas set out in the notes, questions, and introductory essays; to Merritt Fox, Walter Kamiat, Wendy Wiener Katz, Mitt Regan, Steven Salop, Warren Schwartz, John Sheckler, Michelle White, and two anonymous reviewers for comments on the manuscript; and to Pam Irwin, Louise Muse, Andrew Niebler, Martin Saad, John Sheckler, and Wendy Wiener Katz for invaluable clerical and research assistance.

Contents

9 An Application on the Frontier: Family Law, 371

Foundations of the

Economic Approach to Law

1

Methodology of the
Economic Approach

This chapter introduces the basic methods and tools of the economic approach. It focuses on the question, how does economics differ from other ways of thinking about social life and legal institutions? As the sequence of readings suggests, economics is distinctive in at least three ways. First, it offers a particular descriptive theory of human behavior, which it uses to explain past events and predict future ones. This theory forms the basis of what is often called *positive economics*. Second, economics proceeds from a particular set of moral and ethical principles, which it uses to evaluate existing institutions and to recommend reforms. This endeavor is usually called *normative economics*. Third, and more subtly, economics follows a particular methodological program in pursuing its aims—one grounded in the separation of positive and normative issues and focused on the construction of abstract models of empirical reality. All three aspects of the economic approach differ from traditional legal methodology in important respects.

The first reading in this chapter, by Gary Becker, lays out the essential assumptions of positive economics—what we will call the *model of rational choice*. Becker was awarded the Nobel Prize in Economics in 1992 for his work extending the domain of economic analysis to a wide range of human behavior and interactions that had been traditionally considered outside the boundaries of the discipline. The excerpt reprinted here is adapted from the introduction to his book *The Economics of Human Behavior*, in which he presents economic analyses of such diverse social phenomena as crime, racial discrimination, marriage and divorce, and the rearing of children. In Becker's view, the usefulness of economics is not restricted to markets; rather, any aspect of human behavior that arises from the deliberate pursuit of ends is amenable to economic methods. In particular, he argues, all intentional behavior can be understood as *constrained maximization*—people pursuing their goals as best they can subject to the constraints of the external world. Accordingly, the tools that economists have developed to study maximization in the market setting are equally useful outside markets. According to Becker, just three conditions are necessary to make the economic approach relevant: individual maximization, market equilibrium, and stable preferences.

The readings by Jules Coleman and Thomas Schelling outline the basic framework of normative economics: the related concepts of *Pareto efficiency, Pareto superiority,* and *potential Pareto superiority* (also called the *Kaldor-Hicks criterion*). These concepts form the basis for virtually all policy recommendations in law and economics, and underlie the practice of cost-benefit analysis. Coleman, a professional philosopher, analyzes the ethical underpinnings of these concepts, paying particular attention to their relationship to *classical utilitarianism,* the philosophy that one is morally obligated to try to maximize the total amount of happiness in the world. He concludes that although efficiency (and hence cost-benefit analysis) cannot be justified on utilitarian grounds, it might be justifiable on grounds of consent or liberty. Schelling, a professional economist, offers a pragmatic argument in favor of the efficiency criterion. He argues that if a situation is inefficient, then there necessarily exists some potential arrangement, perhaps involving side payments among the parties, that would make everyone concerned better off. Tolerating the inefficient situation, therefore, leaves money on the table.

Finally, the selections by Mark Blaug and Milton Friedman present economics' approach to scientific method, focusing on the field's central methodological distinction: the difference between positive and normative analysis. Blaug explains this distinction and then, drawing on ideas from the philosophy of science, argues that descriptive and ethical issues cannot in principle be separated. In his view, people necessarily accept or reject positive descriptions on grounds that are themselves social conventions. This results in an inevitable interplay between fact and value, as people interpret empirical evidence in light of their ethical values while at the same time revising those values in order to accommodate the practical conditions of the world. Nonetheless, he regards the positive/normative distinction as a useful and healthy methodological approach because, in his words, "our ordinary methods for settling disputes about facts are less divisive than those for settling disputes about values."

Friedman's essay, which is generally regarded by the economics profession as its classic statement of methodology, focuses both on the relationship between positive and normative analysis and on the practice of economic modeling. With regard to the positive/normative distinction, Friedman argues not only that economists should try to keep their positive predictions separate from their normative arguments but that they should concentrate on the former rather than the latter. In his view, public controversies in Western society derive more from people's holding different views about the way the world works than from any fundamental difference in their values. Accordingly, since economists can do little to help resolve value conflicts in any event, they should devote their professional efforts to promoting a positive consensus. The best way to do this, he argues, is to build theories about the world and to test them against the available empirical evidence.

The theory-building aspect of economic method has often met with resistance from lawyers and other noneconomists, for the analytical models that economists use to generate predictions and propose reforms are based on simplified and sometimes counterfactual assumptions about the empirical world. The model of rational choice, for instance, posits that individuals are constantly engaged in the attempt to maximize their utility, whereas we know that actual human beings often behave in ways that are habitual, inconsistent, or even random. Similarly, while the behavior of business organizations arises from a complex interaction among the goals and purposes of their constituents, neoclassical economic theory reduces this interaction to the single-minded pursuit of profit maximization. Does the literal falsity of the assumptions of utility and profit maximization make them an unreliable foundation for modeling?

Friedman's answer is that models should not be judged by the realism of their assumptions. Economics is not an abstract search for truth, he argues, but an applied science, a pragmatic endeavor designed to serve human purposes. Whether a model is a good one, accordingly, depends on how well it serves these purposes. In Friedman's view, it is the accuracy of a model's *predictions* that are important, not the accuracy of its assumptions, for a model that predicts well will help us to plan for the future and manipulate the material world toward our desired ends. To illustrate, the standard economic model of the firm predicts that business enterprises will respond in particular ways to changes in the market and regulatory environment—for example, that they will lay off workers in response to an increase in the minimum wage or will increase their scale of operations in response to the imposition of a fixed tax on the right to do business. If the empirical evidence confirms these predictions, argues Friedman, then the model has been validated for use; if the predictions are disconfirmed, conversely, then it is time to build a new and better model. Since the ultimate test of a model is pragmatic and empirical, criticizing its theoretical assumptions is beside the point.

1.1   BEHAVIORAL PREMISES

The Economic Approach to Human Behavior
GARY BECKER

Economy is the art of making the most of life.

George Bernard Shaw

. . . I believe that what most distinguishes economics as a discipline from other disciplines in the social sciences is not its subject matter but its approach. Indeed, many kinds of behavior fall within the subject matter of several disciplines: for example, fertility behavior is considered part of sociology, anthropology, economics, history, and perhaps even politics. I contend that the economic approach is uniquely powerful because it can integrate a wide range of human behavior.

Everyone recognizes that the economic approach assumes maximizing behavior more explicitly and extensively than other approaches do, be it the utility or wealth function of the household, firm, union, or government bureau that is maximized. Moreover, the economic approach assumes the existence of markets that with varying degrees of efficiency coordinate the actions of different participants—individuals, firms, even nations—so that their behavior becomes mutually consistent. Since economists generally have had little to contribute, especially in recent times, to the understanding of how preferences are formed, preferences are assumed not to change substantially over time, nor to be very different between wealthy and poor persons, or even between persons in different societies and cultures.

Prices and other market instruments allocate the scarce resources within a society and thereby constrain the desires of participants and coordinate their actions. In the economic approach, these market instruments perform most, if not all, of the functions assigned to "structure" in sociological theories.

The preferences that are assumed to be stable do not refer to market goods and services, like oranges, automobiles, or medical care, but to underlying objects of choice that are produced by each household using market goods and services, their own time, and other inputs. These underlying preferences are defined over fundamental aspects of life, such as health, prestige, sensual pleasure, benevolence, or envy, that do not always bear a stable relation to market goods and services. The assumption of stable preferences provides a stable foundation for generating predictions about responses to various changes, and prevents the analyst from succumbing to the temptation of simply postulating the required shift in preferences to "explain" all apparent contradictions to his predictions.

The combined assumptions of maximizing behavior, market equilibrium, and stable preferences, used relentlessly and unflinchingly, form the heart of the economic approach as I see it. They are responsible for the many theorems associated with this approach. For example, that (1) a rise in price reduces quantity demanded, be it a rise in the market price of eggs reducing the demand for eggs, a rise in the "shadow" price of children reducing the demand for children, or a rise in the office waiting time for physicians, which is one component of the full price of physician services, reducing the demand for their services; (2) a rise in price increases the quantity supplied, be it a rise in the market price of beef increasing the number of cattle raised and slaughtered, a rise in the wage rate offered to married women increasing their labor force participation, or a reduction in "cruising" time raising the effective price received by taxicab drivers and thereby increasing the supply of taxicabs; (3) competitive markets satisfy consumer preferences more effectively than monopolistic markets, be it the market for aluminum or the market for ideas . . . ; or (4) a tax on the output of a market reduces that output, be it an excise tax on gasoline that reduces the use of gasoline, punishment of criminals (which is a "tax" on crime) that reduces the amount of crime, or a tax on wages that reduces the labor supplied to the market sector.

The economic approach is clearly not restricted to material goods and wants, nor even to the market sector. Prices, be they the money prices of the market sector or the "shadow" imputed prices of the nonmarket sector, measure the opportunity cost of using scarce resources, and the economic approach predicts the same kind of response to shadow prices as to market prices. Consider, for example, a person whose only scarce resource is his limited amount of time. This time is used to produce various commodities that enter his preference function, the aim being to maximize utility. Even without a market sector, either directly or indirectly, each commodity has a relevant marginal "shadow" price, namely, the time required to produce a unit change in that commodity; in equilibrium, the ratio of these prices must equal the ratio of the marginal utilities. Most importantly, an increase in the relative price of any commodity—i.e., an increase in the time required to produce a unit of that commodity— would tend to reduce the consumption of that commodity.

The economic approach does not assume that all participants in any market necessarily have complete information or engage in costless transactions. Incomplete information or costly transactions should not, however, be confused with irrational or volatile behavior. The economic approach has developed a theory of the optimal or rational accumulation of costly information that implies, for example, greater investment in information when undertaking major than minor decisions—the purchase of a house or entrance into marriage versus the purchase of a sofa or bread. The assumption that information is often seriously incomplete because it is costly to acquire is used in the economic approach to explain the same kind of behavior that is explained by irrational and volatile behavior, or traditional behavior, or "nonrational" behavior in other discussions.

When an apparently profitable opportunity to a firm, worker, or household is not exploited, the economic approach does not take refuge in assertions about irrationality, contentment with wealth already acquired, or convenient ad hoc shifts in values (i.e., preferences). Rather it postulates the existence of costs, monetary or psychic, of taking advantage of these opportunities that eliminate their profitability—costs that may not be easily "seen" by outside observers. Of course, postulating the existence of costs closes or "completes" the economic approach in the same, almost tautological, way that postulating the existence of (sometimes unobserved) uses of energy completes the energy system, and preserves the law of the conservation of energy. Systems of analysis in chemistry, genetics, and other fields are completed in a related manner. The critical question is whether a system is completed in a useful way; the important theorems derived from the economic approach indicate that it has been completed in a way that yields much more than a bundle of empty tautologies in good part because, as I indicated earlier, the assumption of stable preferences provides a foundation for predicting the responses to various changes.

Moreover, the economic approach does not assume that decision units are necessarily conscious of their efforts to maximize or can verbalize or otherwise describe in an informative way reasons for the systematic patterns in their behavior. Thus it is consistent with the emphasis on the subconscious in modern psychology and with the distinction between manifest and latent functions in sociology. . . . In addition, the economic approach does not draw conceptual distinctions between major and minor decisions, such as those involving life and death in contrast to the choice of a brand of coffee; or between decisions said to involve strong emotions and those with little emotional involvement such as in choosing a mate or the number of children in contrast to buying paint; or between decisions by persons with different incomes, education, or family backgrounds.

Indeed, I have come to the position that the economic approach is a comprehensive one that is applicable to all human behavior, be it behavior involving money prices or imputed shadow prices, repeated or infrequent decisions, large or minor decisions, emotional or mechanical ends, rich or poor persons, men or women, adults or children, brilliant or stupid persons, patients or therapists, businessmen or politicians, teachers or students. The applications of the economic approach so conceived are as extensive as the scope of economics in the definition given earlier that emphasized scarce means and competing ends. It is an appropriate approach to go with such a broad and unqualified definition, and with the statement by Shaw that begins this essay. . . .

To convey dramatically the flavor of the economic approach, I discuss briefly three of the more unusual and controversial applications.

Good health and a long life are important aims of most persons, but surely no more than a moment's reflection is necessary to convince anyone that they are not the only aims: somewhat better health or a longer life may be sacrificed because they conflict with other aims. The economic approach implies that there is an "optimal" expected length of life, where the value in utility of an additional year is less than the utility foregone by using time and other resources to obtain that year. Therefore, a person may be a heavy smoker or so committed to work as to omit all exercise, not necessarily because he is ignorant of the consequences or "incapable" of using the information he possesses, but because the lifespan forfeited is not worth the cost to him of quitting smoking or working less intensively. These would be unwise decisions if a long life were the only aim, but as long as other aims exist, they could be informed and in this sense "wise."

According to the economic approach, therefore, most (if not all!) deaths are to some extent "suicides" in the sense that they could have been postponed if more resources had been invested in prolonging life. This not only has implications for the analysis of what are ordinarily called suicides, but also calls into question the common distinction between suicides and "natural" deaths. Once again the economic approach and modern psychology come to similar conclusions since the latter emphasizes that a "death wish" lies behind many "accidental" deaths and others allegedly due to "natural" causes.

The economic approach . . . implies, for example, that both health and medical care would rise as a person's wage rate rose, that aging would bring declining health although expenditures on medical care would rise, and that more education would induce an increase in health even though expenditures on medical care would fall. None of these or other implications are necessarily true, but all appear to be consistent with the available evidence.

[Second, a]ccording to the economic approach, a person decides to marry when the utility expected from marriage exceeds that expected from remaining single or from additional search for a more suitable mate. Similarly, a married person terminates his (or her) marriage when the utility anticipated from becoming single or marrying someone else exceeds the loss in utility from separation, including losses due to physical separation from one's children, division of joint assets, legal fees, and so forth. Since many persons are looking for mates, a market in marriages can be said to exist: each person tries to do the best they can. A sorting of persons into different marriages is said to be an equilibrium sorting if persons not married to each other in this sorting could not marry and make each better off.

Again, the economic approach has numerous implications about behavior that could be falsified. For example, it implies that "likes" tend to marry each other, when measured by intelligence, education, race, family background, height, and many other variables, and that "unlikes" marry when measured by wage rates and some other variables. The implication that men with relatively high wage rates marry women with relatively low wage rates (other variables being held constant) surprises many, but appears consistent with the available data when they are adjusted for the large fraction of married women who do not work. . . . The economic approach also implies that higher-income persons marry younger and divorce less frequently than others, implications consistent with the available evidence . . . but not with common beliefs. Still another implication is that an increase in the relative earnings of wives increases the likelihood of marital dissolution, which partly explains the greater dissolution rate among black than white families.

[Third, the economic approach suggests] that persons only choose to follow scholarly or other intellectual or artistic pursuits if they expect the benefits, both monetary and psychic, to exceed those available in alternative occupations. Since the criterion is the same as in the choice of more commonplace occupations, there is no obvious reason why intellectuals would be less concerned with personal rewards, more concerned with social well-being, or more intrinsically honest than others.

It then follows from the economic approach that an increased demand by different interest groups or constituencies for particular intellectual arguments and conclusions would stimulate an increased supply of these arguments, by the theorem cited earlier on the effect of a rise in "price" on quantity supplied. Similarly, a flow of foundation or government funds into particular research topics, even "ill-advised" topics, would have no difficulty generating proposals for research on those topics. What the economic approach calls normal responses of supply to changes in demand, others may call intellectual or artistic "prostitution" when applied to intellectual or artistic pursuits. Perhaps, but attempts to distinguish sharply the market for intellectual and artistic services from the market for "ordinary" goods have been the source of confusion and inconsistency . . .

I am not suggesting that the economic approach is used by all economists for all human behavior or even by most economists for most. Indeed, many economists are openly hostile to all but the traditional applications. Moreover, economists cannot resist the temptation to hide their own lack of understanding behind allegations of irrational behavior, unnecessary ignorance, folly, ad hoc shifts in values, and the like, which is simply acknowledging defeat in the guise of considered judgment. For example, if some Broadway theater owners charge prices that result in long delays before seats are available, the owners are alleged to be ignorant of the profit-maximizing price structure rather than the analyst ignorant of why actual prices do maximize profits. When only a portion of the variation in earning among individuals is explained, the unexplained portion is attributed to luck or change, not to ignorance of or inability to measure additional systematic components. . . .

Even those believing that the economic approach is applicable to all human behavior recognize that many noneconomic variables also significantly affect human behavior. Obviously, the laws of mathematics, chemistry, physics, and biology have a tremendous influence on behavior through their influence on preference and productions possibilities. That the human body ages, that the rate of population growth equals the birth rate plus the migration rate minus the death rate, that children of more intelligent parents tend to be more intelligent than children of less intelligent parents, that people need to breathe to live, that a hybrid plant has a particular yield under one set of environmental conditions and a very different yield under another set, that gold and oil are located only in certain parts of the world and cannot be made from wood, or that an assembly line operates according to certain physical laws—all these and more influence choices, the production of people and goods, and the evolution of societies.

To say this, however, is not the same as saying that, for example, the rate of population growth is itself "noneconomic" in the sense that birth, migration, and death rates cannot be illuminated by the economic approach, or that the rate of adoption of new hybrids is "noneconomic" because it cannot be explained by the economic approach. Indeed, useful implications about the number of children in different families have been obtained by assuming that families maximize their utility from stable preferences subject to a constraint on their resources and prices, with resources and prices partly determined by the gestation period for pregnancies, the abilities of children, and other noneconomic variables . . .

The heart of my argument is that human behavior is not compartmentalized, sometimes based on maximizing, sometimes not, sometimes motivated by stable preferences, sometimes by volatile ones, sometimes resulting in an optimal accumulation of information, sometimes not. Rather, all human behavior can be viewed as involving participants who maximize their utility from a stable set of preferences and accumulate an optimal amount of information and other inputs in a variety of markets.

If this argument is correct, the economic approach provides a united framework for understanding behavior that has long been sought by and eluded Bentham, Comte, Marx, and others. The reader of the following essays will judge for himself the power of the economic approach.

1.2 NORMATIVE PREMISES

Efficiency, Utility, and Wealth Maximization
JULES COLEMAN

A fully adequate inquiry into the foundations of the economic approach to law would address at least the following four related questions:

1. What is economic efficiency; that is, what does it mean to say that resources are allocated in an economically efficient manner or that a body of law is efficient?
2. Does the principle of efficiency have explanatory merit, that is, can the rules and principles of any or all of the law be rationalized or subsumed under an economic theory of legislation or adjudication?
3. How should law be formulated to promote efficiency; that is, in what ways must legal rights and duties be assigned and enforced so that the rules that assign and enforce them are efficient?
4. Ought the law pursue economic efficiency; that is, to what extent is efficiency a desirable legal value in particular, and a normatively attractive principle in general? . . .

Efficiency and Utility

Economists as well as proponents of the economic analysis of law employ at least four efficiency-related notions, including: (1) productive efficiency, (2) Pareto optimality, (3) Pareto superiority, and (4) Kaldor-Hicks efficiency. . . . In this section of the paper, I want to focus on the analytic and normative relationships between Pareto and Kaldor-Hicks notions of efficiency.

The Pareto Criteria

Resources are allocated in a Pareto-optimal fashion if and only if any further reallocation of them can enhance the welfare of one person only at the expense of another. An allocation of resources is Pareto superior to an alternative allocation if and only if no one is made worse off by the distribution and the welfare of at least one person is improved. These two conceptions of efficiency are analytically related in that a Pareto-optimal distribution has no distributions Pareto superior to it.

Both Pareto concepts express standards for ranking or describing states of affairs. The Pareto-superior criterion relates two states of affairs and says that one is an improvement over the other if at least one person's welfare improves while no one else's welfare is diminished. The optimality standard relates one distribution to all possible distributions and says in effect that no Pareto improvements can be made from any Pareto-optimal state. In addition, Pareto-optimal distributions are Pareto noncomparable; the Pareto-superior standard cannot be employed to choose among them. Another way of putting this last point is to say that the social choice between Pareto-optimal distributions must be made on nonefficiency grounds.

Kaldor-Hicks

Like Pareto superiority, Kaldor-Hicks efficiency is a relational property of states of affairs. One state of affairs (E') is Kaldor-Hicks efficient to another (E) if and only if those whose welfare increases in the move from E to E' could fully compensate those whose welfare diminishes with a net gain in welfare. Under Kaldor-Hicks, compensation to losers is not in fact paid. Were the payment transaction costless and full compensation given to the losers, Kaldor-Hicks distributions would be transformed into Pareto-superior ones. This characteristic of Kaldor-Hicks has led some to refer to it as a "potential Pareto-superior" standard.

Kaldor-Hicks-efficient distributions do not necessarily map onto Pareto-superior distributions. The failure to require compensation has the effect of making some individuals worse off and thus fails to satisfy the requirements of Pareto superiority.

In general, a distribution that is Kaldor-Hicks efficient need not be Pareto optimal either. If a distribution is Kaldor-Hicks efficient, then some individual has been made sufficiently better off so that he could—hypothetically at least—fully compensate those who have been made worse off. It does not follow, however, that from their new relative positions the winners and losers are incapable of further mutual improvement through trade. Thus a Kaldor-Hicks-efficient allocation need be neither Pareto superior nor Pareto optimal, although it may be either or both. . . .

The Pareto Standards and Utilitarianism

Pareto Superiority. The Pareto-superior standard is often thought of as normatively rooted in classical utilitarianism. Briefly, the argument for identifying Pareto efficiency with utilitarianism is as follows: Allocations that are Pareto superior increase at least one person's utility without adversely affecting the utility of another; they produce winners but no losers. Consequently there is no need to compare the relative gains and losses of winners and losers in order to determine if a course of conduct increases total utility. Pareto improvements increase total utility, although not all increases in total utility constitute Pareto improvements. Because the Pareto-superiority criterion appears to obviate the need to make interpersonal utility comparisons in order to determine if a course of conduct increases total utility, it is easy to see why one might be led to consider the justification for pursuing Pareto improvements to be utilitarian.

Aspects of this argument are misleading, others mistaken. First, the Pareto-superior criterion does not eliminate the need for a standard of utility comparison. If the Pareto-superior standard is to be an index of total utility, interpersonal utility comparisons are necessary, since the concept of total utility presupposes the capacity to aggregate individual utility functions, which in turn requires a standard of comparison. Provided such a standard exists, Pareto improvements increase total utility; and because they do, one could argue that the justification for pursuing Pareto improvements relies on its connection to utilitarianism. That would be a mistake.

We presume that the concept of total utility is meaningful, and because we do, we must believe that some standard of interpersonal comparability exists. From this it follows that Pareto improvements increase total utility. Suppose, however, that the very idea of total utility were meaningless. In that case, we should be unable to make sense of claims concerning increases or decreases in total utility. We should, however, still be capable of talking meaningfully about Pareto improvements.

There is an important distinction among individual preference theories, total utility theories, and social welfare theories. Pareto judgments are expressed in terms of orderings of individual preferences rather than in terms of total utility. To say that an action is Pareto superior is to say only that at least one person is higher along his or her preference ranking, while no one else is any worse off with respect to his or hers. The existence of a standard of comparison enables us to bring these distinct judgments about the relative standings of individuals with respect to their preference orderings into a single judgment about total utility. As we have just seen, however, the Pareto judgment has empirical significance independent of its connection to utility.

At bottom is the distinction between the claim that Pareto judgments may warrant claims about utility and the claim that utility justifies the pursuit of Pareto improvements. Those committed to utilitarianism will no doubt embrace the Pareto standard, since they are committed to a conception of the right and the good from which Pareto superiority follows as a particular instance, that is, as one way of promoting utility. But because Pareto judgments have empirical significance apart from their connection to utility it is possible to conceive of alternative justifications for the pursuit of actions and policies that satisfy the Pareto-superior criterion.

For example, one could plausibly argue that the Pareto-superior standard is normatively defensible not because applying it gives an index of total utility, but because rational self-interested persons would consent to its use. That is, few would object to policies that made at least one person better off if doing so never required anyone else to suffer. This line of defense does not rest on the utility of the Pareto standard, but on the fact that rational, self-interested persons would consent to its use.

Alternatively, one might advance a normative defense of Pareto superiority that relies on a libertarian rather than a contractarian or utilitarian political morality.

Exchanges among knowledgeable, rational persons in a free market are generally Pareto superior; rational individuals do not strike bargains with one another unless each perceives it to be in his or her own interest to do so. A successful exchange between such parties is, therefore, one in which the value to each of what he or she relinquishes is perceived as less than the value of what each receives in return. Such exchanges make no individual worse off; often they improve the lot of all concerned. Pareto superiority is connected in this way to the ideal of a free-exchange market.

We can imagine, then, at least three distinct normative defenses for the Pareto-superior standard, only one of which is utilitarian. These are the utilitarian argument, the contractarian or consent argument, and the libertarian argument. The important point for our present purposes is that one can remain unpersuaded by the utilitarian conception of the good and the right, yet stand prepared to endorse certain institutional arrangements that work to at least some person's benefit and to no one's detriment, on the ground either that rational self-interested persons would consent to such arrangements or that certain of these arrangements respect or are required by a deeper commitment to individual liberty or autonomy. . . . [I]t simply does not follow that the normative use of the Pareto-superior criterion must be abandoned even if utilitarianism is wrongheaded.

Pareto Optimality. Let us now turn to Pareto optimality in order to develop its relationship to utilitarianism. To say that resources are distributed in Pareto-optimal fashion is to say that no further distributions are capable of enhancing anyone's welfare without making someone else worse off. Pareto-optimal distributions have no distributions Pareto superior to them. . . .

From the fact that a distribution of resources is Pareto optimal two utility judgments follow: First, every Pareto-optimal distribution contains more total utility than some set of distributions represented by points within the utility-possibility frontier, because every distribution represented by a point on the utility possibility frontier is Pareto superior to some distributions represented by points within the frontier . . . Second, the overall utility of a Pareto-optimal distribution cannot be increased by a Pareto-superior move, because a Pareto-optimal distribution has no distributions Pareto superior to it.

From the fact that a distribution of resources is Pareto optimal it does not follow that the move to it increases utility, because not every move to a Pareto-optimal distribution involves a Pareto improvement. Pareto-optimal distributions may result from Kaldor-Hicks-efficient or non-Pareto-superior moves as well as from Pareto-superior ones. Moreover, no judgments about the relative utility content of the various members of the set of Pareto-optimal distributions is warranted. This is because the set of Pareto-optimal distributions cannot be compared by the Pareto-superiority standard. . . .

Kaldor-Hicks and Utility

Both Pareto-superiority and Pareto-optimality judgments entail certain total utility claims provided the very weak condition that a standard of comparability exists is satisfied. The same cannot be said for Kaldor-Hicks. First, in order to infer from the satisfaction of the Kaldor-Hicks test that there has been a net gain in utility, we need to know whether winners have won more than losers have lost, which requires interpersonal-cardinal comparability. Second, the Kaldor-Hicks test may lead to inconsistent preferences over social states; that is, two states of affairs may be Kaldor-Hicks efficient to one another. This is the Scitovsky paradox. A demonstration of the paradox follows.

Suppose there are two persons, A and B, and two commodities, X and Y. [The figure below] gives the outputs in two states of the economy: E and E'.

In E, Mr. A has two units of X and no units of Y; Ms. B has no units of X and one unit of Y. In E', A has one unit of X and none of Y [; Ms. B has no units of X and two units of Y]. Suppose now that A and B have the following preferences for X and Y: Mr. A prefers one unit of X and one unit of Y to two units of X and no units of Y, which is itself preferred to one unit of X and no units of Y. Ms. B prefers one unit of X and one unit of Y to no units of X and two of Y, which is itself preferred to no units of X and one of Y. E is Kaldor-Hicks efficient to E', and E' is Kaldor-Hicks efficient to E. In E, Mr. A has two units of X and none of Y; were he to give one unit of X to Ms. B, he would be exactly as well off as he was in E'; Ms. B would be better off. She would then have one unit of X and one of Y, which she prefers to both E' and E. Compensation would make her better off and Mr. A no worse off. So E is Kaldor-Hicks efficient to E'.

	E		E'	
	X	Y	X	Y
A	2	0	1	0
B	0	1	0	2

As it stands, Kaldor-Hicks does not provide an adequate efficiency basis for preferring one state of the economy to another. The Kaldor-Hicks criterion may be reformulated so that one state of affairs is Kaldor-Hicks preferable to another if and only if the winners could compensate the losers in going from E to E' but the winners could not compensate the losers in going from E' to E. This eliminates the paradox, but then Kaldor-Hicks will not be transitive.

Consequently, from the satisfaction of the Kaldor-Hicks test it does not follow that there has been a net gain in utility, since (1) one can determine whether Kaldor-Hicks has been satisfied without appealing to any particular standard of interpersonal comparison, but, in cases in which there are both winners and losers, one cannot tell whether total utility has been increased by a Kaldor-Hicks improvement without appealing to such a standard, and (2) E can be Kaldor-Hicks superior to E' and E' to E, whereas E cannot have more utility than E' while E' has more utility than E.

Of the three efficiency-related criteria, only Pareto superiority may be transformed into an index of utility. Utility judgments warranted by Pareto optimality are a consequence of its analytic relationship to Pareto superiority. Kaldor-Hicks efficiency is paradoxical as a standard of utility and therefore cannot be transformed into an index of utility even if interpersonal utility comparisons were possible.

Further, it does not follow from the fact that utilitarianism is mistaken that the normative use of the Pareto standards must be abandoned. Even though Pareto superiority can be employed as a standard of utility, it has content aside from this connection, and its application in social and legal policy matters may be justified on nonutilitarian grounds, in particular on the basis of consent or liberty. If utilitarianism is wrong, a proponent of economic analysis has a choice: He or she can either abandon the normative use of the Pareto criteria in favor of a nonutility-based efficiency criterion or try to construct nonutilitarian normative arguments for the Pareto criteria. . . .

Nonutilitarian Arguments for Paretianism
. . .

One set of arguments for the normative use of the Pareto criteria relies on the fact that exchanges in free markets are generally Pareto superior. Individuals transact when it is in their interest to do so; when each views the transaction as liable to make him or her better off. Moreover, the ultimate outcome of Pareto-superior market behavior is Pareto optimality. Individuals will engage in transactions until it is no longer in the interest of at least one of them to do so. At that point negotiations cease because there are no further mutual gains through trades to be had. In the ideal world of noncoercive markets free from transaction costs and third-party effects, in which individuals are both rational and knowledgeable, the exercise of liberty leads to Pareto-optimal states of affairs through a series of Pareto-superior exchanges.

These considerations provide the bases for a number of related arguments for pursuing Pareto improvements. One argument . . . , emphasizes that in exercising their liberty individuals promote efficiency. Consequently, the pursuit of Pareto efficiency is justified not because of its relationship to net utility but because noncoercive market behavior is efficient. The moral value we attach to individual autonomy is transferred to the pursuit of efficiency.

This line of argument fails for a number of reasons. First, not all markets are Pareto efficient. For example, lack of adequate information may transform free choice into something less than Pareto-superior action. Further, it is at least plausible that some individuals acting freely make themselves worse off; freedom does not necessarily ensure increased happiness. This much we know.

One could respond by saying that choices based on insufficient information are not totally free. To the extent that market failures or inefficiencies result from inadequately informed choice, they are less than fully free. Consequently, an action that is not Pareto superior must be the result of a choice that is less than fully free. . . .

A second, more telling objection to the autonomy argument is that pursuing Pareto superiority or optimality may require intervention in the exchange market—at both the individual and institutional levels. Where the conditions of competitive equilibrium are not satisfied, a market will not secure an efficient outcome. To secure the efficient outcome, the political order must intervene in the market. In sum, not every free exchange is Pareto superior, and pursuing efficiency may sometimes require abandoning noncoercive markets. Consequently, the argument from autonomy is both too weak and too strong to justify adequately the pursuit of Pareto improvements. . . .

Conclusion

. . .

Suppose that none of my objections to the various normative strategies for defending the pursuit of Pareto efficiency are persuasive and that Paretianism constitutes an adequate moral theory. . . . It still would not follow that judges in particular cases had the authority to resolve disputes on the basis of efficiency criteria. It does not follow from the fact that in general we ought to pursue efficiency that every actor or agent, regardless of his or her institutional role and circumstance, has the obligation or authority to promote efficiency. An argument that judicial behavior should be structured by efficiency considerations requires a further theory of institutional competence. The question is whether judges have the authority to seize upon a private dispute framed by and in terms of the litigants' interests as an opportunity to promote desirable social policies, for example, efficiency and distributional justice.

The alternative and I believe commonsense view is that the responsibility of a judge is to determine which of the litigants in a dispute has the relevant legal right. It may be that in determining which party is entitled to a decision in its favor the judge must take account of efficiency considerations. He or she might hold that an individual who suffers harm caused by the unreasonably risky conduct of another is entitled to compensation. Following Learned Hand, he or she might then go so far as to provide an economic analysis of fault or negligence. The fact that an economic argument is relevant to determining fault, for example, does not show that liability based on fault is justified on efficiency grounds, or that in awarding compensation to the plaintiff in a particular negligence case the judge is following an economic theory of adjudication. It is perfectly possible to describe the judicial behavior as follows: The judge is trying to determine which party has a right to the decision; the theory of legal rights he or she believes is correct has as a consequence that individuals who suffer from the economically inefficient conduct of others are entitled to compensation. Consequently, determining whether a particular plaintiff is entitled to recompense would require some sort of economic argument.

The difference between proponents of economic analysis and critics such as Dworkin and myself—on this question at least—is that we have different theories of institutional competence generally, and of adjudication in particular. The advantage of our way of looking at the problem is this: We draw a distinction between what, on balance, it is morally right or defensible to do and what a particular person, occupying an institutional role, has authority to do. Even were we to agree that pursuing efficiency would be morally defensible, we should argue that it is a further question whether judges have the authority to do so. In addition, we argue that adjudication primarily—or always—concerns rights rather than the promotion of some useful social policy while at the same time it provides a substantial and meaningful role for economic argument. The economic theory of adjudication, because it views private disputes in terms of global interests and social goals, explains the reference to eco-

nomic argument in adjudication at the expense of the commonsense view of law as
concerned with rights and obligations—rights and obligations it is the court's duty to
enforce rather than create.

Economic Reasoning and the Ethics of Policy

THOMAS SCHELLING

The Ethics of Policy

What I mean by the ethics of policy is the relevant ethics when we try to think disin-
terestedly about rent control, minimum wages, Medicaid, food stamps, safety regulations, cigarette taxes, or the financing of Social Security.

Farmers have an interest in price supports, laundry operators in minimum-wage
laws, doctors in the financing of Medicare, and electric utilities in clean-air regulation; and until my youngest child is safely overage I shall have a personal interest in
the draft. When people take sides on a leash law we don't expect them to argue it the
way they discuss the space shuttle. I want to define the ethics of policy as what we
try to bring to bear on those issues in which we do not have a personal stake.

It is hard to find issues that are absolutely unsoiled with personal interest. On abortion and capital punishment our personal ethics usually dominate. Food stamps affect
us, whether we qualify or not, because they cost money. Someone meticulously interested in his own welfare could find at least a minuscule personal interest in a U.N.
program for alphabet reform.

Still, most of us on many issues want to think and to talk as though we are not interested parties. We want to discuss welfare and national defense and school construction and unemployment benefits and automobile-mileage standards as though
we were not personally involved. There will be an unmistakable element of social
obligation; nobody can discuss income-tax rates or welfare levels without a partici-
patory awareness that the poor, the unfortunate, the disadvantaged, and the otherwise
deserving have some legitimate claim on those among us who can afford to help. But
although few issues are without financial impact somewhere, and most big issues involve big amounts of money, we can often confine our personal stake to an aggregate
and nonspecific social obligation. Our position in the income scale affects our conditioning as well as our reasoning, but beyond that we can try to be neutral, removed,
vicarious, impartial, judicious. . . .

The Ethics of Pricing

My students always like gasoline rationing. They believe in it on ethical principles.
(They say they do, and they sound as if they do.) Evidently the principles lie deeper

Thomas Schelling, Economic Reasoning and the Ethics of Policy, 63 *The Public Interest* (1981), pp. 37–61.
Copyright © 1981 National Affairs, Inc. Reprinted with permission.

than rationing itself; the students must have some notion of what happens with rationing and without it, or with some specific alternative, and they must have a preference about the outcome. Students know that there are gainers and losers; their ethics appear to relate to who gains, who loses, and how much.

I can talk most of them out of it. It takes longer than fifty minutes, and I never try it if I have only a single class hour. They probably distrust my ethical principles and think I do not care about what they care about, or care as much. They very likely think my ethics are "process oriented" and the free market enchants me, while they are "consequences oriented" and don't like the results.

Time permitting, I accomplish their conversion in two stages. I warn them in advance that I am going to show them that if they like rationing there is something they should like better. I join them in believing the free market needn't be let alone, but I do propose what is sometimes called "rationing by the purse." I suggest we let the price of gasoline rise until there is no shortage, and capture the price increase with a tax. Because that looks hard on the poor my students do not like it.

The first step in subverting their ethical preference is to propose that under any system of rationing that they might devise—and I take a little while to show them that it is not easy to design a "fair" system of rations—people should be encouraged to buy and sell ration coupons. This proposal has little appeal. The rich will obviously burn more than their share of gas, the poor being coerced by their very poverty into releasing coupons for the money they so desperately need. But eventually students recognize that the poor, because they are poor, would like the privilege of turning their coupons into money. Where gas coupons can only provide them gas at a discount, transferable coupons can buy milk at a discount. If it is unfair that the poor cannot drive as much as the rich, it is the poverty that is unfair, not the gasoline system.

The principle established, we observe that coupons are worth cash, whether you buy them, sell them, or merely turn in your own at the local service station. If gas at the pump is $1.25 and coupons sell for 75 cents, the net price of gasoline is $2.00; anyone who gets ten gallons' worth of coupons from the Department of Energy is getting a clumsy equivalent of $7.50 cash. The station that sells ten gallons receives $12.50 in money and $7.50 in coupons that could have been traded for cash. What we have is a 75-cent tax payable at the pump in special money, and a cash disbursement to motorists paid in this special money. We could just as well do it all without the coupons.

There is more to it than that, but the "more" usually does not involve much ethics. It isn't that we resolved an ethical issue. We merely lifted the veil of money and discovered that the ethical issue we thought was there was not. Or, perhaps better, the ethical issue that we associated with rationing was tangential to that procedure. Whatever the compensatory principle is that appeals to the students' sense of fairness, there are many procedures that can achieve it, some better than others, rationing neither worst nor best; and once it is all converted to money, it is easier to see what some of the alternatives are and whether they are ethically superior. Superficially it may seem wrong to give gas coupons to people who don't drive; but if the gasoline is taxed instead and the proceeds rebated to the public, we can judge the ethics of alternative distributions of the proceeds, not just those based on drivers' licenses and car registrations.

Persuasion is a little harder with rent control, partly because students do not like landlords. We try to see whether there might be something better, even in principle; whether people seeking apartments are losers under rent control; whether some nonevictable tenants in rent-controlled apartments would like to cash in their precious property right but are locked in because their claim is to a specific apartment. Usually by the time we have identified all the interested parties and the likely magnitudes of their interests, and have considered a few alternative ways to accomplish the things rent control is intended to accomplish, the liveliness of the issues is undiminished but the ethical loading has mostly evaporated.

I dislike "counting coup" over vanquished students in order to display, and to hope you are impressed with, some of the ways that economics can contribute to the clarification of ethical issues. But at least the claim for economics is modest: it often helps to diagnose misplaced identification of an ethical issue. And it does this solely by helping to identify what is happening. It is not clarifying ethics; it is only clarifying economics. . . .

The Clash between Equity and Incentives

Policy issues are preponderantly concerned with helping, in compensatory fashion, the unfortunate and the disadvantaged. We have welfare for those who cannot work, unemployment benefits for people out of jobs, disability benefits for the disabled, hospital care for the injured and the ill, disaster relief for the victims of floods, income tax relief for the victims of accidental loss, and rescue services for people who find themselves in danger. Social Security is based on the premise that people will arrive at post-working age with inadequate savings to live on.

An unsympathetic way to restate this is that a preponderance of government policies have the purpose of rewarding people who get into difficulty. People are paid handsomely for losing their jobs; if you smash your car the IRS will share the cost of a new one; and if your injury requires hospitalization you can stay in an air-conditioned room as long as the doctor certifies that you will recover better if you don't go home. By treating the absence of a "man in the house" as a special grievance for a woman with dependent children, families have even received a bonus for fathers' leaving home.

There is no getting away from it. Almost any compensatory program directed toward a condition over which people have any kind of control, even remote and probabilistic control, reduces the incentive to stay out of that condition and detracts from the urgency of getting out of it. It is a rare ameliorative program that has no visible way, by its influence on behavior, to affect the likelihood or the duration or the severity of the circumstances it is intended to ameliorate. And most commonly—not always but most commonly—the effect on behavior is undesired and in the wrong direction.

To keep the issue in perspective we can observe that private insurance, even the informal kind that allows us to ask for help when we run out of gas, can have the same adverse influence on behavior. People more willingly drive on slippery roads the more nearly complete their collision coverage; back doors are unlocked if the homeowner's policy is liberal in its provisions for burglary. I am more indulgent of my sore throat if my employer provides an ample quota of sick days.

There is no use denying it in defense of social programs. As is usually the case with important issues, principles conflict. On the one hand, we want to treat unemployment as a collective liability, sustaining the family at public expense when working members lose their livelihoods. And on the other, we want not to induce people to get conveniently disemployed or to feel no need when unemployed to seek work vigorously. What helps toward one objective hurts toward the other. Offering 90 percent of normal pay can make unemployment irresistible for some, and even a net profit for those who can moonlight or work around the home. Providing only 40 percent over a protracted period makes living harsher than we want it to be. There is nothing to do but compromise. But a compromise that makes unemployment a grave hardship for some makes it a pleasant respite for others, and we cannot even be comfortable with the compromise.

Decent welfare in a high-income state is bound to be at a higher level than in a low-income state. It induces migration. Even if we favor migration, the state that finds more and more migrants on its welfare rolls did not intend to reduce the poverty of other states by helping any and all who could get up and move. But to provide an unattractive level of benefits would condemn the intended beneficiaries to a level of living below what their home state wanted to provide them. Again two principles conflict. . . .

I do not know whether one of the principles, helping the disadvantaged, should be considered ethical and the other, not letting them get away with it, not ethical. Much of the discussion about welfare rights, about not proportioning medical care to the ability to pay, and about not producing a "work ethic" by threatening the unemployed with their families' starvation, is in an ethical mode. To a lesser extent, ethical considerations are evoked over the encouragement of malingering, rewarding those who beat the system, or inducing dependence on the state. Once it is recognized, however, that two principles conflict, that two desiderata point in opposite directions and neither is so overwhelming that the other can be ignored, that both objectives have merit, and even that there is no ideal compromise because there is an diversified population at risk, the ethical contents of the principles begin to seem tangential to the inescapable problem of locating an acceptable compromise.

It is a universal problem. It won't go away. It can't be neglected. It isn't even unique to public policy. The word "compromise" has those two different meanings. Compromising a principle sounds wrong. Compromising between principles is all right.

Valuing the Priceless

Among the poignant issues that policy has to face, explicitly or by default, are some that seem to pit finite cost against infinite value. What is it worth to save a life? How much to spend on fair trial to protect the innocent against false verdicts? What limits to put on the measures, some costly in money and some in anguish, to extend the lives of people who will die soon anyway or whose lives, in someone's judgment, are not worth preserving?

These issues are ubiquitous. They arise in designing a national health program. They are directly involved in decisions for traffic lights, airport safety, medical research, fire and Coast Guard protection, and the safety of government employees. They are implicitly involved in regulation for occupational safety or safe water supplies, in building codes and speed laws, even helmets for motorcyclists—because somebody has to pay the costs.

It is characteristic of policymakers, especially at the federal level, that they usually think of themselves as making decisions that affect others, not themselves. Hurricane and tornado warnings are for those living where hurricanes and tornados strike; mine safety is a responsibility of legislators and officials above-ground concerning the lives of people who work underground. Policies toward the senile, the comatose, the paralyzed, and the terminally ill are deliberated by people who are none of the above. Occasionally the legislator debating a 55-mile speed limit pauses to think whether the benefits in safety to his own family will be worth the added driving time, but if he or she is conscientious even that calculation may be surreptitious.

The situation is different when a small community considers a mobile cardiac unit or a new fire engine. The question then is not what we ought to spend to save someone else's life but what we can afford to make our lives safer. Spending or stinting on the lives of others invites moral contemplation; budgeting my expenditures for my own benefit, alone or with neighbors for the school safety program, is less a moral judgment than a consumer choice, a weighing of some reduction in risk against the other things that money will buy.

There is a suggestion here. Maybe we can reduce the unmanageable moral content of that paternalistic decision at the national level by making it more genuinely vicarious. Instead of asking what society's obligation to them is, we should ask how they would want us to spend their money. In deciding how much to require people to spend on their own seat belts, smoke alarms, fire extinguishers, and lightning rods, it is easier to be vicarious and it is legitimate to get our bearings by reflecting on how much we might reasonably spend on our own safety. The question still may not be easy, but is less morally intimidating.

Surely, if we are all similarly at risk and in like economic circumstances, this would be the way to look at it, whether for the town bandstand or the town ambulance. On a national scale it is less transparently so, but nevertheless so, that we should want our appropriations committees to think of themselves as spending our money in our behalf. We want them neither to skimp where it really counts nor to go overboard to prepare at great expense—our expense—for the remotest of dangers. We want them to be thinking not about what concern the government owes its citizens for their safety but how much of our own money we taxpayers want spent for our safety.

With that perspective it is remarkable how quickly the issue, now collectively self-regarding instead of other-regarding, drops the ethical content that was only a construct of the initial formulation. We can still find ethical issues, but not the one that seemed so central. . . .

The "Something Better" Approach

What the reader will have noticed both with gas rationing and with [expenditures on] safety is a technique that economics commonly employs in addressing whether a particular condition or policy or program has virtue. That technique is to explore whether, in respect of alternative outcomes or consequences, some alternative policy or condition or program technique is "better." And "better" has a particular definition: superior, as an outcome, for everyone involved or, somewhat less ambitiously, for all the identifiable interests. Of gasoline rationing we explore whether there is something better, something that meets whatever objectives rationing was supposed to fulfill and does a little more besides, or meets some of them more amply, or achieves the same results at lower costs to someone concerned. To find something better does not necessarily mean that rationing is not among the better policies, only that it is still inferior to some identifiable alternative. Sometimes, but not always, it is possible to measure or estimate a lower or upper bound to the magnitude of the superiority. And sometimes if an alternative is better for not quite everybody and disadvantageous to some, we can find a way to estimate the extent of disadvantage, or put an upper bound on it. . . .

Let me close this part of the discussion by reiterating with emphasis two points. First, this line of reasoning attempts, when honestly done, to reshuffle the consequences by rearranging proposed programs and comparing alternatives, leaving intact the original weighing system by which the outcomes for different people, or different interests, were to be evaluated. It explores alternative consequences, assessing those consequences from the points of view of all the affected parties, to see whether, whatever the proposal is or the situation being evaluated, there is "something better." It is therefore of limited, but genuine, usefulness. And the second point is that there is no mystery, nothing that cannot be penetrated by a responsible policymaker, one who is willing to make some effort to discover whether indeed there is something better.

Unfortunately, economists use the term "efficiency" to describe this process, and often distinguish between considerations of "equity" and "efficiency." The word "efficiency" sounds more like engineering than human satisfaction; and if I tell you that it is not "efficient" to put the best runway lights at the poorer airport, you are likely to think you know exactly what I mean and not like it, perhaps also not liking me. If I tell you that "not efficient" merely means that I can think of something better— something potentially better from the points of view of all parties concerned—you can at least be excused for wondering why I use "efficient" in such an unaccustomed way. The only explanation I can think of is that economists talk mainly to each other. . . .

The Market Ethic

Nothing distinguishes economists from other people as much as a belief in the market system, or what some call the free market. A perennial difficulty in dealing with economics and policy is the inability of people who are not economists, and some who are, to ascertain how much of an economist's confidence in the way markets work is faith and how much is analysis and observation. How much is due to the economist's observing the way markets work and judging actual outcomes, and how much is a belief that the process is right and just? (Or right, if occasionally unjust; or right, and justice is indeterminate.)

The problem is compounded because some economists do identify markets with freedom of choice, or construe markets as processes that yield returns that are commensurate with an individual's deserts. A conclusion that arises in the analysis of a perfectly working competitive market is that people who work for hire are paid amounts equivalent to their marginal contributions to the total product, to the difference it makes if one's contribution is withdrawn while the rest of the system continues. An ethical question is whether one's marginal product constitutes an appropriate rate of remuneration. Critics of the theory, however, typically direct their energies toward the empirical issue, arguing that actual markets work differently. Nevertheless, there are economists who have given considerable thought to the matter who find that a system that distributes the fruits of economic activity in accordance with marginal contributions, be they contributions in effort or ideas or property, is ethically attractive, and others who have given considerable thought to it and find that such a system has great practical merit but little ethical claim. Most of these others believe there is a need for policies to readjust the results.

There is even an important socialist school of market economics, pioneered by Abba Lerner and Oscar Lange in the 1930s, that asserts that pricing in a socialist economy should mimic the pricing of a perfectly competitive free market, that such an economy would be least wasteful of resources, and that extramarket income transfers should compensate for any results that one does not like.

And there is a large body of professional opinion among economists, perhaps more among older than among younger ones, to the effect that markets left to themselves may turn in a pretty poor performance, but not nearly as poor when left alone as when tinkered with, especially when the tinkering is simplistically done or done cleverly to disguise the size and distribution of the costs or losses associated with some "innocuous" favoritism.

Whether or not an economist shares the ethic or the ideology that values the working of the market system for its own sake (or that identifies it not only with personal freedom but as personal freedom), most professional economists accept certain principles that others, if not the economists themselves, would recognize to have ethical content.

An example is incentives. Economists see economic incentives operating everywhere; they find nothing offensive or coercive about the responses of people to economic opportunities and sanctions; they have no interest in overcoming or opposing incentives for the sake of victory over an enemy; and they have a predilection toward tilting incentives and augmenting and dampening and restructuring incentives and even inventing incentives, to induce people to behave in ways that are collectively more rewarding or less frustrating. You can usually tell an economist from a noneconomist by asking whether at the peak season for tourism and camping there should be substantial entrance fees at the campgrounds of national parks.

A related touchstone of market economics is the idea that most people are better at spending their own money than somebody else is at spending it for them. Sometimes this is directly elevated into an ethical principle: the consumer's right to make his own mistakes. But usually it is simply that giving a poor family a shopping cart filled from the shelves of a supermarket is not as good as giving them the money and the cart and letting them do their own shopping. The idea is that they will get more for your money if they get to spend it. A given amount of your money will do more good for the family from the family's point of view if it is spent the way they want it spent.

Economists have a long checklist of exceptions to this principle, exceptions from the point of view of that family's welfare and from other points of view, but generally the economist thinks the burden of proof belongs on those who want to give food stamps or subway tokens or eyeglasses to the poor and the elderly, not money. Proof may not be hard to come by; but the burden, for most economists, should be on those who don't trust the efficacy of money. It often sounds like an ethical principle. Maybe it is.

There have recently been proposals to compensate poor families in cash for the exact amount by which their heating bills, at deregulated prices, exceed what the same amount of fuel would have cost at regulated prices. The question was raised why they shouldn't merely receive, unconditionally, an amount of money estimated in advance by that formula. The retort was that, being poor, they couldn't be trusted to spend the money on heating fuel. They might spend it on something else!

This is the point at which most economists can only shake their heads slowly.

1.3 MODELS AND MODELLING

The Distinction between Positive and Normative Economics

MARK BLAUG

The distinction between positive and normative economics, between "scientific" economics and practical advice on economic policy questions, is now 150 years old, going back to the writings of Nassau Senior and John Stuart Mill. Somewhere in the latter half of the nineteenth century, this familiar distinction in economics became entangled, and almost identified with, a distinction among philosophical positivists between "is" and "ought," between facts and values, between supposedly objective, declarative statements about the world and prescriptive evaluations of states of the

Mark Blaug, The Distinction Between Positive and Normative Economics, from *The Methodology of Economics: or, How Economists Explain* (Cambridge: Cambridge University Press, 1980), pp. 112–134. Copyright © 1980 Cambridge University Press. Reprinted with permission.

world. Positive economics was now said to be about facts and normative economics about values. . . .

It was David Hume in his *Treatise of Human Nature* who long ago laid down the proposition that "one cannot deduce ought from is," that purely factual, descriptive statements by themselves can only entail or imply other factual, descriptive statements and never norms, ethical pronouncements, or prescriptions to do something. This proposition has been aptly labeled "Hume's guillotine," implying as it does a watertight logical distinction between the realm of facts and the realm of values [see figure below].

Hume's Guillotine: Equivalent Antonyms	
positive	normative
is	ought
facts	values
objective	subjective
descriptive	prescriptive
science	art
true/false	good/bad

But how do we tell whether a given utterance is an is-statement or an ought-statement? It is clearly not to be decided by whether the sentence containing the statement is or is not grammatically formulated in the indicative mood, because there are sentences in the indicative mood, like "murder is a sin," which are thinly disguised ought-statements dressed up as is-statements. Nor is it decided by the fact that people agree more readily to is-statements than to ought-statements, since it is easy to see that there is far less agreement, say, about the factual proposition that the universe originated without supernatural intervention in a big bang eons ago than about the normative proposition that, say, we should not eat babies. An is-statement is simply one that is either materially true or false: it asserts something about the state of the world—that it is such and such, and not otherwise—and we can employ interpersonally testable methods to discover whether it is true or false. An ought-statement expresses an evaluation of the state of the world—it approves or disapproves, it praises or condemns, it extols or deplores—and we can only employ arguments to persuade others to accept it.

Surely, it will be objected, the normative proposition that we should not eat babies can likewise be tested by interpersonally testable methods, say, by a political referendum? But all that a political referendum can establish is that all of us agree that eating babies is wrong; it cannot establish that it is wrong. But it will again be objected, this is just as true of every interpersonally testable verification or falsification of an is-statement. Ultimately, a factual, descriptive is-statement is held to be true because we have agreed among ourselves to abide by certain "scientific" rules that instruct us to regard that statement as true, although it may in fact be false. . . . Moral judgments are usually defined as prescriptions enjoining a certain kind of behavior, which everyone is supposed to comply with in the same circumstances. But are assertions about facts not exactly the same kind of judgments, enjoining certain kinds of attitude rather than certain kinds of behavior? . . .

Methodological Judgments versus Value Judgments

Nagel seeks to protect Hume's guillotine against precisely this sort of objection by drawing a distinction in social science between two types of value judgments—characterizing value judgments and appraising value judgments. Characterizing value judgments involve the choice of subject matter to be investigated, the mode of investigation to be followed and the criteria for judging the validity of the findings, such as adherence to the canons of formal logic, the selection of data in terms of reliability, explicit prior decisions about levels of statistical significance, et cetera; in short, everything that we have earlier called methodological judgments. Appraising value judgments, on the other hand, refer to evaluative assertions about states of the world, including the desirability of certain kinds of human behavior and the social outcomes that are produced by that behavior; thus, all statements of the "good society" are appraising value judgments. Science as a social enterprise cannot function without methodological judgments, but it can free itself, at least in principle, Nagel contends, of any commitment to appraising or normative value judgments.

At a deep philosophical level, this distinction is perhaps misleading. Ultimately, we cannot escape the fact that all nontautological propositions rest for their acceptance on the willingness to abide by certain rules of the game, that is, on judgments that we players have collectively adopted. An argument about facts may appear to be resolvable by a compelling appeal to so-called objective evidence, whereas an argument about moral values can only be resolved by a hortatory appeal to the emotions, but at bottom both arguments rest on certain definite techniques of persuasion, which in turn depend for their effectiveness on shared values of one kind or another. But at the working level of a scientific inquiry, Nagel's distinction between methodological and normative judgments is nevertheless real and significant.

Every economist recognizes that there is a world of difference between the assertion that there is a Phillips curve, a definite functional relationship between the level of unemployment and the rate of change in wages or prices, and the assertion that unemployment is so deplorable that we ought to be willing to suffer any degree of inflation to get rid of it. When an economist says that every individual should be allowed to spend his income as he or she likes, or that no able person is entitled to the support of others, or that governments must offer relief to the victims of inexorable economic forces, it is not difficult to see that he or she is making normative value judgments. There are long-established, well-tried methods for reconciling different methodological judgments. There are no such methods for reconciling different normative value judgments—other than political elections and shooting it out at the barricades. It is this contrast in the methods of arbitrating disagreements that gives relevance to Nagel's distinction.

We have overstated the case in suggesting that normative judgments are the sort of judgments that are never amenable to rational discussion designed to reconcile whatever differences there are between people. Even if Hume is right in denying that "ought" can be logically deduced from "is," and of course "is" from "ought," there is no denying that "oughts" are powerfully influenced by "ises" and that the values we hold almost always depend on a whole series of factual beliefs. This indicates how a rational debate on a disputed value judgment can proceed: we pose alternative factual circumstances and ask, should these circumstances prevail, would you be willing to abandon your judgment? A famous and obvious example is the widespread value judgment that economic growth, as measured by real national income, is always desirable; but is it, we might ask, even if it made the bottom quartile, decile, quintile of the size distribution of personal incomes absolutely worse off ? Another example is the frequently expressed value judgment that capital punishment is always wrong. But if there were incontrovertible evidence that capital punishment deterred potential murderers, we might ask, would you still adhere to your original opinion? And so on.

In thinking along these lines, we are led to a distinction between "basic" and "nonbasic" value judgments, or what I would prefer to call pure and impure value judgments. . . . So long as a value judgment is nonbasic or impure, a debate on value judgments can take the form of an appeal to facts, and that is all to the good because our standard methods for settling disputes about facts are less divisive than those for settling disputes about values. It is only when we finally distill a pure value judgment— think of a strict pacifist opposition to any and all wars, or the assertion that "I value this for its own end"—that we have exhausted all the possibilities of rational analysis and discussion. There is hardly any doubt that most value judgments that are expressed on social questions are highly impure and hence perfectly amenable to the attempt to influence values by persuading the parties holding them that the facts are other than what they believe them to be. . . .

The Methodology of Positive Economics

MILTON FRIEDMAN

The Relation between Positive and Normative Economics

Confusion between positive and normative economics is to some extent inevitable. The subject matter of economics is regarded by almost everyone as vitally important to himself and within the range of his own experience and competence; it is the source of continuous and extensive controversy and the occasion for frequent legislation. Self-proclaimed "experts" speak with many voices and can hardly all be regarded as disinterested; in any event, on questions that matter so much, "expert" opinion could hardly be accepted solely on faith even if the "experts" were nearly unanimous and clearly disinterested. The conclusions of positive economics seem to be, and are, immediately relevant to important normative problems, to questions of what ought to be

Milton Friedman, The Methodology of Positive Economics, from *Essays in Positive Economics* (Chicago: University of Chicago Press, 1953), pp. 3–43. Copyright © 1953 University of Chicago Press. Reprinted with permission.

done and how any given goal can be attained. Laymen and experts alike are inevitably tempted to shape positive conclusions to fit strongly held normative preconceptions and to reject positive conclusions if their normative implications—or what are said to be their normative implications—are unpalatable.

Positive economics is in principle independent of any particular ethical position or normative judgements. As Keynes says, it deals with "what is," not with "what ought to be." Its task is to provide a system of generalizations that can be used to make correct predictions about the consequences of any change in circumstances. Its performance is to be judged by the precision, scope, and conformity with experience of the predictions it yields. In short, positive economics is, or can be, an "objective" science, in precisely the same sense as any of the physical sciences. Of course, the fact that economics deals with the interrelations of human beings, and that the investigator is himself part of the subject-matter being investigated in a more intimate sense than in the physical sciences, raises special difficulties in achieving objectivity at the same time that it provides the social scientist with a class of data not available to the physical scientist. But neither the one nor the other is, in my view, a fundamental distinction between the two groups of sciences.

Normative economics and the art of economics, on the other hand, cannot be independent of positive economics. Any policy conclusion necessarily rests on a prediction about the consequences of doing one thing rather than another, a prediction that must be based—implicitly or explicitly—on positive economics. There is not, of course, a one-to-one relation between policy conclusions and the conclusions of positive economics; if there were, there would be no separate normative science. Two individuals may agree on the consequences of a particular piece of legislation. One may regard them as desirable on balance and so favor the legislation; the other, as undesirable and so oppose the legislation.

I venture the judgement, however, that currently in the Western world, and especially in the United States, differences about economic policy among disinterested citizens derive predominantly from different predictions about the economic consequences of taking action—differences that in principle can be eliminated by the progress of positive economics—rather than from fundamental differences in basic values, differences about which men can ultimately only fight. An obvious and not unimportant example is minimum-wage legislation. Underneath the welter of arguments offered for and against such legislation there is an underlying consensus on the objective of achieving a "living wage" for all, to use the ambiguous phrase so common in such discussions. The difference of opinion is largely grounded on an implicit or explicit difference in predictions about the efficacy of this particular means in furthering the agreed-on end. Proponents believe (predict) that legal minimum wages diminish poverty by raising the wages of those receiving less than the minimum wage as well as of some receiving more than the minimum wage without any counterbalancing increase in the number of people entirely unemployed or employed less advantageously than they otherwise would be. Opponents believe (predict) that legal minimum wages increase poverty by increasing the number of people who are unemployed or employed less advantageously and that this more than offsets any favorable effect on the wages of those who remain employed. Agreement about the economic consequences of the legislation might not produce complete agreement about its desirability, for differences might still remain about its political or social consequences; but, given agreement on objectives, it would certainly go a long way towards producing consensus.

Closely related differences in positive analysis underlie divergent views about the appropriate role and place of trade unions and the desirability of direct price and wage controls and of tariffs. Different predictions about the importance of so-called "economies of scale" account very largely for divergent views about the desirability or necessity of detailed government regulation of industry and even of socialism rather than private enterprise. And this list could be extended indefinitely. Of course, my judgement that the major differences about economic policy in the Western world are of this kind is itself a "positive" statement to be accepted or rejected on the basis of empirical evidence.

If this judgement is valid, it means that a consensus on "correct" economic policy depends much less on the progress of normative economics proper than on the progress of a positive economics yielding conclusions that are, and deserve to be, widely accepted. It means also that a major reason for distinguishing positive economics sharply from normative economics is precisely the contribution that can thereby be made to agreement about policy.

Positive Economics

The ultimate goal of a positive science is the development of a "theory" or "hypothesis" that yields valid and meaningful (i.e., not truistic) predictions about phenomena not yet observed. Such a theory is, in general, a complex intermixture of two elements. In part, it is a "language" designed to promote "systematic and organized methods of reasoning." In part, it is a body of substantive hypotheses designed to abstract essential features of complex reality. . . .

Viewed as a body of substantive hypotheses, theory is to be judged by its predictive power for the class of phenomena which it is intended to "explain." Only factual evidence can show whether it is "right" or "wrong" or, better, tentatively "accepted" as valid or "rejected." As I shall argue at greater length below, the only relevant test of the validity of a hypothesis is comparison of its predictions with experience. The hypothesis is rejected if its predictions are contradicted ("frequently" or more often than predictions from an alternative hypothesis); it is accepted if its predictions are not contradicted; great confidence is attached to it if it has survived many opportunities for contradiction. Factual evidence can never "prove" a hypothesis; it can only fail to disprove it, which is what we generally mean when we say, somewhat inexactly, that the hypothesis has been "confirmed" by experience.

To avoid confusion, it should perhaps be noted explicitly that the "predictions" by which the validity of a hypothesis is tested need not be about phenomena that have not yet occurred, that is, need not be forecasts of future events; they may be about phenomena that have occurred but observations on which have not yet been made or are not known to the person making the prediction. For example, a hypothesis may imply that such and such must have happened in 1906, given some other known circumstances. If a search of the records reveals that such and such did happen, the prediction is confirmed; if it reveals that such and such did not happen, the prediction is contradicted. . . .

In so far as a theory can be said to have "assumptions" at all, and in so far as their "realism" can be judged independently of the validity of predictions, the relation between the significance of a theory and the "realism" of its "assumptions" is almost the opposite of that suggested by the view under criticism. Truly important and significant hypotheses will be found to have "assumptions" that are widely inaccurate descriptive representations of reality, and, in general, the more significant the theory, the more unrealistic the assumptions (in this sense). The reason is simple. A hypothesis is important if it "explains" much by little, that is, if it abstracts the common and crucial elements from the mass of complex and detailed circumstances surrounding the phenomena to be explained and permits valid predictions on the basis of them alone. To be important, therefore, a hypothesis must be descriptively false in its assumptions; it takes account of, and accounts for, none of the many other attendant circumstances, since its very success shows them to be irrelevant for the phenomena to be explained.

To put this point less paradoxically, the relevant question to ask about the "assumptions" of a theory is not whether they are descriptively "realistic," for they never are, but whether they are sufficiently good approximations for the purpose in hand. And this question can be answered only by seeing whether the theory works, which means whether it yields sufficiently accurate predictions. The two supposedly independent tests thus reduce to one test.

The theory of monopolistic and imperfect competition is one example of the neglect in economic theory of these propositions. The development of this analysis was explicitly motivated, and its wide acceptance and approval largely explained, by the belief that the assumptions of "perfect competition" or "perfect monopoly" said to underlie neoclassical economic theory are a false image of reality. And this belief was itself based almost entirely on the directly perceived descriptive inaccuracy of the assumptions rather than on any recognized contradiction of predictions derived from neoclassical economic theory. . . .

Can a Hypothesis Be Tested by the Realism of Its Assumptions?

We may start with a simple physical example, the law of falling bodies. It is an accepted hypothesis that the acceleration of a body dropped in a vacuum is a constant— g, or approximately 32 feet per second per second on the earth—and is independent of the shape of the body, the manner of dropping it, etc. This implies that the distance travelled by a falling body in any specified time is given by the formula $s = \frac{1}{2} gt^2$, where s is the distance travelled in feet and t is time in seconds. The application of this formula to a compact ball dropped from the roof of a building is equivalent to saying that a ball so dropped behaves *as if* it were falling in a vacuum. Testing this hypothesis by its assumptions presumably means measuring the actual air pressure and deciding whether it is close enough to zero. At sea-level the air pressure is about 15 pounds per square inch. Is 15 sufficiently close to zero for the difference to be judged insignificant? Apparently it is, since the actual time taken by a compact ball to fall from the roof of a building to the ground is very close to the time given by the formula. Suppose, however, that a feather is dropped instead of a compact ball. The formula then gives wildly inaccurate results. Apparently, 15 pounds per square inch is significantly different from zero for a feather but not for a ball. Or, again, suppose the formula is applied to a ball dropped from an airplane at an altitude of 30,000 feet. The air pressure at this altitude is decidedly less than 15 pounds per square inch. Yet, the actual time of fall from 30,000 feet to 20,000 feet, at which point the air pressure is still much less than at sea-level, will differ noticeably from the time predicted by the formula—much more noticeably than the time taken by a compact ball to fall from the roof of a building to the ground. According to the formula, the velocity of the ball should be gt and should therefore increase steadily. In fact, a ball dropped at 30,000 feet will reach its top velocity well before it hits the ground. And similarly with other implications of the formula.

The initial question of whether 15 is sufficiently close to zero for the difference to be judged insignificant is clearly a foolish question by itself. Fifteen pounds per square inch is 2,160 pounds per square foot, or 0.0075 tons per square inch. There is no possible basis for calling these numbers "small" or "large" without some external standard of comparison. And the only relevant standard of comparison is the air pressure for which the formula does or does not work under a given set of circumstances. But this raises the same problem at a second level. What is the meaning of "does or does not work"? Even if we could eliminate errors of measurement, the measured time of fall would seldom if ever be precisely equal to the computed time of fall. How large must the difference between the two be to justify saying that the theory "does not work"? Here there are two important external standards of comparison. One is the accuracy achievable by an alternative theory with which this theory is being compared and which is equally acceptable on all other grounds. The other arises when there exists a theory that is known to yield better predictions but only at a greater cost. The gains from greater accuracy, which depend on the purpose in mind, must then be balanced against the costs of achieving it. . . .

A largely parallel example involving human behavior has been used elsewhere by Savage and me. Consider the problem of predicting the shots made by an expert billiard player. It seems not at all unreasonable that excellent predictions would be yielded by the hypothesis that the billiard player made his shots *as if* he knew the complicated mathematical formulas that would give the optimum directions of travel, could estimate accurately by eye the angles, etc., describing the location of the balls, could make lightning calculations from the formulas, and could then make the balls travel in the direction indicated by the formulas. Our confidence in this hypothesis is not based on the belief that billiard players, even expert ones, can or do go through the process described; it derives rather from the belief that, unless in some way or other they were capable of reaching essentially the same result, they would not in fact be *expert* billiard players.

It is only a short step from these examples to the economic hypothesis that under a wide range of circumstances individual firms behave as if they were seeking rationally to maximize their expected returns (generally if misleadingly called "profits") and had full knowledge of the data needed to succeed in this attempt; *as if,* that is, they knew the relevant cost and demand functions, calculated marginal cost and marginal revenue from all actions open to them, and pushed each line of action to the point at which the relevant marginal cost and marginal revenue were equal. Now, of course, businessmen do not actually and literally solve the system of simultaneous equations in terms of which the mathematical economist finds it convenient to express this hypothesis, any more than leaves or billiard players explicitly go through complicated mathematical calculations or falling bodies decide to create a vacuum. The billiard player, if asked how he decides where to hit the ball, may say that he "just figures it out" but then also rubs a rabbit's foot just to make sure; and the businessman may well say that he prices at average cost, with of course some minor deviations when the market makes it necessary. The one statement is about as helpful as the other, and neither is a relevant test of the associated hypothesis.

Confidence in the maximization-of-returns hypothesis is justified by evidence of a very different character. This evidence is in part similar to that adduced on behalf of the billiard-player hypothesis—unless the behavior of businessmen in some way or other approximated behavior consistent with the maximization of returns, it seems unlikely that they would remain in business for long. Let the apparent immediate determinant of business behavior be anything at all—habitual reaction, random chance, or whatnot. Whenever this determinant happens to lead to behavior consistent with rational informed maximization of returns, the business will prosper and acquire resources with which to expand; whenever it does not, the business will tend to lose resources and can be kept in existence only by the addition of resources from outside. The process of "natural selection" thus helps to validate the hypothesis—or, rather, given natural selection, acceptance of the hypothesis can be based largely on the judgement that it summarizes appropriately the conditions for survival.

An even more important body of evidence for the maximization-of-returns hypothesis is experience from countless applications of the hypothesis to specific problems and the repeated failure of its implications to be contradicted. This evidence is extremely hard to document; it is scattered in numerous memorandums, articles, and monographs concerned primarily with specific concrete problems rather than with submitting the hypothesis to test. Yet the continued use and acceptance of the hypothesis over a long period, and the failure of any coherent, self-consistent alternative to be developed and be widely accepted, is strong indirect testimony to its worth. The evidence *for* a hypothesis always consists of its repeated failure to be contradicted, continues to accumulate so long as the hypothesis is used, and by its very nature is difficult to document at all comprehensively. It tends to become part of the tradition and folklore of a science revealed in the tenacity with which hypotheses are held rather than in any textbook list of instances in which the hypothesis has failed to be contradicted. . . .

Conclusion

Economics as a positive science is a body of tentatively accepted generalizations about economic phenomena that can be used to predict the consequences of changes in circumstances. Progress in expanding this body of generalizations, strengthening our confidence in their validity, and improving the accuracy of the predictions they yield is hindered not only by the limitations of human ability that impede all search for knowledge but also by obstacles that are especially important for the social sciences in general and economics in particular, though by no means peculiar to them. Familiarity with the subject-matter of economics breeds contempt for special knowledge about it. The importance of its subject matter to everyday life and to major issues of public policy impedes objectivity and promotes confusion between scientific analysis and normative judgement. The necessity of relying on uncontrolled experience rather than on controlled experience makes it difficult to produce dramatic and clear-cut evidence to justify the acceptance of tentative hypotheses. Reliance on uncontrolled experience does not affect the fundamental methodological principle that a hypothesis can be tested only by the conformity of its implications or predictions with observable phenomena; but it does render the task of testing hypotheses more difficult and gives greater scope for confusion about the methodological principles involved. More than other scientists, social scientists need to be self-conscious about their methodology. . . .

One confusion that has been particularly rife and has done much damage is confusion about the role of "assumptions" in economic analysis. A meaningful scientific hypothesis or theory typically asserts that certain forces are, and other forces are not, important in understanding a particular class of phenomena. It is frequently convenient to present such a hypothesis by stating that the phenomena it is desired to predict behave in the world of observation *as if* they occurred in a hypothetical and highly simplified world containing only the forces that the hypothesis asserts to be important. In general, there is more than one way to formulate such a description—more than one set of "assumptions" in terms of which the theory can be presented. The choice among such alternative assumptions is made on the grounds of the resulting economy, clarity, and precision in presenting the hypothesis; their capacity to bring indirect evidence to bear on the validity of the hypothesis by suggesting some of its implications that can be readily checked with observation or by bringing out its connection with other hypotheses dealing with related phenomena; and similar considerations.

Such a theory cannot be tested by comparing its "assumptions" directly with "reality." Indeed, there is no meaningful way in which this can be done. Complete "realism" is clearly unattainable, and the question whether a theory is realistic "enough" can be settled only by seeing whether it yields predictions that are good enough for the purpose in hand or that are better than predictions from alternative theories. Yet the belief that a theory can be tested by the realism of its assumptions independently of the accuracy of its predictions is widespread and the source of much of the perennial criticism of economic theory as unrealistic. Such criticism is largely irrelevant, and, in consequence, most attempts to reform economic theory that it has stimulated have been unsuccessful. . . .

NOTES AND QUESTIONS

1. Each of Becker's three central assumptions—maximization, equilibrium, and stable preferences—is open to criticism as a description of empirical reality. Market equilibrium, for

instance, is ordinarily taken to imply that individuals behave atomistically—that is, that they take the constraints they face and the actions of others as given. This rules out strategic behavior intended to influence other actors or the state. Much of the modern microeconomics literature, however, is devoted to studying such behavior. Chapter 4 of this reader discusses the application of this literature to law. The assumption of individual maximization is more universally accepted by the economics profession, but it has also been questioned. Chapter 7 of this reader discusses criticisms of this assumption and offers an alternative account of human decisionmaking premised on imperfect rationality.

2. As Becker points out, the assumption of stable preferences stands on a different footing than his two other fundamental assumptions. Specifically, he sees stable preferences less as a description of reality than as a necessary condition for having any predictive theory at all. In his view, if preferences can change arbitrarily, then it is impossible to rule out any human behavior whatsoever, making the theory empirically untestable and hence empty. Still, it is apparent that preferences do change in the real world. While economists have tended to leave the study of preference formation to other social scientists, there is some economic literature on the subject, particularly with reference to phenomena such as advertising, addiction, and acquired tastes. Does the existence of such phenomena undermine the relevance of the model of rational choice? See the notes to chapter 8, infra, as well as George Stigler and Gary Becker, De Gustibus Non Es Disputandum, 64 *American Economic Review* 76 (1977) [arguing that acquired tastes can be seen as a deliberate investment in the productivity of future consumption and addiction as a deliberate decision to favor present pleasures over future pains.]

3. Some economists have disagreed with Becker's conclusion that the model of rational choice adequately explains all purposive human action. Amartya Sen, for instance, has argued that many important decisions are better understood as a form of commitment, that is, as a choice to do something even though it works against one's interests or preferences. Examples of commitment might include paying taxes even when one is unlikely to be audited, voting even though there is no likelihood of influencing the outcome of the election, or keeping a deathbed promise. While it is theoretically possible to interpret commitment as a special type of maximization, Sen argues that such an interpretation fails to account for the emotional salience of such choices or for the special problems of self-control and temptation they raise. See Amartya Sen, Rational Fools: A Critique of the Behavioral Foundations of Economic Theory, 6 *Philosophy and Public Affairs* 317 (1977). Do you agree with this criticism? Are there goals that do not fit within the model of rational choice, and can you think of any that are important in the design of legal institutions?

4. The model of rational choice has also been subject to the criticism that certain alternatives are *incommensurable*—that they are so different in kind that rational comparisons (and hence tradeoffs) among them are not possible. Examples might include material goods versus health and safety, the rights of the accused versus the interests of potential crime victims, or distributional equity versus economic efficiency. In this view, decisions among such goods when they are in conflict can be made only through the exercise of private or collective will. Accordingly, the best we can do is to ensure that decisions are made through an open process of conversation and deliberation. For an elaboration of this critique, see Elizabeth Anderson, *Value in Ethics and Economics* (Cambridge, Mass.: Harvard University Press, 1993); for a discussion of applications to law and legal institutions, see Cass Sunstein, Incommensurability and Valuation in Law, 92 *Michigan Law Review* 779 (1994). A related critique is that even if it is possible to make such choices rationally, the psychological and social trauma we suffer from doing so is so great that we are better off referring such decisions to an arational but impersonal process such as a lottery. See Guido Calabresi and Philip Bobbitt, *Tragic Choices* (New York: Norton, 1978).

5. Coleman's analysis focuses on the differences rather than the similarities between classical utilitarianism and efficiency. It remains the case, however, that efficiency, Pareto

superiority, and the Kaldor-Hicks criterion all derive from the utilitarian moral tradition and in important respects reflect its world view. For instance, both efficiency and utilitarianism incorporate a value that might be called normative individualism: the idea that the interest of society is reducible to the interests of its individual members. Similarly, both utilitarians and advocates of economic efficiency tend to support the value of consumer sovereignty—the idea that individuals are ordinarily the best judges of their own interests. Accordingly, many of the standard philosophical criticisms of classical utilitarianism will also apply to modern normative economics. Some of these criticisms are presented in chapter 8, infra; for a survey of others, see Amartya Sen and Bernard Williams, eds., *Utilitarianism and Beyond* (New York: Cambridge University Press, 1982). For a discussion of the historical development of Pareto efficiency as a criterion distinct from classical utilitarianism, see Robert Cooter and Peter Rappoport, Were the Ordinalists Wrong About Welfare Economics? 22 *Journal of Economic Literature* 507 (1984). For a more detailed exposition of the efficiency concept, see generally Harvey Rosen, *Public Finance,* 3d ed. (Homewood, Ill.: Richard D. Irwin, Inc. 1992), chap. 4 ("Tools of Normative Analysis").

6. Schelling's pragmatic defense of the Kaldor-Hicks criterion would likely be endorsed by most contemporary economists. His argument does assume, however, that there exists some way to share the gains from an improvement in efficiency among the original winners and losers. In many instances, this will be straightforward (for instance, the gains from efficient terms in a labor contract can be shared by adjusting the contract wage); but in many instances such side payments will be infeasible, especially when there are many gainers and losers from a proposed change in public policy. In such cases, there will be a tradeoff between efficiency and fair distribution; and economics offers no way to resolve the tradeoff. Instead, the answer must come from societal values. For a discussion of the assumptions of efficiency analysis, with particular attention to the practical problems of side payments, see A. Mitchell Polinsky, Economic Analysis as a Potentially Defective Product: A Buyer's Guide to Posner's "Economic Analysis of Law," 87 *Harvard Law Review* 1655 (1974). For a general discussion of the tension between efficiency and distributional equity, see Arthur Okun, *Equality and Efficiency: The Big Tradeoff* (Washington, D.C.: The Brookings Institution, 1975). In your view, is it reasonable to focus on efficiency alone when analyzing the rules of private law? For arguments that it is, see Richard Craswell, Passing on the Costs of Legal Rules: Efficiency and Distribution in Buyer-Seller Relationships, 43 *Stanford Law Review* 361 (1991); Louis Kaplow and Steven Shavell, Why the Legal System Is Less Efficient than the Income Tax in Redistributing Income, 23 *Journal of Legal Studies* 667 (1994).

7. Consider Schelling's proposal, advocated by a variety of economists across the political spectrum, that government support for the needy take the form of cash payments, rather than allotments of particular goods such as food or housing or vouchers tied to their purchase. Schelling argues that a system of exclusive cash payments would be more efficient, given consumers' natural incentives to spend cash on those goods that best satisfy their needs. An obvious counterargument, of course, is paternalism—that the needy do not know their own interests. Can you think of any additional arguments against Schelling's proposal that do not rely on paternalism? See, for example, Steven Kelman, A Case for In-kind Transfers, 2 *Economics and Philosophy* 55 (1986).

8. The most controversial aspect of Friedman's essay is his claim that an economic model's assumptions are entirely irrelevant. His defense of unrealistic assumptions, however, can be criticized as being both too broad and too narrow. It is too broad in that he overstates the precision with which theoretical predictions can be empirically tested. In actual practice, we often lack the evidence to decide definitively between competing models on the basis of past predictive success. If we wish our future predictions to be as reliable as possible, therefore, we should not ignore any information we may have about the empirical realism of our assumptions.

Conversely, Friedman's conception of the purposes of modeling is too narrow. Economists and lawyers are not merely interested in prediction. They also construct theoretical models for pedagogical or mnemonic purposes, to catalog the existing stock of knowledge, to organize research programs for the future, and to establish a benchmark for normative discussion. They may make unrealistic assumptions for expositional convenience, to abstract from aspects of a problem that are not under consideration at the moment in order better to focus on the one that is, or as part of a deliberate reductio ad absurdum. Whether a particular model is a good one, therefore, depends on which of these purposes is the goal at hand and how well the model promotes them. For a theoretical survey of such purposes, see Allan Gibbard and Hal Varian, Economic Models, 75 *Journal of Philosophy* 664 (1978). For a more practically oriented or "cookbook" discussion, see Edith Stokey and Richard Zeckhauser, *A Primer for Policy Analysis* (New York: W. W. Norton, 1978), chap. 2 ("Models: A General Discussion").

9. An alternate justification of economic methodology can be found in Donald McCloskey, *The Rhetoric of Economics* (Madison: University of Wisconsin Press, 1985) [see especially chap. 1, "The Poverty of Economic Modernism"]. McCloskey, a critic of what he refers to as "modernist" scientific method, contends that economics is ultimately an interpretive discipline, employing methods of rhetoric and persuasion similar to the humanities. In this view, economic models are essentially extended metaphors; like metaphors, they help bring out essential truths about aspects of the world that otherwise might escape our notice. Accordingly, they should be judged by the same criteria as other literary devices and figures of speech. While McCloskey defends much of the actual practice of economic modeling, he argues that economists should be receptive to verbal models as well as quantitative ones, and should consider a wider variety of empirical evidence, including opinion surveys and introspection.

10. Do you agree with Friedman that contemporary public controversies derive more from people's holding different views about the way the world works than from fundamental difference in values? If so, what does this imply for the professional tasks of economists and lawyers? Do you think that progress on understanding positive issues will help lead to social consensus on normative ones, or does positive analysis merely divert attention away from normative conflict? See the critiques in chapter 8, infra.

Two Competing Economic Models of Law

This chapter introduces two competing approaches to the economic analysis of law. For most applications in law and economics, one finds differing and sometimes contradictory arguments put forward in the scholarly literature. These arguments often imply different policy conclusions, making it difficult for students of the subject to form a coherent view of its overall approach. The apparent contradiction is resolved, however, once we recognize that economics has traditionally accommodated two competing schools of thought regarding the merits of public control over resource allocation and that many contributors to law and economics draw on both traditions in their writings. One of these paradigms takes the position that in the absence of legal constraints, rational actors will find it in their self-interest to find and exploit all opportunities for mutually beneficial exchange. I refer to this view as the *model of cooperation.* The other paradigm starts with the observation that self-interested individuals will act in socially beneficial ways only in ideal circumstances; thus leaving substantial scope for legal rules and institutions to promote efficient exchange. I refer to this view as the *model of market failure.* The readings in this chapter identify and compare the basic claims and arguments of these two paradigms.

The branch of economics dealing with efficient resource allocation is known as *welfare economics,* and dates back to Adam Smith's famous argument that the competitive market acts as an invisible hand allocating all resources to their highest and best use. The "invisible hand" metaphor, however, assumes that resources are allocated in markets where everyone has full information and an equal opportunity to bid and where no one acts strategically in an attempt to influence the outcome. If these conditions are not satisfied, there is no guarantee of efficiency; instead, the result is what economists call market failure. For the purposes of law and economics, the most important category of market failure is *externality:* any time an individual can obtain goods without bidding for them explicitly or implicitly on the market, he has an incentive to consume excessively. This occurs when goods are allocated entirely outside the market, as in the case of most environmental resources; it can also occur when extramarket activities spill over to influence the value of marketable resources, as in the case of many traditional common-law nuisances. Economists' usual policy recommendation for dealing with externality is to create a quasi-market through government intervention. The externality should be *internalized,* that is, brought within the scope of market calculations, by taxing or subsidizing its producer in an amount equal to the economic value of any spillover. Such state taxes or subsidies are commonly referred to as *Pigouvian,* in honor of their most influential proponent, the twentieth-century English economist A. C. Pigou.

The readings in this chapter by Guido Calabresi and Robert Cooter apply this model to the economic analysis of law; both argue, in the Pigouvian tradition, that state intervention is needed to promote an efficient allocation of resources. Calabresi, in what is generally regarded as one of the two seminal articles in the field of law and economics (the other being Coase's "The Problem of Social Cost"), suggests that the law of torts should be understood as a regulatory regime for the control of externalities. The legal system manages to internalize costs not by using taxes or subsidies, however, but through damage awards imposed by common-law courts—what might be called a system of Pigouvian liability. Not only is this the major purpose of the tort system, in Calabresi's view, but it helps explain many of the common law's traditional limits on liability as well as the law's move away from negligence and toward strict liability in the middle decades of the twentieth century.

Cooter's article extends this argument to include liability based on fault, generalizing the concept of externality to what he calls the *model of precaution*. He suggests that legal doctrines across a variety of fields should be understood as regulating decisions about whether and how much to take care. For instance, a motorist hurrying to an important appointment must decide how fast and how safely to drive; a commercial supplier of goods must decide how large a stock of goods to keep in inventory. Because the benefits of precaution are often external, Pigouvian liability may be needed to encourage an efficient amount of it. The efficient rule of law, however, is complicated by the fact that both parties to an interaction usually can take precautions to lessen the severity and probability of a undesirable outcome; for example, both motorists and pedestrians have some influence on the risk of traffic accidents. As a result, it is necessary to provide incentives to both sides or, in Cooter's words, "double responsibility on the margin." He suggests that apparently diverse rules in tort, property, and contract are designed with the common purpose of providing such double responsibility.

The model of cooperation is illustrated here by two articles by its major exponent, Ronald Coase. Though many surveys of the economic analysis of law begin with Coase's work, his ideas are better presented after the model of market failure, because properly understood, they are intended as a critique of the Pigouvian tradition. The essence of this critique is that it is a fallacy to regard government intervention as the presumptive solution to imperfections in the private market. Instead, government, the market, and the private business firm should be seen as alternate institutional arrangements for dealing with the problem of economic organization. In Coase's view, the choice among these various institutions should come down to their comparative abilities to conserve on the organizational and administrative costs of exchange—what he called *transaction costs.*

The concept of transaction costs, for which Coase was awarded the Nobel Prize in Economics in 1991, is introduced here by his classic article, "The Nature of the Firm." In this article, Coase argues that business firms exist in order to conserve on the costs of transacting through the market, which include, inter alia, the expense of keeping in touch with a variety of potential suppliers and customers and of drafting and enforcing formal contracts. Because administering a business enterprise through a system of centralized command entails its own transaction costs, however, the optimal size and scope of the firm is determined by the rate at which the costs of administrative coordination grow relative to the benefits of central planning as the enterprise increases in scale.

In "The Problem of Social Cost," Coase then applies this general theory of economic organization to the specific problem of externality, using the legal category of nuisance as an example. In Coase's framework, externality is merely one of the transaction costs of using the market. The fact that markets suffer from such costs, however, is not a sufficient argument for abandoning them in favor of a cumbersome government bureaucracy with inefficiencies of its own. Moreover, private individuals may have better incentives to devise a solution to the externality problem, since they profit from finding an efficient outcome while bureaucrats do not. In any event, public regulatory agencies have no inherent advantage over markets in dealing with externalities. Where transaction costs are least, rather, is an empirical question.

Coase develops this argument through a rhetorical device: a reductio ad absurdum that has come to be called the *Coase theorem.* Suppose, he hypothesizes, that there were no transaction costs associated with private exchange. In that case the victim of a nuisance would be willing to pay its perpetrator to have it removed. If the amount the victim were willing to pay to have the nuisance removed exceeded the perpetrator's price to remove it, the two sides would enter into a bargain to do just that. If, conversely, the victim's reservation price were less than the perpetrator's, the nuisance would remain, but in either case the outcome would be efficient. In effect, the opportunity for bargaining would work to internalize the externality. Accordingly, Coase argues that in the absence of transaction costs it would not matter whether or not the government imposed a Pigouvian tax, or whether or not courts imposed liability. Whatever the rule of law, the parties would have the same incentive to bargain their way to an efficient outcome and the allocation of resources would be the same.

The lessons of the Coase theorem have been widely misunderstood. Coase argues against state intervention as the preferred response to market failure, it is true, and he does suggest that the cost of the former is often greater than the costs of the latter, but he does not claim that market transaction costs are zero or that laissez-faire is always justified. Rather, his criticism is that it was misleading for the Pigouvians to argue for state intervention by stressing the transaction costs of the market while ignoring the transaction costs of government institutions. His fundamental thesis is that comparative transaction costs are more important than the presence of an externality and that the efficient allocation of resources ultimately will depend more on such transaction costs than on the nominal rules of the legal regime. The excerpt from Harold Demsetz's article "When Does the Rule of Liability Matter?" explains and develops this thesis in greater detail, concluding that legal rules matter most when the transaction costs of private exchange are high.

2.1   THE MODEL OF MARKET FAILURE

Some Thoughts on Risk Distribution and the Law of Torts

GUIDO CALABRESI

In their excellent new casebook on torts, Professors Gregory and Kalven state that "the central policy issue in tort law is whether the principal criterion of liability is to be based on individual fault or on a wide distribution of risk and loss." And so, I suppose, it is. But to say "risk distribution" is really to say very little. Indeed, under the heading "risk distribution" have come the most diverse schemes for allocating losses, schemes that have almost nothing to do with each other.

The reason for the difficulty is, presumably, that while many people have talked about "risk distribution" and some have even used it as a basis for proposed modifications in the law of torts, few have in recent years attempted to examine in any depth just what it is they are striving for when they say "distribute losses." They could mean one of three things. Do they wish as broad a spreading of all losses, both interpersonally and intertemporally, as is possible? Or do they want the burden of losses to be borne by those classes of people "most able" to pay? Or do they seek something

Guido Calabresi, Some Thoughts on Risk Distribution and the Law of Torts, 70 *Yale Law Journal,* pp. 499–553. Copyright © 1961 Yale Law Journal Company and Fred B. Rothman & Company. Reprinted with permission.

entirely different—that those "enterprises" which give rise to a loss "should" bear the burden, whether or not this accomplishes the prior two aims ? The answer, I suppose, is that some times they mean each of these things, and at other times all of them. Unfortunately, these goals are not always consistent with each other. They are, moreover, supported by quite different ethical and economic postulates—postulates of quite varied acceptability. To decide when and how we wish to distribute losses we must, therefore, examine the theoretical justifications of each of these three positions. This article takes some first steps in that direction. . . .

Enterprise Liability—or the Allocation of Resources Justification

"Activities should bear the costs they engender"; "it is only fair that an industry should pay for the injuries it causes." "Enterprise liability"—the notion that losses should be borne by the doer, the enterprise, rather than distributed on the basis of fault—is usually explained in such terms. A statement of this kind is generally followed by an additional one which implies that the enterprise can pass the loss on to the consumers in price rises, and that therefore enterprise liability is really a form of "risk spreading." It is, of course, true that enterprise liability sometimes does spread losses; it is equally true, however, that sometimes it does not. . . . And since risk spreading is not always a valid justification for enterprise liability we are at the moment less concerned with the risk spreading potential of enterprise liability than with whether another, more general, justification exists for the "should" in the phrase "an enterprise should bear its costs."

The problem of this "should" and what it means is analogous to the problem of why workmen's compensation should be limited to injuries arising out of or in the course of employment, and why master-servant liability should be limited to those acts which are in some sense within the scope of employment. If the "should" were merely a way of saying, "because this is a handy way of spreading losses through the price mechanism to a broad group of people—the consumers," one would wonder why workmen's compensation or master-servant liability should be so limited. And, indeed, writers have long wondered why. Some have answered directly that there is no logical reason for limiting liability to injuries related to employment. Others have said about the same thing, but have masked their answer by stating that some "innate sense of fairness" justifies the limitations. What that "fairness" is, unfortunately, is never clearly explained.

But the "should" is used so often that one suspects it must have a more clearly defined justification than some vague sense of fairness. And indeed it does; though it is a justification that only some of us would accept, and which, strangely enough, has been all but ignored in tort law in recent years. That justification can be called the "allocation of resources" justification. At its base are certain fundamental ethical postulates. One of these, perhaps the most important, is that by and large people know what is best for themselves. If people want television sets, society should produce television sets; if they want licorice drops, then licorice drops should be made. And, the theory continues, in order for people to know what they really want they must know the relative costs of producing different goods. The function of prices is to reflect the actual costs of competing goods, and thus to enable the buyer to cast an informed vote in making his purchases.

An example may help clear the mind a bit. Assume two different societies, Athens and Sparta: in Sparta all accident costs are borne by the state and come out of general taxes; in Athens accident costs are in some way or another charged to the doer. C. J. Taney, a business man in Athens, has one car, but he wants to buy another. The cost of owning a second, used, car would come to about $200 a year, plus an addition to his insurance bill of another $200. The cost of train fares, the occasional taxis he would need to use to be as comfortable without the car, and other forms of entertainment which make up for the car, come to about $250. Contrasting the $400 additional car cost with the $250 expense of riding in trains and taxis, he decides to forgo the car. If C. J. lived in Sparta, on the other hand, he would have to pay a certain sum in taxes to cover the general accident program. He could not avoid this cost whatever he bought. As a result, the comparative cost of buying a car and going by taxi in Sparta would be $200 per year for the car as contrasted with $250 for train and taxi fares. Chances are Taney would buy the car. In purchasing a second car the Sparta C. J. is not made to pay the full $400 that it costs. And in fact, he must pay part of that cost whether or not he buys one. He will, therefore, buy a car. If he alone had to carry the full burden of a second car, he would use trains and taxis, spending the money saved on something else—a TV set or a rowboat.

One need not imagine that any of us sit around at home thinking about relative costs of different goods and the relative pleasures derived from them for the theory to make sense. The fact is that if the cost of all auto accidents were suddenly to be paid out of a general government fund the expense of owning a car would be a lot lower than it is now since people would no longer need to worry about buying insurance; the result would be that some people would buy more cars. Perhaps they would be teenagers who can afford $100 for an old jalopy, but who cannot afford—or whose fathers cannot afford—the insurance. Or they might be people who could buy a second car so long as no added insurance was involved. In any event, the demand for cars would increase, and, therefore, so would the number of cars produced. Indeed, the effect would be the same as if the government suddenly chose to pay the cost of steel used by car makers, and to raise the money out of taxes. In each case the objection would be the same. In each, an economist would say, resources are misallocated in that goods are produced which the purchaser would not want if he really had to pay the full extent of their cost to society—their cost, whether in terms of the physical components of the item or of the expense of accidents associated with its production and use.

The resource-allocation theory is not, however, without its limitations. A primary difficulty with it involves the existence of monopoly power. . . . And since monopoly distorts allocation of resources, any system of loss allocation based on this theory must take this possible bias into account.

But forgetting for a moment the problems monopoly brings, the most desirable system of loss distribution under a strict resource-allocation theory is one in which the prices of goods accurately reflect their full cost to society. The theory therefore requires, first, that the cost of injuries should be borne by the activities which caused them, whether or not fault is involved, because, either way, the injury is a real cost of those activities. (It is because of this nonfault basis, of course, that "enterprise liability" is often lumped together with other nonfault systems of loss allocation under the general heading, "risk distribution.") Second, the theory requires that among the several parties engaged in an enterprise the loss should be placed on the party which is most likely to cause the burden to be reflected in the price of whatever the enterprise sells.

But which is that party? Is it the worker who has been injured, or his employer; is it the depositor whose check is forged, or the bank; is it the pedestrian, or the driver of the car that hit him? . . . [A] pedestrian—even if tempted to buy accident insurance because of the risk of being hit by a car—would not be able to make this part of the price of cars. As a result, car buyers would have no reason not to buy cars, even though their purchases raised the cost of pedestrian auto insurance. In fact, they would be in the same situation as C. J. Taney in Sparta for whom the real cost of a car is not reflected in its purchase price. Were the risk of accident put instead on the car owner as driver, this added cost would be reflected in the real expense of owning a car and would affect purchases. Secondly, in the real world not all parties evaluate losses equally, or are equally likely to insure. Before workmen's compensation the individual worker simply did not evaluate the risk of injury to be as great as it actually was. He took his chances; and even if he did not wish to take his chances, the fact that other workmen took a chance forced him to do the same, or to starve. The result—apart from some individual tragedies—was that wages and prices in certain industries simply did not reflect the losses those industries caused. Finally, insurance may cost one party less than it costs another. If that is so, the proper party to bear the risk is the party whose insurance costs are lower. For only then are the true costs of injuries, and not some false costs of more expensive insurance, reflected in price.

Effect of Monopoly Power on the Allocation of Resources Justification

The foregoing analysis of the resource-allocation and loss-distribution theories is clearly valid only in the absence of monopoly power altogether, or where a similar degree of monopoly power exists in all industries. But since in the American economy monopoly power in fact varies enormously from industry to industry, the difficult question of whether these theories are equally justified in the presence of monopolies is crucial. As noted, this is because the relatively monopolistic seller charges a price which is higher in relation to his costs than that charged by the relatively competitive seller; he thereby causes a shift in choices away from monopoly goods, less of which are demanded than would be justified by their true costs. It might be argued, therefore, that charging a monopolistic producer with all his accident costs would frequently do nothing to correct the distortion, and where the accident costs of the monopolist were relatively high, might actually increase that distortion. Thus, at first glance at least, it might appear that accident costs should be charged to competitive industries in order to induce them to charge more and produce less, while to counteract the monopolist's relative under-production these costs should not be placed on monopolistic industries. Some of the reasons for the undesirability of such a system will appear later in the discussion of the "deep pocket" or "let the rich man pay" side of what is called risk distribution. For the moment, however, it is enough to note that while the allocation-of-resources theory may be strong enough to justify some modifications in the way losses are allocated, it is not strong enough to justify modifications which run counter to basic political beliefs in our society—like the belief that monopolists should be treated worse than small competitors, or at least not better.

Fortunately for the allocation-of-resources theory, more careful analysis destroys much of the theoretical validity of this "subsidize monopoly" argument. In the first place, the allocation-of-resources theory is of primary importance in situations involving two or more products which can to some significant extent substitute for each other. C. J. Taney is faced with the alternative of using aluminum and steel in making widgets. Suppose that one of these can only be produced with a high accident cost, while the other involves few accidents indeed. Taney's choice between the two metals will be influenced by their relative prices, and these will be influenced by whether or not the accident cost is charged to the metal-producing industries. From the standpoint of resource allocation, the fact that both steel and aluminum have a high degree of monopoly power when compared to corner hash-houses is quite irrelevant. The choice is between steel and aluminum, not between these and fried clams. Putting accident costs on corner hash-houses and not on steel and aluminum plants might help counter a minor misallocation of purchases between metals and clams. But this adjustment would create a major resource misallocation between steel and aluminum, the prices of which would not reflect their relative costs because of the difference in accident rates in the two industries. In America, industries producing goods which can to some real extent substitute for each other have, by and large, similar degrees of monopoly power. Hence, a system of loss allocations which charged all industries with their accident costs would be a pretty good one from the standpoint of resource allocation, even though monopoly power differs greatly in the economy as a whole. . . .

The traditional or "marginal" theory [of pricing] assumes that the seller is less concerned with the average cost of production than with what the last units he produces cost him relative to what they bring in. So long as it costs him less to produce the last 1, or the last 1,000, widgets than he will make when he sells these additional widgets, he will produce them. If, on the other hand, he believes that the effect of producing more widgets will be to increase his costs more than his revenue, he will not produce them. At this equilibrium point profits would be maximized and losses minimized. Assuming that increases in widget production are accompanied by higher accident costs, a seller who is saddled with accident costs will produce fewer widgets and charge a higher price for each one. He will do this because the point at which producing more widgets will increase his total costs at a higher rate than his total revenues will have been shifted back by the fact that while the additional revenue derived from producing an extra 1,000 widgets is unchanged, it now costs more to produce the last thousand widgets than it did before.

But instead of varying according to output, the tort liability costs may be constant regardless of production volume. They may in effect be a tax for entering the industry. An example of such fixed costs would be the lump sum damages awarded to neighboring property owners if a factory creates a nuisance. Once the payment—say $10,000—is made, it makes no difference whether one makes 1 or 1,000,000 widgets. In this situation, though profits will decrease by the amount of the damages, the placing of accident costs on the industry will not affect price or output at all, unless someone who used to produce widgets decides to produce them no longer. The increase in total cost caused by the production of the last 1,000 widgets will, like the increase in revenue due to their sale, be unchanged by the fact that it cost $10,000 to get into the business in the first place. Since the equilibrium point at which profits are maximized remains the same, no amount of output changing will tend to mitigate this loss or return profits to their previous levels. In short, Taney may decide that because of the added $10,000 cost of being in the widget business—a cost he incurs whether he produces one or one million widgets—it is no longer worth his while to manufacture widgets. But if he decides it is still profitable, he will sell as much, and at the same price, as before. . . .

If one accepts the traditional "marginal" theory the result will be the same if accident costs vary fairly continuously with output, but will be quite different where they are either fixed or are subject to change in large lumps only. In the latter cases the immediate effect on prices and output will be nil and there will, therefore, be no immediate effect on allocation of resources one way or another. There is, however, a markedly different secondary effect in monopolistic and competitive firms under this kind of cost burden. And this secondary effect, without favoring monopolies, promotes a favorable resource allocation in the difficult case where a relatively monopolistic industry competes with an industry that is relatively competitive.

Taney makes widgets. Widget-making is a highly competitive business, and Taney is barely able to make a go of it. Suddenly he is slapped with the requirement that he pay for accidents caused by widget-making. Assume that insurance costs are such that he will be charged the same whatever he produces, so long as he produces at all. Insuring will drive him out of business; failure to insure will ultimately drive him, or an unlucky competitor who had accidents, out too. Fewer widget-makers will remain, output will be lower, and that output will now sell at a higher price, one sufficient to cover accident as well as other costs. Were Taney in a monopolistic industry he would also have suffered a decrease in profits from the fact that he now had to bear accident costs, and since these accident costs did not vary with output he could not pass any part of them on to the consumers through output and price shifts. But, in all probability, he would still be making enough after his decrease in profits to make staying in the business worthwhile. His extra, or monopoly, profits would have been cut, but he would still be surviving. So would the few others in his industry. Output would therefore remain the same and so would price. In short, if the theory is accurate, competitive industries would ultimately react to increases in fixed costs by losing some firms; monopolistic industries, on the other hand, would be unaffected in their size and output, although their extra profits would decrease. The net result would be a relatively higher price and lower output in the competitive industry, a desirable result from the standpoint of allocation of resources. . . .

Some Tentative Conclusions

We are now in a better position to understand what may be meant when it is said that masters "should" be liable for the torts of their servants, but should "only" be liable for them if they occur in the scope of the servants' employment. Similarly, we can now understand the "arising out of or in the course of employment" limitation on workmen's compensation. More detailed analysis of the specific legal doctrines of workmen's compensation, respondeat superior, and independent contractor will have to wait until we have discussed the other elements in what is called risk distribution—the other pieces of our puzzle. But it is not difficult to see that whatever the other elements in risk distribution will show, allocation of resources gives quite substantial support to doctrines which rely essentially on an enterprise concept of scope of liability.

Proper resource allocation militates strongly against allocating to an enterprise costs not closely associated with it—"liability should be limited to injuries arising out of or in the course of employment." But it also militates for allocating to an enterprise all costs that are within the scope of that enterprise. "The enterprise is held liable for the injuries even though no fault on its part can be shown." Not charging an enterprise with a cost which arises from it leads to an understatement of the true cost of producing its goods; the result is that people purchase more of those goods than they would want if their true cost were reflected in price. On the other hand, placing a cost not related to the scope of an enterprise on that enterprise results in an overstatement of the costs of those goods, and leads to their underproduction. Either way the postulate that people are by and large best off if they can choose what they want, on the basis of what it costs our economy to produce it, would be violated. . . .

We can also begin to see why strict fault liability had such a strong vogue from the middle to the end of the 19th century. Many factors were involved, of course. Not the least among them is the fact that the justifications for the risk spreading and "let the rich man pay" elements in risk distribution were not such as would commend themselves to a 19th century Weltanschaung. But, on the other hand, the allocation-of-resources theory would seem to fit in with the 19th century approach to output and production as much, if not more, than with the 20th century one. Why then did it play so small a role in the choice of a system of loss allocation?

Perhaps the answer can be found in the rather peculiar state of industry at the time. In the early days of the industrial revolution many industries were operating on a decreasing cost basis. That is, if an industry could expand sufficiently its costs would fall as a result of that expansion. It is an interesting fact that in cases where an industry is operating on a decreasing cost basis a subsidy to that industry will probably help, rather than hinder, proper allocation of resources.

An example may help. Widget-maker Taney has such high costs in making widgets that he must sell them at a price which only the rich can afford. As a result he makes few widgets. If he could sell at a lower price, however, many more people would want to buy widgets. If he could reach this greater level of production his costs would be sufficiently low to enable him, after a time, to meet them all and to sell all the widgets he produced. Taney, however, cannot just start producing at this much higher output, if for no other reason than that he would go broke, selling widgets so cheaply, before his costs would drop. If he had a subsidy, however—if he did not for some years have to meet all his costs—he would be able to establish himself at the higher output, and in the long run all people would be better off. There would be a widget in every pot, as well as in every garage.

If this was the situation of most American industry in the nineteenth century—and the fact that high tariffs were being justified even by "free trade" economists at the time, on just this ground, indicates that it was—then an argument could be made that proper "long run" allocation of resources required that industry be spared from paying hidden accident costs—at least unless other factors like fault were involved. I do not suggest, of course, that 19th century judges made the transfer to fault liability on the basis of this rather complicated economic theory. But their statements that nonfault liability would deprive our land of the benefits and promises of industrial expansion may represent a rough-and-ready, noneconomist's, way of recognizing the fact that industry was simply not ready to bear all of its costs, and that the country would in the long run be better off if it did not. To this extent these phrasings are no different from those of modern writers who, conditions having changed, say without further analysis that enterprises "should" bear all the accident costs they cause, regardless of fault.

Of course, the fact that a subsidy may have made sense does not suggest that the injured worker should have been the one to pay the subsidy. Today we would be inclined to have the subsidy come out of taxes—either a general tax, or one on those who benefit most from the innovation. This is, however, giving a 20th century answer to a 19th century question. For, the reason we quite properly find the idea of workers subsidizing industrial expansion intolerable is because we are wedded to "risk spreading" and "deep pocket" notions, and these are notions which did not appeal especially to the 19th century mind. (In addition, industry itself would have borne a heavy part of the burden of taxes; subsidization through taxation might, therefore, have discouraged industrial expansion in the same way as nonfault liability.)

. . .

Some General Observations

At the end of this rather long analysis we might well consider some of the consequences of the different justifications for nonfault distribution of losses, and some inconsistent results these theories would seem to require. Perhaps the most dramatic inconsistencies exist between some of the requirements of the allocation-of-resources justification and the loss-spreading justification. The treatment of losses which are definitely caused by enterprises, but which could not be foreseen by those enterprises—and which are therefore probably not insured against—is a strong example. Unless they were covered by a general state social-insurance program, such losses would be unlikely to be thoroughly spread. They would be left either on the injured parties or on the enterprises which engendered them. . . .

Insofar as resource allocation is concerned, such losses are just as truly costs of producing particular goods as are more foreseeable risks. Lack of foreseeability makes it somewhat more difficult to include these costs in the price of the item produced, but does not make it impossible. Industries with more than their share of unforeseeable losses—and, as a result, more than their share of bad years or failures—get reputations for being risky. Fewer firms enter such industries and, over time, higher prices prevail. Thus, the desired allocation effect is accomplished. Higher prices do not mean, however, that any substantial loss spreading occurs. They only mean that entrepreneurs in such industries make greater profits, subject to the danger that, when the risk strikes, one of them may be so severely damaged that he will never recover his losses, or that he may be wiped out altogether. In such cases undesirable secondary social and economic losses would, of course, follow.

None of this would occur if instead of being handled by a system of enterprise liability risks of all injuries were covered by a general state accident program. But neither would these losses be reflected in prices under such a scheme. Advocates of allocation-of-resources enterprise liability would argue further that though such secondary losses seem harsh, they are a necessary part of any free enterprise system. Entrepreneurs always take "uninsurable risks"—indeed, the danger of going into business, which, many economists say, is the very source of "profits" in business, as distinguished from mere payments for labor or for use of capital, is just such a risk. And advocates of enterprise liability would say that this is merely another indication of how enterprise liability is really the "free enterprise" way of allocating losses, as against more collectivist social insurance plans.

Of course, it is true that enterprise liability must ultimately be supported primarily on a free enterprise argument. Though as a system of loss-spreading enterprise liability has some merits, it is still relatively inefficient. In the first place, we are not prepared to charge enterprises with losses which are not readily assignable to some specific activity. And, of course, many such losses do exist. If risk spreading is really important, these general losses of living would in themselves require some kind of social insurance. Enterprise liability may be similarly inefficient where the cost of collecting the loss from the enterprise is very large either in terms of court costs, or lawyers' fees. (In such situations, neither would there be an allocation-of-resources justification for placing these losses on any activity. Indeed, the justification would run the other way. A greater misallocation is caused by incurring the avoidable costs of trying to allocate the loss than by leaving it where it falls and letting the price of the product involved understate its true costs.) At best, then, if risk spreading is deemed crucial, enterprise liability could do only part of the job; the other part would have to be filled in by some social insurance scheme.

In the second place, even in the area where enterprise liability does play its part, it would in all probability be a far less thorough risk spreader than a social insurance plan. We have seen that the danger of creating sick industries and the possibility of driving out small competitors—at least during a transition period to enterprise liability—indicate that harmful secondary economic and social effects may well occur with enterprise liability, while they could be avoided under general social insurance. This is not to say that enterprise liability would do a bad job of spreading losses; it is only to say that social insurance probably would do a better one.

Similarly, from the point of view of the "deep pocket" justification, social insurance would probably be preferable to enterprise liability. It is true that in the long run enterprise liability promises either wide loss spreading or—by and large—a tax on monopoly. But it does not tax all monopolies equally; nor does it tax wealthy men who are not monopolists. The taxing system—with all its weaknesses—is far more refined in taking from the rich and giving to the poor than enterprise liability could ever be. For all these reasons, many writers who have been concerned primarily with risk spreading or "deep pocket" have tended to view enterprise liability as, at best, a half way house on the road to social insurance.

One can argue with this position by raising questions about the actual costs of running a program of social insurance, and by suggesting that, in view of those costs, enterprise liability does what it does in the way of risk spreading pretty cheaply. Or one can go back to questions of deterrence, and to some of the other justifications for fault liability, and see whether they do not form some justification for enterprise liability as against social insurance. But the first of these approaches is not really subject to proof, and the second, though potentially fruitful, is really outside the scope of this article, since it would involve a thorough discussion of the role fault plays in our system of loss allocation.

On the basis of the discussion in this article, however, enterprise liability is superior to social insurance in that it promotes proper allocation of resources. And the importance of allocation of resources increases to the extent that we value free enterprise. Therefore, so long as our society remains committed to free enterprise, enterprise liability is unlikely to be relegated to the role of a stop-gap measure on the road to social insurance. . . .

Of course, if the costs of administering enterprise liability prove exorbitant, or if damages rise out of all proportion to the injuries sustained—if, in other words, the amount charged to the industry becomes much greater than the loss caused—it will be difficult to make out a case for enterprise liability on resource-allocation grounds. And we may look for an increased trend toward social insurance. Similarly, if we become more concerned with the elimination of any possible economic dislocation, and if at the same time—for the two are quite consistent—we become increasingly disenchanted with production in accordance with the apparent desires of consumers, then social insurance is bound to increase in importance. But if these things don't happen, there is every reason to think that we shall try to combine broader enterprise liability—in which risk spreading, loss allocation, and deep pocket values are synthesized—with limited social insurance programs, paid out of progressive taxes, to cover those losses which are too general to be assigned to any single activity or group of activities. . . .

Unity in Tort, Contract, and Property: The Model of Precaution

ROBERT COOTER

Forms of Precaution

Even when necessary or unavoidable, an accident, breach of contract, taking, or nuisance causes harm. The affected parties, however, can usually take steps to reduce the probability or magnitude of the harm. The parties to a tortious accident can take precautions to reduce the frequency or destructiveness of accidents. In contract, the promisor can take steps to avoid breach, and the promisee, by placing less reliance

Robert Cooter, Unity in Tort, Contract and Property: The Model of Precaution, 73 *California Law Review,* pp. 1–51 (1985). Copyright © The California Law Review, Inc. Reprinted with permission.

on the promise, can reduce the harm caused by the promisor's breach. Similarly, for governmental takings of private property, the condemnor can conserve on its need for private property, while property owners can reduce the harm they suffer by avoiding improvements whose value would be destroyed by the taking. Finally, the party responsible for a nuisance can abate; furthermore, the victim can reduce his exposure to harm by avoiding the nuisance.

Generalizing these behaviors, I extend the ordinary meaning of the word "precaution" and use it as a term of art in this article to refer to any action that reduces harm. Thus the term "precaution" includes, for example, prevention of breach and reduced reliance on promises, conservation of the public need for private property and limited improvement of private property exposed to the risk of a taking, and abatement and avoidance of nuisances. These examples are, of course, illustrative, not exhaustive.

The Paradox of Compensation

When each individual bears the full benefits and costs of his precaution, economists say that social value is internalized. When an individual bears part of the benefits or part of the costs of his precaution, economists say that some social value is externalized. The advantage of internalization is that the individual sweeps all of the values affected by his actions into his calculus of self-interest, so that self-interest compels him to balance all the costs and benefits of his actions. According to the marginal principle, social efficiency is achieved by balancing all costs and benefits. Thus, the incentives of private individuals are socially efficient when costs and benefits are fully internalized, whereas incentives are inefficient when some costs and benefits are externalized.

In situations when both the injurer and the victim can take precaution against the harm, the internalization of costs requires both parties to bear the full cost of the harm. To illustrate, suppose that smoke from a factory soils the wash at a commercial laundry, and the parties fail to solve the problem by private negotiation. One solution is to impose a pollution tax equal to the harm caused by the smoke. The factory will bear the tax and the laundry will bear the smoke, so pollution costs will be internalized by both of them, as required for social efficiency. In general, when precaution is bilateral, the marginal principle requires both parties to be fully responsible for the harm. The efficiency condition is called double responsibility at the margin.

One problem with the combination of justice and efficiency, however, is that compensation in its simplest form is inconsistent with double responsibility at the margin. In the preceding example, justice may require the factory not only to pay for harm caused by the smoke, but also to compensate the laundry for that harm. Compensation, however, permits the laundry to externalize costs, thereby compromising efficiency. Thus, a paradox results: If the factory can pollute with impunity, harm is externalized by the factory; if the factory must pay full compensation, harm is externalized by the laundry; if compensation is partial, harm is partly externalized by the factory and partly externalized by the laundry. Assigning full responsibility for the injury to one party or parceling it out between the parties cannot fully internalize costs for both of them. Thus, there is no level of compensation that achieves double responsibility at the margin. In technical terms, when efficiency requires bilateral precaution, strict liability for any fraction of the harm, from zero percent to 100 percent, is inefficient.

Rules that combine compensation for harm with incentives for efficient precaution are therefore patently difficult to formulate. The problem confronted in this part of the article is to explain how the law combines compensation with double responsibility at the margin. The law has evolved three distinct mechanisms for achieving this end, which I will sketch by reference to the law of torts, contracts, and property.

Accidents

Assume that Xavier and Yvonne are engaged in activities that sometimes result in accidents. If an accident occurs, Yvonne's property is damaged and Xavier's is not. For this reason I will call Xavier the injurer and Yvonne the victim, regardless of who is at fault. The probability that an accident will occur depends on the precautions taken by both of them, which are costly. The relationship between harm and precaution is easy to visualize in concrete cases. Drawing on a famous example, suppose that Xavier operates a railroad train that emits sparks that sometimes set fire to Yvonne's cornfield. Xavier can reduce the harm to the corn by installing spark arresters, by running the trains more slowly, or by running fewer trains. In a like manner, Yvonne can reduce the harm by planting her corn farther from the tracks, by planting cabbage instead of corn, or by leaving the fields fallow.

There are two rules that assign liability without regard to fault. The first of these is a rule of no liability, which means that courts will not redistribute the cost of accidents. Under such a rule, the victim bears the full cost of accidents. The second rule is strict liability, which means that the injurer must compensate the victim whenever an accident occurs. The rule used by the courts for allocating accident costs will determine Xavier's and Yvonne's incentives for precaution. . . .

The desirability of no liability or strict liability can be evaluated from the viewpoint of economic efficiency. The measure of social costs in the simple model of precaution is the sum of the parties' costs of precaution and the expected cost of harm. Efficient levels of precaution minimize the social costs of accidents. For most accidents, precaution is bilateral in the sense that social efficiency requires both injurer and victim to take at least some precaution. The rule of no liability and the rule of strict liability with perfect compensation both lack incentives for one of the parties to take precaution, so these rules cannot be efficient for accidents that are bilateral in this sense.

A similar statement is true when compensation is imperfect rather than perfect. Compensation is less than perfect if the victim would prefer no accident to an accident with (imperfect) compensation. Under a rule of strict liability with less than perfect compensation, the injurer externalizes the uncompensated portion of the harm and the victim externalizes the compensated portion of the harm. Since neither of them internalizes the full cost of harm, both have inadequate incentives for precaution. Thus, when efficiency requires bilateral precaution, rules of no liability or strict liability provide inadequate incentives for precaution, regardless of the level of compensation.

This is an instance of the paradox of compensation. Nonetheless, the paradox can be resolved by adopting fault rules that assign responsibility for harm according to the fault of the parties. To illustrate, a simple negligence rule requires the victim to be compensated by the injurer if, and only if, the latter is at fault. Under a simple negligence rule, Xavier will satisfy the legal standard in order to avoid liability. Thus, if the legal standard corresponds to the efficient level of precaution, Xavier's precaution will be efficient. Since Yvonne knows that she bears residual responsibility, she internalizes the costs and benefits of precaution; therefore, her incentives are efficient. Thus, if the legal standard of fault corresponds to the efficient level of care, both parties will take efficient precaution.

Like the tax solution, a simple negligence rule creates a condition in which each party bears the cost of the harm caused by a small decrease in his precaution. The injurer responds by minimally fulfilling the legal standard of care, so that even a small reduction in his care will cause him to be liable. Absent that reduction in care by the injurer, however, the victim will be responsible. Thus, each party bears the full cost of the increase in harm caused by the decrease in his precaution. This is double responsibility at the margin.

The same method of reasoning can be used to show that efficient incentives for precaution are created by fault rules other than simple negligence, such as negligence with contributory negligence, strict liability with contributory negligence, or comparative negligence. Under any fault rule, the injurer can escape responsibility by satisfying the legal standard, so an efficient legal standard will cause his behavior to be efficient. Similarly, the victim's precaution will be efficient because he bears residual responsibility and thus internalizes the costs and benefits of precaution. So long as the legal standards correspond to efficient precaution, all such rules create double responsibility at the margin. Thus the particular rule can be chosen that best accords with the requirements of just compensation. . . .

Breach of Contract

Yvonne and Xavier enter into a contract in which Yvonne pays for Xavier's promise to deliver a product in the future. There are certain obstacles to Xavier's performance that might arise, and if severe obstacles materialize, Xavier will not be able to deliver the product as promised. The probability of timely performance depends in part on Xavier's efforts to prevent such obstacles from arising. These efforts are costly.

One purpose of contracting is to give Yvonne confidence that Xavier's promise will be performed, so that she can rely upon his promise. Reliance on the contract increases the value to Yvonne of Xavier's performance. However, reliance also increases the loss suffered in the event of breach. The more the promisee relies, therefore, the greater the benefit from performance and the greater the harm caused by breach.

To make this description concrete, suppose that Xavier is a builder who signs a contract to construct a store for Yvonne by the first of September. Many events could jeopardize timely completion of the building; for example, the plumbers union may strike, the city's inspectors may be recalcitrant, or the weather may be inclement. Xavier can increase the probability of timely completion by taking costly measures, such as having the plumbers work overtime before their union contract expires, badgering the inspectors to finish on time, or rescheduling work to complete the roof before the rainy season arrives. Yvonne, on the other hand, must order merchandise for her new store in advance if she is to open with a full line on the first of September. If she orders many items for September delivery and the store is not ready for occupancy, she will have to place the goods in storage, which is costly. The more merchandise she orders, the larger her profit will be in the event of performance, and the larger her loss in the event of nonperformance.

As thus described, the structure of the contractual model is similar to the model developed for tortious accidents. The precaution taken by the potential tortfeasor against accidents parallels the steps taken by the promisor to avoid obstacles to performance. The parallel between the tort victim and the promisee, however, is more subtle. More precaution by the tort victim is like less reliance by the contract promisee, because each action reduces the harm caused by an accident or a breach. Therefore, the tort victim's precaution against accidents and the contract promisee's reliance upon the contract are inversely symmetrical.

If Xavier does not perform, then a court must decide whether a breach has occurred or whether nonperformance is excused by circumstances. Among the excuses that the law recognizes are: that the quality of assent to the contract was too low due to mistake, incapacity, duress, or fraud; that the terms of the contract were unconscionable; or that performance was impossible or commercially impractical. If the court narrowly construes excuses, usually finding nonperformance to be a breach, then Xavier will usually be liable. If the court construes excuses broadly, usually finding nonperformance to be justified, then Xavier will seldom be liable.

The incentive effects of a broader or narrower construction of excuses are similar to the effects of strict liability and no-liability rules in tort. If defenses are narrowly construed and perfect expectation damages are awarded for breach, the promisee will rely as if performance were certain. Specifically, Yvonne will order a full line of merchandise as if the store were certain to open on the first of September. A promisee's reliance to the same extent as if performance were certain corresponds to a tort victim's failure to take precaution against harm.

A broad construction of excuses has the symmetrically opposite effect: the promisor expects to escape liability for harm caused by his breach, so he will not undertake costly precautions to avoid nonperformance. Specifically, if Xavier is unconcerned about his reputation or the possibility of future business with Yvonne, and if nonperformance due to a plumber's strike, recalcitrant inspectors, or inclement weather will be excused, say, on grounds of impossibility, then Xavier will not take costly precautions against these events. The promisor's lack of precaution against possible obstacles to performance corresponds to the injurer's lack of precaution against tortious accidents.

As explained, the narrow and broad constructions of excuses for breach of contract affect behavior in ways that parallel no liability and strict liability in tort. Furthermore, the effects of these constructions on cost internalization and efficiency are also parallel. Specifically, if excuses are broadly construed, allowing the promisor to avoid responsibility for breach regardless of his precaution level, the promisor will externalize some of the costs of breach. As a result, his incentives to take precaution against the events that cause him to breach are insufficient relative to the efficient level. If, on the other hand, excuses are narrowly construed and full compensation is available for breach, the promisee can externalize some of the costs of reliance. Insofar as the promisee can transfer the risk of reliance to the promisor, her incentives are insufficient to provide efficient reliance and, therefore, reliance will be excessive.

To illustrate, social efficiency requires Xavier to hire the plumbers to work overtime if the additional cost is less than the increase in Yvonne's expected profits caused by the higher probability of timely completion. Suppose, however, that there are circumstances in which tardiness will be excused regardless of whether or not Xavier hired the plumbers to work overtime. Suppose for example that inclement weather excuses tardiness on grounds of impossibility. In the event inclement weather provides Xavier with an excuse, the extra cost of hiring the plumbers to work overtime, which is valuable to Yvonne, has no value to Xavier. Anticipating this eventuality, Xavier may not hire the plumbers to work overtime, even though social efficiency may require him to do so.

Social efficiency also requires Yvonne to restrain her reliance in light of the objective probability of breach. To be more precise, social efficiency requires her to order additional merchandise until the resulting increase in profit from anticipated sales in the new store, discounted by the probability that Xavier will finish the store on time, equals the cost of storing the goods, discounted by the probability that Xavier will finish the new store late. Suppose, however, that Xavier must compensate Yvonne for her storage costs in the event that the goods must be stored. From a self-interested perspective, Yvonne has no incentive to restrain her reliance in these circumstances. Anticipating this possibility, instead of weighting the cost of storage by the objective probability of breach, Yvonne will weight it by the probability of breach without compensation. Since in this example the probability of breach is greater than the probability of breach without compensation, the weight Yvonne gives to the possibility of storage cost is too small. Therefore, her reliance will be excessive and thus inefficient.

In general, the possibility of successful excuses may externalize the costs of not taking precaution, so that the promisor takes too little precaution and the probability of breach is excessive. Similarly, the possibility of compensation may externalize the costs of reliance, so the promisee relies too heavily and the harm that materializes in the event of breach is excessive. This is an aspect of the paradox of compensation that arises in tort with respect to no liability and strict liability. As with tort law, contract law has a solution to the paradox, but the contract solution is different from the tort solution. To illustrate the characteristic remedy in contracts, consider the liquidation of damages. If the contract stipulates damages for breach, requiring Xavier to remit, say, $200 per day for late completion, then the promisor will have a material incentive to prevent breach. Specifically, Xavier may find that paying the plumbers to work overtime is cheaper than running the risk of late completion. If the promisee receives the stipulated damages as compensation, then the level of her compensation is independent of her level of reliance, so she has a material incentive to restrain her reliance. Specifically, if Yvonne receives $200 per day in damages for late completion whether or not she orders the bulky merchandise, she may avoid the risk of bearing storage costs by not ordering it.

Like a negligence rule in tort, liquidation of damages in a contract imposes double responsibility at the margin: the promisor is responsible for the stipulated damages and the victim is responsible for the actual harm. By adjusting the level of stipulated damages, efficient incentives can be achieved for both parties. Stipulated damages are efficient when they equal the loss that the victim would suffer from breach if her reliance were efficient. To illustrate, assume that efficient reliance requires Yvonne to order the compact merchandise and not the bulky merchandise. Furthermore, assume that if Yvonne orders the compact merchandise she will lose $200 in profits for each day that Xavier is late in completing the new store. Under these assumptions, liquidating damages at $200 per day for late completion provides efficient incentives for both Xavier and Yvonne. . . .

There are other doctrinal approaches to damages that have similar effects. For example, suppose that Xavier fails to complete the building on the first of September as promised, and Yvonne has to rent temporary space elsewhere. The court might award damages based in part on the additional rent, if it finds Yvonne's calculation of lost profits too speculative. If damages are based on the additional rent, and if the additional rent varies less than Yvonne's profits with respect to her reliance, then her incentive to overrely is reduced. As another example, failure to perform on a franchise agreement may result in an award of damages equal to the profit of similar franchise establishments, but not the "speculative profits" lost by the particular plaintiff. The general point of these two examples is that if compensation is restricted to nonspeculative damages, and if nonspeculative damages vary less with respect to reliance than the actual harm, then restricting compensation to nonspeculative damages reduces the incentive to overrely. . . .

Takings

Xavier is a government official whose wall is covered by a large map with a thick blue line across it. The blue line represents a proposed government project, such as a highway, park, sewer line, or the boundaries of a neighborhood being downzoned. Yvonne is contemplating major improvements on her property, which is located on the blue line. If the government carries out its plan, it will either take Yvonne's title or an easement in her land, or restrict her ownership rights by regulation. By so doing, the government project will also destroy the value of Yvonne's proposed improvements. In brief, Yvonne's improvements will be valuable if the government project is abandoned and valueless if the project is carried out. Yvonne's improvements are therefore analogous to reliance in the contract example: the more she invests, the greater the benefit if there is no taking and the larger the loss if there is.

There are several ways in which a dispute could arise between Xavier and Yvonne. Xavier might take the property and offer compensation that Yvonne considers inadequate. Xavier might regulate the property and offer no compensation. Or Xavier might threaten to take the property without actually taking it, thus eroding its value. Furthermore, the law offers Yvonne several remedies; for example, the courts may overturn the regulation or award her damages in the dispute. . . .

The incentive effects of takings and regulation are like the incentive effects of strict liability and no liability in tort, or like the narrow and broad construction of excuses for breach of contract. Moreover, the effects upon cost internalization and efficiency are also similar. When government action is likely to be judged a taking, private property owners externalize the risk associated with improvements the value of which may be destroyed by the government action. Instead of restraining investment in light of the objective probability of a taking, private investors have incentives to invest excessively relative to the socially efficient level.

Social efficiency requires Yvonne to scale down or delay her planned improvements in light of the probability that Xavier's study will recommend downzoning. To be more precise, social efficiency requires her to make additional improvements until the resulting increase in her profits when there is no government action, multiplied by the probability of no government action, equals the loss in profits when there is government action, multiplied by the probability of government action. When the government action is likely to be judged a regulation it is in Yvonne's self-interest to make this calculation, so her incentives for investment will be efficient. However, when the government action is likely to be judged a taking with full compensation, Yvonne will give insufficient weight to her loss in profits in the event of government action. Therefore, she will invest excessively.

Similarly, when government action is likely to be judged a taking, the government internalizes the cost of its actions and thus restrains its taking of private property. On the other hand, when government action is likely to be judged a mere regulation, the government lacks material incentives to conserve its use of valuable private property rights. In general, the possibility that government action will be judged to be a mere regulation externalizes government costs, resulting in excessive government action. The possibility that takings will be fully compensated externalizes the risk to private individuals that government will destroy the value of private improvements, resulting in excessive private investment. . . .

The solution to this problem in property law is similar to the solution in contracts. The contract remedy for breach is to liquidate damages so that the injurer is liable for liquidated damages and the victim for actual damages. This outcome is achieved in the condemnation setting when the government purchases an option from the property owner. An option entitles the government to buy the property at any time within a prescribed interval at a price specified in the contract. If government buys the property on which it holds an option, then it is liable for the price and the property owner is liable for any actual loss.

For example, if Xavier buys an option to purchase Yvonne's property for one million dollars, Yvonne must sell on demand at that price whether or not she has made improvements on her property. If the option is exercised, Xavier will bear the stipulated cost of one million dollars and Yvonne will bear the loss on any improvements. The stipulated price will provide an incentive for Xavier to conserve on taking private property, and the noncompensability of the actual cost of improvements will provide an incentive for Yvonne to restrain her investments, thus resolving the problem of efficient incentives.

There is, however, an important practical difference between liquidating contract damages and the use of governmental options to acquire private property rights. Liquidating damages involves adding an additional clause to the primary contract; the parties must come together and negotiate the primary contract anyway. When the government purchases an option to buy private property, however, it must enter into negotiations that would not otherwise occur. The transaction costs of buying and selling options will often outweigh the gains from correcting efficiency incentives for investment. . . .

Summary . . .

Tort, contract, and property law all allocate the cost of harm. For many types of harm, efficiency requires precaution by both injurer and victim. Incentives for precaution are efficient when both parties are responsible for the harm caused by their marginal reductions in precaution (double responsibility at the margin). An absolute rule, such as strict liability, erodes the victim's incentives for precaution. Conversely, a rule of no liability erodes the injurer's incentives for precaution. There are at least three mechanisms in the common law that combine compensation and incentives for efficient precaution.

The first mechanism is the fault rule. When an injurer satisfies the legal standard of care, a small reduction in his precaution will make him negligent. Thus, he is liable for the harm resulting from marginal reductions in his precaution. Furthermore, when the injurer satisfies the legal standard of care, the victim bears residual responsibility and is responsible for harm resulting from any reduction in her own precaution. Thus, a negligence rule encourages double responsibility at the margin—the injurer takes efficient precaution to avoid legal responsibility and the victim takes efficient precaution because she bears residual responsibility.

The second common law mechanism is invariant damages, as exemplified by liquidation of damages provisions in the law of contracts. A liquidation clause stipulates a dollar amount to be paid as compensation in the event of breach. When damages are liquidated, the breaching party is responsible for the stipulated damages and the victim of the breach is responsible for actual damages. Thus, invariant damages encourage double responsibility at the margin—the promisor balances the cost of precaution against the stipulated damages and the victim balances the benefits of reliance against the potential loss.

The third mechanism is the coercive order from a court, such as an injunction against a nuisance. Economists are unenthusiastic about coercive orders for reasons that are developed at length in the economic critique of regulation. However, the right to an injunction may have desirable economic effects if it is used as a bargaining chip rather than actually exercised. Unlike coercive orders, bargaining solutions have desirable economic properties. The right to obtain an injunction may enable nuisance victims to achieve adequate compensation by private agreement with the injurer, and the parties to the bargain will desire to make its terms efficient.

Fault rules are prominent in tort law, invariant damages are frequently found in contracts, and injunctions are a common remedy in property law; however, each of the three mechanisms can be found in other branches of the law. For example, workers' compensation law stipulates damages for accidents, the purchase of an option entitling the state to buy private property effectively stipulates damages for a taking, and specific performance provides an injunctive remedy for breaches of contract. . . .

2.2   THE MODEL OF PRIVATE COOPERATION

The Nature of the Firm

RONALD COASE

Our task is to attempt to discover why a firm emerges at all in a specialized exchange economy. The price mechanism (considered purely from the side of the direction of resources) might be superseded if the relationship which replaced it was desired for its own sake. This would be the case, for example, if some people preferred to work under the direction of some other person. Such individuals would accept less in order to work under someone, and firms would arise naturally from this. But it would appear that this cannot be a very important reason, for it would rather seem that the opposite tendency is operating if one judges from the stress normally laid on the advantage of "being one's own master." Of course, if the desire was not to be controlled but to control, to exercise power over others, then people might be willing to give up something in order to direct others; that is, they would be willing to pay others more than they could get under the price mechanism in order to be able to direct them. But this implies that those who direct pay in order to be able to do this and are not paid to direct, which is clearly not true in the majority of cases. Firms might also exist if purchasers preferred commodities which are produced by firms to those not so produced; but even in spheres where one would expect such preferences (if they exist) to be of negligible importance, firms are to be found in the real world. Therefore there must be other elements involved.

The main reason why it is profitable to establish a firm would seem to be that there is a cost of using the price mechanism. The most obvious cost of "organizing" production through the price mechanism is that of discovering what the relevant prices are. This cost may be reduced but it will not be eliminated by the emergence of specialists who will sell this information. The costs of negotiating and concluding a separate contract for each exchange transaction which takes place on a market must also be taken into account. Again, in certain markets, e.g., produce exchanges, a technique is devised for minimizing these contract costs; but they are not eliminated. It is true that contracts are not eliminated when there is a firm, but they are greatly reduced. A factor of production (or the owner thereof) does not have to make a series of contracts with the factors with whom he is co-operating within the firm, as would be necessary, of course, if this co-operation were a direct result of the working of the price mechanism. For this series of contracts is substituted one. At this stage, it is important to note the character of the contract into which a factor enters that is employed within a firm. The contract is one whereby the factor, for a certain remuneration (which may be fixed or fluctuating), agrees to obey the directions of an entrepreneur within certain limits. The essence of the contract is that it should only state the limits to the powers of the entrepreneur. Within these limits, he can therefore direct the other factors of production.

There are, however, other disadvantages—or costs—of using the price mechanism. It may be desired to make a long-term contract for the supply of some article or service. This may be due to the fact that if one contract is made for a longer period instead of several shorter ones, then certain costs of making each contract will be avoided. Or, owing to the risk attitude of the people concerned, they may prefer to make a long- rather than a short-term contract. Now, owing to the difficulty of forecasting, the longer the period of the contract is for the supply of the commodity or service, the less possible and indeed the less desirable it is for the person purchasing to specify what the other contracting party is expected to do. It may well be a matter of indifference to the person supplying the service or commodity which of several courses of action is taken, but not to the purchaser of that service or commodity. But the purchaser will not know which of these several courses he will want the supplier to take. Therefore, the service which is being provided is expressed in general terms, the exact details being left until a later date. All that is stated in the contract is the limits to what the person supplying the commodity or service is expected to do. The details of what the supplier is expected to do are not stated in the contract but are decided later by the purchaser. When the direction of resources (within the limits of the contract) becomes dependent on the buyer in this way, that relationship which I term a "firm" may be obtained. A firm is likely, therefore, to emerge in those cases where a very short-term contract would be unsatisfactory. It is obviously of more importance in the case of services—labour—than it is in the case of the buying of commodities. In the case of commodities, the main items can be stated in advance and the details which will be decided later will be of minor significance.

We may sum up this section of the argument by saying that the operation of a market costs something and that, by forming an organization and allowing some authority (an "entrepreneur") to direct the resources, certain marketing costs are saved. The entrepreneur has to carry out his function at less cost, taking into account the fact that he may get factors of production at a lower price than the market transactions which he supersedes, because it is always possible to revert to the open market if he fails to do this. . . .

Another factor that should be noted is that exchange transactions on a market and the same transactions organized within a firm are often treated differently by governments or other bodies with regulatory powers. If we consider the operation of a sales tax, it is clear that it is a tax on market transactions and not on the same transactions organized within the firm. Now since these are alternative methods of "organization"—by the price mechanism or by the entrepreneur—such a regulation would bring into existence firms which otherwise would have no raison d'être. It would furnish a reason for the emergence of a firm in a specialized exchange economy. Of course, to the extent that firms already exist, such a measure as a sales tax would merely tend to make them larger than they would otherwise be. Similarly, quota schemes, and methods of price control which imply that there is rationing and which do not apply to firms producing such products for themselves, by allowing advantages to those who organize within the firm and not through the market, necessarily encourage the growth of firms. But it is difficult to believe that it is measures such as those mentioned in this paragraph which have brought firms into existence. Such measures would, however, tend to have this result if they did not exist for other reasons.

These, then, are the reasons why organizations such as firms exist in a specialized exchange economy in which it is generally assumed that the distribution of resources is "organized" by the price mechanism. A firm, therefore, consists of the system of relationships which comes into existence when the direction of resources is dependent on an entrepreneur.

The approach which has just been sketched would appear to offer an advantage, in that it is possible to give a scientific meaning to what is meant by saying that a firm gets larger or smaller. A firm becomes larger as additional transactions (which could be exchange transactions coordinated through the price mechanism) are organized by the entrepreneur, and it becomes smaller as he abandons the organization of such transactions. The question which arises is whether it is possible to study the forces which determine the size of the firm. Why does the entrepreneur not organize one less transaction or one more? . . . Why, if by organizing one can eliminate certain costs and in fact reduce the cost of production, are there any market transactions at all? Why is not all production carried on by one big firm? There would appear to be certain possible explanations.

First, as a firm gets larger, there may be decreasing returns to the entrepreneur function, that is, the costs of organizing additional transactions within the firm may rise. Naturally, a point must be reached where the costs of organizing an extra transaction within the firm are equal to the costs involved in carrying out the transaction in the open market or to the costs of organizing by another entrepreneur. Second, it may be that, as the transactions which are organized increase, the entrepreneur fails to place the factors of production in the uses where their value is greatest, that is, fails to make the best use of the factors of production. Again, a point must be reached where the loss through the waste of resources is equal to the marketing costs of the exchange transaction in the open market or to the loss if the transaction was organized by another entrepreneur. Finally, the supply price of one or more of the factors of production may rise, because the "other advantages" of a small firm are greater than those of a large firm. Of course, the actual point where the expansion of the firm ceases might be determined by a combination of the factors mentioned above. The first two reasons given most probably correspond to the economists' phrase of "diminishing returns to management." . . .

Up to now it has been assumed that the exchange transactions which take place through the price mechanism are homogeneous. In fact, nothing could be more diverse than the actual transactions which take place in our modern world. This would seem to imply that the costs of carrying out exchange transactions through the price mechanism will vary considerably, as will the costs of organizing these transactions within the firm. It seems therefore possible that, quite apart from the question of diminishing returns, the costs of organizing certain transactions within the firm may be greater than the costs of carrying out the exchange transactions in the open market. This would necessarily imply that there were exchange transactions carried out through the price mechanism; but would it mean that there would have to be more than one firm? Clearly not, for all those areas in the economic system where the direction of resources was not dependent directly on the price mechanism could be organized within one firm. The factors which were discussed earlier would seem to be the important ones, though it is difficult to say whether "diminishing returns to management" or the rising supply price of factors is likely to be the more important. Other things being equal, therefore, a firm will tend to be larger:

a. the less the costs of organizing and the slower these costs rise with an increase in the transactions organized;
b. the less likely the entrepreneur is to make mistakes and the smaller the increase in mistakes with an increase in the transactions organized;
c. the greater the lowering (or the less the rise) in the supply price of factors of production to firms of larger size.

Apart from variations in the supply price of factors of production to firms of different sizes, it would appear that the costs of organizing and the losses through mistakes will increase with an increase in the spatial distribution of the transactions organized, in the dissimilarity of the transactions, and in the probability of changes in the relevant prices. As more transactions are organized by an entrepreneur, it would appear that the transactions would tend to be either different in kind or different in place. This furnishes an additional reason why efficiency will tend to decrease as the firm gets larger. Inventions which tend to bring factors of production nearer together, by lessening spatial distribution, tend to increase the size of the firm. Changes like the telephone and the telegraph, which tend to reduce the cost of organizing spatially, will tend to increase the size of the firm. All changes which improve managerial technique will tend to increase the size of the firm. . . .

The Problem of Social Cost

RONALD COASE

The Problem To Be Examined

This paper is concerned with those actions of business firms which have harmful effects on others. The standard example is that of a factory, the smoke from which has harmful effects on those occupying neighbouring properties. The economic analysis of such a situation has usually proceeded in terms of a divergence between the private and social product of the factory, in which economists have largely followed the treatment of Pigou in *The Economics of Welfare*. The conclusions to which this kind of analysis seems to have led most economists is that it would be desirable to make the owner of the factory liable for the damage caused to those injured by the smoke; or to place a tax on the factory owner varying with the amount of smoke produced and equivalent in money terms to the damage it would cause; or, finally, to exclude

Ronald Coase, "The Problem of Social Cost," 3 *Journal of Law and Economics* (1960). Copyright © 1960 University of Chicago Law School. Reprinted with permission.

the factory from residential districts (and presumably from other areas in which the emission of smoke would have harmful effects on others). It is my contention that the suggested courses of action are inappropriate in that they lead to results which are not necessarily, or even usually, desirable.

The Reciprocal Nature of the Problem

The traditional approach has tended to obscure the nature of the choice that has to be made. The question is commonly thought of as one in which A inflicts harm on B and what has to be decided is, How should we restrain A? But this is wrong. We are dealing with a problem of a reciprocal nature. To avoid the harm to B would be to inflict harm on A. The real question that has to be decided is, Should A be allowed to harm B or should B be allowed to harm A? The problem is to avoid the more serious harm. I instanced in my previous article the case of a confectioner, the noise and vibrations from whose machinery disturbed a doctor in his work. To avoid harming the doctor would be to inflict harm on the confectioner. The problem posed by this case was essentially whether it was worth while, as a result of restricting the methods of production which could be used by the confectioner, to secure more doctoring at the cost of a reduced supply of confectionery products. Another example is afforded by the problem of straying cattle which destroy crops on neighbouring land. If it is inevitable that some cattle will stray, an increase in the supply of meat can only be obtained at the expense of a decrease in the supply of crops. The nature of the choice is clear: meat or crops. What answer should be given is, of course, not clear unless we know the value of what is obtained as well as the value of what is sacrificed to obtain it. To give another example, George J. Stigler instances the contamination of a stream. If we assume that the harmful effect of the pollution is that it kills the fish, the question to be decided is, Is the value of the fish lost greater or less than the value of the product which the contamination of the stream makes possible? It goes almost without saying that this problem has to be looked at in total and at the margin.

The Pricing System with Liability for Damage

I propose to start my analysis by examining a case in which most economists would presumably agree that the problem would be solved in a completely satisfactory manner: when the damaging business has to pay for all damage caused and the pricing system works smoothly (strictly this means that the operation of a pricing system is without cost).

A good example of the problem under discussion is afforded by the case of straying cattle which destroy crops growing on neighbouring land. Let us suppose that a farmer and a cattle-raiser are operating on neighbouring properties. Let us further suppose that, without any fencing between the properties, an increase in the size of the cattle-raiser's herd increases the total damage to the farmer's crops. What happens to the marginal damage as the size of the herd increases is another matter. This depends on whether the cattle tend to follow one another or to roam side by side, on whether they tend to be more or less restless as the size of the herd increases, and on other similar factors. For my immediate purpose, it is immaterial what assumption is made about marginal damage as the size of the herd increases.

To simplify the argument, I propose to use an arithmetical example. I shall assume that the annual cost of fencing the farmer's property is $9 and that the price of the crop is $1 per ton. Also, I assume that the relation between the number of cattle in the herd and the annual crop loss is as follows:

Number in Herd (steers)	Annual Crop Loss (tons)	Crop Loss per Additional Steer (tons)
1	1	1
2	3	2
3	6	3
4	10	4

Given that the cattle-raiser is liable for the damage caused, the additional annual cost imposed on the cattle-raiser if he increased his herd from, say, 2 to 3 steers is $3, and in deciding on the size of the herd, he will take this into account along with his other costs. That is, he will not increase the size of the herd unless the value of the additional meat produced (assuming that the cattle-raiser slaughters the cattle) is greater than the additional costs that this will entail, including the value of the additional crops destroyed. Of course, if, by the employment of dogs, herdsmen, aeroplanes, mobile radio, and other means, the amount of damage can be reduced, these means will be adopted when their cost is less than the value of the crop which they prevent being lost. Given that the annual cost of fencing is $9, the cattle-raiser who wished to have a herd with 4 steers or more would pay for fencing to be erected and maintained, assuming that other means of attaining the same end would not do so more cheaply. When the fence is erected, the marginal cost due to the liability for damage becomes zero, except to the extent that an increase in the size of the herd necessitates a stronger and therefore more expensive fence because more steers are liable to lean against it at the same time. But, of course, it may be cheaper for the cattle-raiser not to fence and to pay for the damaged crops, as in my arithmetical example, with 3 or fewer steers.

It might be thought that the fact that the cattle-raiser would pay for all crops damaged would lead the farmer to increase his planting if a cattle-raiser came to occupy the neighbouring property. But this is not so. If the crop was previously sold in conditions of perfect competition, marginal cost was equal to price for the amount of planting undertaken, and any expansion would have reduced the profits of the farmer. In the new situation, the existence of crop damage would mean that the farmer would sell less on the open market, but his receipts for a given production would remain the same since the cattle-raiser would pay the market price for any crop damaged. Of course, if cattle-raising commonly involved the destruction of crops, the coming into existence of a cattle-raising industry might raise the price of the crops involved and farmers would then extend their planting. But I wish to confine my attention to the individual farmer.

I have said that the occupation of a neighbouring property by a cattle-raiser would not cause the amount of production, or perhaps more exactly the amount of planting, by the farmer to increase. In fact, if the cattle-raising has any effect, it will be to decrease the amount of planting. The reason for this is that, for any given tract of land, if the value of the crop damaged is so great that the receipts from the sale of the undamaged crop are less than the total costs of cultivating that tract of land, it will be profitable for the farmer and the cattle-raiser to make a bargain whereby that tract of land is left uncultivated. This can be made clear by means of an arithmetical example. Assume initially that the value of the crop obtained from cultivating a given tract of land is $12 and that the cost incurred in cultivating this tract of land is $10, the net gain from cultivating the land being $2. I assume for purposes of simplicity that the farmer owns the land. Now assume that the cattle-raiser starts operations on the neighbouring property and that the value of the crops damaged is $1. In this case $11 is obtained by the farmer from sale on the market and $1 is obtained from the cattle-raiser for damage suffered and the net gain remains $2. Now suppose that the cattle-raiser finds it profitable to increase the size of his herd, even though the amount of damage rises to $3; which means that the value of the additional meat production is greater than the additional costs, including the additional $2 payment for damage. But the total payment for damage is now $3. The net gain to the farmer from cultivating the land is still $2. The cattle-raiser would be better off if the farmer would agree not to cultivate his land for any payment less than $3. The farmer would be agreeable to not cultivating the land for any payment greater than $2. There is clearly room for a mutually satisfactory bargain which would lead to the abandonment of cultivation. But the same argument applies not only to the whole tract cultivated by the farmer but also to any subdivision of it. Suppose, for example, that the cattle have a well-defined route, say, to a brook or to a shady area. In these circumstances, the amount of damage to the crop along the route may well be great; and if so, it could be that the farmer and the cattle-raiser would find it profitable to make a bargain whereby the farmer would agree not to cultivate this strip of land.

But this raises a further possibility. Suppose that there is such a well-defined route. Suppose further that the value of the crop that would be obtained by cultivating this strip of land is $10 but that the cost of cultivation is $11. In the absence of the cattle-raiser, it could well be that if the strip was cultivated, the whole crop would be destroyed by the cattle. In this case, the cattle-raiser would be forced to pay $10 to the farmer. It is true that the farmer would lose $1. But the cattle-raiser would lose $10. Clearly this is a situation which is not likely to last indefinitely since neither party would want this to happen. The aim of the farmer would be to induce the cattle-raiser to make a payment in return for an agreement to leave this land uncultivated. The farmer would not be able to obtain a payment greater than the cost of fencing off this piece of land nor so high as to lead the cattle-raiser to abandon the use of the neighbouring property. What payment would in fact be made would depend on the shrewdness of the farmer and the cattle-raiser as bargainers. But as the payment would not be so high as to cause the cattle-raiser to abandon this location and as it would not vary with the size of the herd, such an agreement would not affect the allocation of resources but would merely alter the distribution of income and wealth between the cattle-raiser and the farmer.

I think it is clear that if the cattle-raiser is liable for damage caused and the pricing system works smoothly, the reduction in the value of production elsewhere will be taken into account in computing the additional cost involved in increasing the size of the herd. This cost will be weighed against the value of the additional meat production and, given perfect competition in the cattle industry, the allocation of resources in cattle-raising will be optimal. . . .

The Pricing System with No Liability for Damage

I now turn to the case in which, although the pricing system is assumed to work smoothly (that is, costlessly), the damaging business is not liable for any of the damage which it causes. This business does not have to make a payment to those damaged by its actions. I propose to show that the allocation of resources will be the same in this case as it was when the damaging business was liable for damage caused. As I showed in the previous case that the allocation of resources was optimal, it will not be necessary to repeat this part of the argument.

I return to the case of the farmer and the cattle-raiser. The farmer would suffer increased damage to his crop as the size of the herd increased. Suppose that the size of the cattle-raiser's herd is three steers (and that this is the size of the herd that would be maintained if crop damage was not taken into account). Then the farmer would be willing to pay up to $3 if the cattle-raiser would reduce his herd to two steers, up to $5 if the herd were reduced to one steer, and up to $6 if cattle-raising was abandoned. The cattle-raiser would therefore receive $3 from the farmer if he kept two steers instead of three. This $3 foregone is therefore part of the cost incurred in keeping the third steer. Whether the $3 is a payment which the cattle-raiser has to make if he adds the third steer to his herd (which it would be if the cattle-raiser was liable to the farmer for damage caused to the crop) or whether it is a sum of money which he would have received if he did not keep a third steer (which it would be if the cattle-raiser was not liable to the farmer for damage caused to the crop) does not affect the final result. In both cases $3 is part of the cost of adding a third steer, to be included along with the other costs. If the increase in the value of production in cattle-raising through increasing the size of the herd from two to three is greater than the additional costs that have to be incurred (including the $3 damage to crops), the size of the herd will be increased. Otherwise, it will not. The size of the herd will be the same whether the cattle-raiser is liable for damage caused to the crop or not.

It may be argued that the assumed starting point—a herd of three steers—was arbitrary. And this is true. But the farmer would not wish to pay to avoid crop damage which the cattle-raiser would not be able to cause. For example, the maximum annual payment which the farmer could be induced to pay could not exceed $9, the annual cost of fencing. And the farmer would only be willing to pay this sum if it did not reduce his earnings to a level that would cause him to abandon cultivation of this particular tract of land. Furthermore, the farmer would only be willing to pay this amount if he believed that, in the absence of any payment by him, the size of the herd maintained by the cattle-raiser would be four or more steers. Let us assume that this is the case. Then the farmer would be willing to pay up to $3 if the cattle-raiser would reduce his herd to three steers, up to $6 if the herd were reduced to two steers, up to $8 if one steer only were kept, and up to $9 if cattle-raising were abandoned. It will be noticed that the change in the starting point has not altered the amount which would accrue to the cattle-raiser if he reduced the size of his herd by any given amount. It is still true that the cattle-raiser could receive an additional $3 from the farmer if he agreed to reduce his herd from three steers to two and that the $3 represents the value of the crop that would be destroyed by adding the third steer to the herd. Although a different belief on the part of the farmer (whether justified or not) about the size of the herd that the cattle-raiser would maintain in the absence of payments from him may affect the total payment he can be induced to pay, it is not true that this different belief would have any effect on the size of the herd that the cattle-raiser will actually keep. This will be the same as it would be if the cattle-raiser had to pay for damage caused by his cattle, since a receipt forgone of a given amount is the equivalent of a payment of the same amount.

It might be thought that it would pay the cattle-raiser to increase his herd above the size that he would wish to maintain once a bargain had been made, in order to induce the farmer to make a larger total payment. And this may be true. It is similar in nature to the action of the farmer (when the cattle-raiser was liable for damage) in cultivating land on which, as a result of an agreement with the cattle-raiser, planting would subsequently be abandoned (including land which would not be cultivated at all in the absence of cattle-raising). But such manoeuvres are preliminaries to an agreement and do not affect the long-run equilibrium position, which is the same whether or not the cattle-raiser is held responsible for the crop damage brought about by his cattle.

It is necessary to know whether the damaging business is liable or not for damage caused, since without the establishment of this initial delimitation of rights there can be no market transactions to transfer and recombine them. But the ultimate result (which maximizes the value of production) is independent of the legal position if the pricing system is assumed to work without cost.

The Problem Illustrated Anew

The harmful effects of the activities of a business can assume a wide variety of forms. An early English case concerned a building which, by obstructing currents of air, hindered the operation of a windmill. A recent case in Florida concerned a building which cast a shadow on the cabana, swimming pool, and sunbathing areas of a neighbouring hotel. The problem of straying cattle and the damaging of crops which was the subject of detailed examination in the two preceding sections, although it may have appeared to be rather a special case, is in fact but one example of a problem which arises in many different guises. . . .

Let us first reconsider the case of *Sturges v. Bridgman.* In this case, a confectioner (in Wigmore Street) used two mortars and pestles in connection with his business (one had been in operation in the same position for more than sixty years and the other for more than twenty-six years). A doctor then came to occupy neighbouring premises (in Wimpole Street). The confectioner's machinery caused the doctor no harm until, eight years after he had first occupied the premises, he built a consulting room at the end of his garden right against the confectioner's kitchen. It was then found that the noise and vibration caused by the confectioner's machinery made it difficult for the doctor to use his new consulting room. "In particular . . . the noise prevented him from examining his patients by auscultation for diseases of the chest. He also found it impossible to engage with effect in any occupation which required thought and attention." The doctor therefore brought a legal action to force the confectioner to stop using his machinery. The courts had little difficulty in granting the doctor the injunction he sought. "Individual cases of hardship may occur in the strict carrying out of the principle upon which we found our judgment, but the negation of the principle would lead even more to individual hardship, and would at the same time produce a prejudicial effect upon the development of land for residential purposes."

The court's decision established that the doctor had the right to prevent the confectioner from using his machinery. But, of course, it would have been possible to modify the arrangements envisaged in the legal ruling by means of a bargain between the parties. The doctor would have been willing to waive his right and allow the machinery to continue in operation if the confectioner would have paid him a sum of money which was greater than the loss of income which he would suffer from having to move to a more costly or less convenient location, from having to curtail his activities at this location, or (and this was suggested as a possibility) from having to build a separate wall which would deaden the noise and vibration. The confectioner would have been willing to do this if the amount he would have had to pay the doctor was less than the fall in income he would suffer if he had to change his mode of operation at this location, abandon his operation, or move his confectionery business to some other location. The solution of the problem depends essentially on whether the continued use of the machinery adds more to the confectioner's income than it subtracts from the doctor's. But now consider the situation if the confectioner had won the case. The confectioner would then have had the right to continue operating his noise- and vibration-generating machinery without having to pay anything to the doctor. The boot would have been on the other foot: the doctor would have had to pay the confectioner to induce him to stop using the machinery. If the doctor's income would have fallen more through continuance of the use of this machinery than it added to the income of the confectioner, there would clearly be room for a bargain whereby the doctor paid the confectioner to stop using the machinery. That is to say, the circumstances in which it would not pay the confectioner to continue to use the machinery and to compensate the doctor for the losses that this would bring (if the doctor had the right to prevent the confectioner's using his machinery) would be those in which it would be in the interest of the doctor to make a payment to the confectioner which would induce him to discontinue the use of the machinery (if the confectioner had the right to operate the machinery). The basic conditions are exactly the same in this case as they were in the example of the cattle which destroyed crops. With costless market transactions, the decision of the courts concerning liability for damage would be without effect on the allocation of resources. It was of course the view of the judges that they were affecting the working of the economic system—and in a desirable direction. Any other decision would have had "a prejudicial effect upon the development of land for residential purposes," an argument which was elaborated by examining the example of a forge operating on a barren moor which was later developed for residential purposes. The judges' view that they were settling how the land was to be used would be true only in the case in which the costs of carrying out the necessary market transactions exceeded the gain which might be achieved by any arrangement of rights. And it would be desirable to preserve the areas (Wimpole Street or the moor) for residential or professional use (by giving non-industrial users the right to stop the noise, vibration, smoke, etc., by injunction) only if the value of the additional residential facilities obtained was greater than the value of cakes or iron lost. But of this the judges seem to have been unaware. . . .

The Cost of Market Transactions Taken into Account

The argument has proceeded up to this point on the assumption . . . that there were no costs involved in carrying out market transactions. This is, of course, a very unrealistic assumption. In order to carry out a market transaction, it is necessary to discover who it is that one wishes to deal with, to inform people that one wishes to deal and on what terms, to conduct negotiations leading up to a bargain, to draw up the contract, to undertake the inspection needed to make sure that the terms of the contract are being observed, and so on. These operations are often extremely costly, sufficiently costly at any rate to prevent many transactions that would be carried out in a world in which the pricing system worked without cost.

In earlier sections, when dealing with the problem of the rearrangement of legal rights through the market, I argued that such a rearrangement would be made through the market whenever this would lead to an increase in the value of production. But this assumed costless market transactions. Once the costs of carrying out market transactions are taken into account, it is clear that such a rearrangement of rights will only be undertaken when the increase in the value of production consequent upon the rearrangement is greater than the costs which would be involved in bringing it about. When it is less, the granting of an injunction (or the knowledge that it would be granted) or the liability to pay damages may result in an activity being discontinued (or may prevent its being started) which would be undertaken if market transactions were costless. In these conditions, the initial delimitation of legal rights does have an effect on the efficiency with which the economic system operates. One arrangement of rights may bring about a greater value of production than any other. But unless this is the arrangement of rights established by the legal system, the costs of reaching the same result by altering and combining rights through the market may be so great that this optimal arrangement of rights, and the greater value of production which it would bring, may never be achieved. . . .

It is clear that an alternative form of economic organization which could achieve the same result at less cost than would be incurred by using the market would enable the value of production to be raised. . . . [T]he firm represents such an alternative to organizing production through market transactions. Within the firm, individual bargains between the various co-operating factors of production are eliminated and for a market transaction is substituted an administrative decision. The rearrangement of production then takes place without the need for bargains among the owners of the factors of production. A landowner who has control of a large tract of land may devote his land to various uses, taking into account the effect that the interrelations of the various activities will have on the net return of the land, thus rendering unnecessary bargains between those undertaking the various activities. Owners of a large building or of several adjoining properties in a given area may act in much the same way. In effect, based upon our earlier terminology, the firm would acquire the legal rights of all the parties, and the rearrangement of activities would not follow on a rearrangement of rights by contract but as a result of an administrative decision as to how the rights should be used.

It does not, of course, follow that the administrative costs of organizing a transaction through a firm are inevitably less than the costs of the market transactions which are superseded. But where contracts are peculiarly difficult to draw up and an attempt to describe what the parties have agreed to do or not to do (for example, the amount and kind of a smell or noise that they may make or will not make) would necessitate a lengthy and highly involved document, and where, as is probable, a long-term contract would be desirable, it would be hardly surprising if the emergence of a firm or the extension of the activities of an existing firm was not the solution adopted on many occasions to deal with the problem of harmful effects. This solution would be adopted whenever the administrative costs of the firm were less than the costs of the market transactions that it supersedes and the gains which would result from the rearrangement of activities greater than the firm's costs of organizing them. . . .

But the firm is not the only possible answer to this problem. The administrative costs of organizing transactions within the firm may also be high, and particularly so when many diverse activities are brought within the control of a single organization. In the standard case of a smoke nuisance, which may affect a vast number of people engaged in a wide variety of activities, the administrative costs might well be so high as to make any attempt to deal with the problem within the confines of a single firm impossible. An alternative solution is direct governmental regulation. Instead of instituting a legal system of rights which can be modified by transactions on the market, the government may impose regulations which state what people must or must not do and which have to be obeyed. Thus, the government (by statute or perhaps more likely through an administrative agency) may, to deal with the problem of smoke nuisance, decree that certain methods of production should or should not be used (for example, that smoke-preventing devices should be installed or that coal or oil should not be burned) or may confine certain types of business to certain districts (zoning regulations).

The government is, in a sense, a super-firm (but of a very special kind) since it is able to influence the use of factors of productions by administrative decision. But the ordinary firm is subject to checks in its operations because of the competition of other firms which might administer the same activities at lower cost, and also because there is always the alternative of market transactions against organization within the firm if the administrative costs become too great. The government is able, if it wishes, to avoid the market altogether, which a firm can never do. The firm has to make market agreements with the owners of the factors of production that it uses. Just as the government can conscript or seize property, so it can decree that factors of production should only be used in such-and-such a way. Such authoritarian methods save a lot of trouble (for those doing the organizing). Furthermore, the government has at its disposal the police and the other law enforcement agencies to make sure that its regulations are carried out.

It is clear that the government has powers which might enable it to get some things done at a lower cost than could a private organization (or at any rate one without special governmental powers). But the governmental administrative machine is not itself costless. It can, in fact, on occasion be extremely costly. Furthermore, there is no reason to suppose that the restrictive and zoning regulations, made by a fallible administration subject to political pressures and operating without any competitive check, will necessarily always be those which increase the efficiency with which the economic system operates. Furthermore, such general regulations which must apply to a wide variety of cases will be enforced in some cases in which they are clearly inappropriate. From these considerations it follows that direct governmental regulations will not necessarily give better results than leaving the problem to be solved by the market or the firm. But equally, there is no reason why, on occasion, such governmental administrative regulation should not lead to an improvement in economic efficiency. This would seem particularly likely when, as is normally the case with the smoke nuisance, a large number of people is involved and when therefore the costs of handling the problem through the market or the firm may be high.

There is, of course, a further alternative, which is to do nothing about the problem at all. And given that the costs involved in solving the problem by regulations issued by the governmental administrative machine will often be heavy (particularly if the costs are interpreted to include all the consequences which follow from the government engaging in this kind of activity), it will no doubt be commonly the case that the gain which would come from regulating the actions which give rise to the harmful effects will be less than the costs involved in governmental regulation.

The discussion of the problem of harmful effects in this section (when the costs of market transactions are taken into account) is extremely inadequate. But at least it has made clear that the problem is one of choosing the appropriate social arrangement for dealing with the harmful effects. All solutions have costs, and there is no reason to suppose that governmental regulation is called for simply because the problem is not well handled by the market or the firm. Satisfactory views on policy can only come from a patient study of how, in practice, the market, firms, and governments handle the problem of harmful effects. Economists need to study the work of the broker in bringing parties together, the effectiveness of restrictive covenants, the problems of the large-scale real-estate development company, the operation of governmental zoning, and other regulating activities. It is my belief that economists, and policymakers generally, have tended to overestimate the advantages which come from governmental regulation. But this belief, even if justified, does not do more than suggest that governmental regulation should be curtailed. It does not tell us where the boundary line should be drawn. This, it seems to me, has to come from a detailed investigation of the actual results of handling the problem in different ways. But it would be unfortunate if this investigation were undertaken with the aid of a faulty economic analysis. . . .

2.3   COMPARING THE MODELS: TRANSACTION COSTS

When Does the Rule of Liability Matter?

HAROLD DEMSETZ

. . . The questions with which we shall be concerned are whether and under what conditions a legal decision about liability affects the uses to which resources will be put and the distribution of wealth between owners of resources. If ranchers are held liable for the damage done by their cattle to corn fields, how will the outputs of meat and corn be affected? If drivers or pedestrians, alternatively, are held liable for automobile-pedestrian accidents, how will the accident rate be affected? What implications for extortion (an extreme form of wealth redistribution) are found in the decision about who is liable for damages?

. . .

The problem of "extortion" is part of the larger problem of wealth redistribution that may accompany a change in the rule of liability. Our concern here is with situations in which such a redistribution takes place. However, it should be noted that, when there is no restriction on contracting, a change in the rule of liability need not be accompanied by wealth redistribution. If owners of firms are made liable for industrial accidents, for example, then the equilibrium wage will move downward to reflect the shifting of this explicit cost from workers to employers. Employers no longer will need to cover the cost of industrial accidents in the wages they pay since this cost will be paid by them in the form of industrial accident insurance or self-insurance required by the new rule of liability. The general effect of shifting accident liability directly to firms will be merely to change the classification, not the amount, of remuneration. What under no employer liability were simply wages become under employer liability wages plus accident benefits. No redistribution of wealth accompanies the change in liability. Workers who, when they had to bear the cost of accidents directly, received $X in wages will, under the new rule of liability, receive part of the $X in the form of accident compensation and the remainder in wages, but there will be no change in their total income after taking account of expected accident costs under the two systems.

This holds strictly only if workers and employers are allowed to enter into voluntary contractual arrangements for reshifting the explicit cost back to workers, a matter that need not be discussed in detail here. If such agreements are disallowed by the law—i.e., if the costs of making such agreements is prohibitively high because of their illegality—then some wealth may be redistributed from those workers who would have found it advantageous to self-insure to workers who find it advantageous to buy insurance; such a law would force workers, in the wage reductions they must

Harold Demsetz, When Does the Rule of Liability Matter?, 1 *Journal of Legal Studies* (1972), pp. 13–28.

accept, to purchase insurance for industrial accidents from their employers. The problem of "extortion" arises when a change in liability gives rise to a redistribution in wealth. In the farmer-rancher case, the relative values of nearby farm and ranchlands will be changed when the rule of liability is altered. Under one rule of liability, with farmers required to bear the cost of crop damage, farmers will need to pay ranchers to reduce herd size; under the other rule ranchers will have to pay farmers for damages or for any alteration in the quantity of corn grown nearby. The change in the direction of payments must affect the rents that can be collected by owners of these lands and thus the market values of these lands.

In these cases the owner of the specialized resource, ranchland or farmland, that is not required to bear the cost of the interaction may threaten to increase the intensity of the interaction in an attempt to get his neighbor to pay him a larger sum than would ordinarily be required to obtain his cooperation in adjusting the intensity of the interaction downward. The owner of ranchland, if he is not liable for crop damage done by straying cattle, might, in the absence of a neighboring farmer, raise only 1,000 head of cattle. With proximity between farming and ranching, a neighboring owner of farmland might be willing to pay the rancher the sum required to finance a 200-head reduction in herd size. However, if the owner of ranchland threatens to raise 1,500 head, he may be able to secure more than this sum from the farmer because of the additional crop damage that would be caused by the larger herd size. With or without this "extortion" threat, the size of the herd will be reduced to 800 because that is the size, by assumption, that maximizes the total value of both activities. Given the interrelationship between the two activities, that is the herd size that will maximize the return to the farmer and, indirectly, the sum available for possible transfer to the rancher. What is at issue is the sharing of this maximum return.

To the extent that there exist alternative farm sites, the ability of the owner of ranchland to make such a threat credible is compromised. Competition among such owners will reduce the payment that farmers make to ranchers to that sum which is just sufficient to offset the revenue forgone by ranchers when herd size is reduced. No rancher could succeed in a threat to increase herd size above normal numbers because other ranchers would be willing to compete to zero the price that farmers are asked to pay to avoid abnormally large herd sizes. Abnormally large herd size, in itself, will generate losses to owners of ranchland and, for this reason, competition among such owners will reduce the price that owners of farmland must pay to avoid such excessive herd sizes to zero.

But if a ranchland owner has a locational monopoly, in the sense that there are no alternative sites available to farmers, then the rancher may succeed in acquiring a larger sum from his neighboring farmer in order to avoid abnormally large herd sizes. The acquisition of a larger sum by the owner of ranchland generally will require him to incur some cost to make his threat credible, perhaps by actually beginning to increase herd size beyond normal levels. If the cost of making this threat credible is low relative to the sum that available for transfer from the owner of farmland, the rancher will be in a good position to accomplish the transfer. The sum available for transfer will be the amount by which the value of the neighboring land when used as farmland exceeds its value in the next best use. If the rancher were to demand a larger payment from his neighbor, the neighboring land would be switched to some other use.

The temptation to label such threats extortion or blackmail must be resisted by economists for these are legal and not economic distinctions. The rancher merely attempts to maximize profits. If his agreements with neighboring farmers are marketed in competition with other ranchers, profit maximization constrained by competition implies that an agreement to reduce herd size can be purchased for a smaller payment than if effective competition in such agreements is absent. The appropriate economic label for this problem is nothing more nor less than monopoly. It takes on the cast of such legal classifications as extortion only because the context seems to be one where the monopoly return is received by threatening to produce something that is not wanted—excessively large herds. The conventional monopoly problem involves a reduction or a threat to reduce the output of a desired good. In the unconventional monopoly problem presented here, there is a threat to increase herd size beyond desirable levels. But this difference is superficial. The conventional monopoly problem can be viewed as one in which the monopolist produces more scarcity than is desired, and the unconventional monopoly problem discussed here can be considered one in which the monopolist threatens to produce too small a reduction in crop damage. Any additional sum that the rancher succeeds in transferring to himself from the farmer is correctly identified as a monopoly return.

The temptation to resolve this monopoly problem merely by reversing the rule of liability must be resisted. Should the liability rule be reversed and the owner of ranchland now be held liable for damage done by his cattle to surrounding crops, the specific monopoly problem that we have been discussing would be resolved. But if the farmer enjoys a locational monopoly such that the rancher has nowhere else to locate, the shoe will now be on the other foot. The farmer can threaten to increase the number of bushels of corn planted, and hence the damage for which the rancher will be liable, unless the rancher pays the farmer a sum greater than would be required under competitive conditions. The potential for monopoly and the wealth redistribution implied by monopoly is present in principle whether or not the owner of ranchland is held liable for damages. Both the symmetry of the problem and its disappearance under competitive conditions refute the allegation that Coase's analysis implicitly endorses the use of resources in undesirable activities.

. . .

The costly interaction between farming and ranching is not properly attributed to the actions of either party individually, being "caused," instead, by resource scarcity, the scarcity of land and fencing materials. If transaction cost is negligible, it would seem that the choice of liability rule cannot depend on who "causes" the damage since both jointly do, or on how resource allocation will be altered, since no such alteration will take place, but largely on judicial or legislative preferences with regard to wealth distribution.

Once significant transacting or negotiating cost is admitted into the analysis, the choice of liability rule will have effects on resource allocation, and it no longer follows that wealth distribution is the main or even an important consideration in choosing the liability rule. The assumption of negligible transacting cost can be only a beginning to understanding the economic consequences of the legal arrangements that underlie the operations of the economy, but little more can be done here than to illustrate the nature of the considerations.

The most obvious effect of introducing significant transacting cost is that negotiations will not be consummated in those situations where the expected benefits from exchange are less than the expected cost of exchanging. Exchange opportunities will be exploited only up to the point where the marginal gain from trade equals the marginal cost of trade. Of course, there is nothing necessarily inefficient in halting exchange at this point. If this were all that could be said on the subject, there would be little more to do than call the reader's attention to the similar analytical roles of transport cost in international trade and transacting cost in exchange generally. But there is more to say.

Significant transacting cost implies that the rule of liability generally will have allocative effects (as Coase recognizes). Consider the problem of liability for automobile-pedestrian accidents. To the extent that "accident" has any economic meaning it must mean that circumstances are such that voluntary negotiations between the driver and the pedestrian are prohibitively costly in many driving situations. The parties to an accident, either because of the speed with which the accident occurs or because of a failure to notice the presence of a competing claimant for the right-of-way, cannot conclude an agreement over the use of the right-of-way at costs that are low enough, ex ante, to make the effort worthwhile.

Partly as a consequence of the costliness of such negotiations, rules of the road are developed. Speed limits, traffic signals, and legal constraints on passing are substituted for the development of saleable private rights. In a specific case it may be possible to assign private rights to use the road in a way that makes the exchange of these rights feasible, but, in general, if these rules make economic sense it is precisely because the cost of transacting is expected to be too high in most cases to warrant the development of saleable private rights to the use of roads.

The practicality of such rules is not an argument for or against government action, but a rationale for the substitution of rules for negotiation. The use of rules to eliminate costly negotiations can be found in the management of privately owned parking lots and toll roads as well as in those that are publicly owned.

Such rules notwithstanding, accidents do take place. Assuming that the cost of transacting is too high to make negotiated agreements practical in such cases, we can compare the effect on resource allocation of the rule of liability that is chosen. If drivers are held liable in automobile-pedestrian accidents, the incentives for pedestrians to be careful about how and where they cross streets will be reduced. The incentives for drivers to be careful will be increased. Indeed, if each pedestrian could be guaranteed full compensation for all financial, physical, and psychological costs suffered in an accident, then pedestrians would become indifferent between being struck by an auto and not being struck. Drivers, however, would actively seek to avoid accidents since they would always be liable, whereas if it were possible to have a system of complete and full pedestrian liability it would be the drivers who became indifferent between accidents and no accidents and it would be the pedestrians who actively sought to avoid accidents.

In a regime in which transacting cost was zero, either system of liability would generate the same accident-avoiding behavior, as the Coase analysis suggests. With driver liability, drivers would themselves avoid accidents or, if such avoidance could be purchased at lower cost from pedestrians, drivers would pay pedestrians to avoid accidents. Under a scheme of pedestrian liability it would be the pedestrians who took direct action to avoid accidents or indirect action by paying drivers to avoid accidents. Under either rule of liability those accidents are avoided for which the accident cost exceeds the least cost method of avoiding accidents, where the least cost is the lesser of either the driver or pedestrian cost of avoiding accidents. Both rules of liability, assuming zero transacting cost, yield the same accident rate and the same accident-avoiding behavior. The effect of switching from one rule of liability to another is limited to wealth redistribution.

In a situation in which transacting cost is prohibitively high, driver liability leads to the avoidance only of those accidents for which the cost of avoidance to the driver is less than the expected accident cost, and pedestrian liability leads to the avoidance of only those accidents for which the cost of avoidance to the pedestrian is less than the expected accident cost. In general, the accident rate that results will differ under these two systems since the cost of avoiding accidents will not be the same for drivers and pedestrians. Both systems will lead to higher accident rates than would be true if transacting costs were zero. The effect of positive transacting cost is to raise the cost of avoiding accidents through the foreclosure of the use of possibly cheaper cost-avoidance techniques when these can be employed only by the other party to the accident. A similar conclusion can be reached for all liability problems when transacting cost is prohibitive and when the law cannot particularize the rule of liability to take account of who is the least-cost damage avoider in every instance.

One liability rule may be superior to another if transacting costs are more than negligible precisely because the difficulty of avoiding costly interactions is not generally the same for the interacting parties. It may be less costly for pedestrians to avoid accidents or for farmers to relocate their crops than it is for drivers to avoid accidents or ranchers to reduce the number of cattle they raise. If information about this were known, it would be possible for the legal system to improve the allocation of resources by placing liability on that party who in the usual situation could be expected to avoid the costly interaction most cheaply.

The use of words such as "blame," "responsible," and "fault" must be treated with care by the economist because they have no useful meanings in an economic analysis of these problems other than as synonyms for the party who could have most easily avoided the costly interaction. Whether the interaction problem involves crop damage, accidents, soot, or water pollution, the qualitative relationship between the interacting parties is symmetrical. It is the joint use of a resource, be it geographic location, air, or water that leads to these interactions. It is the demand for scarce resources that leads to conflicting interests.

The legal system does produce rules for determining prima facie "fault," but in this context "fault" means only according to some acceptable and applicable legal precedent. In an accident involving a rear-end collision, the court generally will place the burden for proving the absence of negligence on the party driving the following car. If a car strikes a person running across a fenced expressway at night the burden of proving the absence of negligence is likely to be placed on the pedestrian. In treating such cases differently, the law bases its decisions on acceptable and appropriate precedents, but the acceptability of these precedents should not be confused with the morality of the interacting parties. A deeper analysis of these precedents may reveal that they generally make sense from the economic viewpoint of placing the liability on that party who can, at least cost, reduce the probability of a costly interaction happening. Less care need be taken by the driver of the following car in a rear-end collision than would need to be taken by the lead driver to avoid the accident, and less care is needed by a pedestrian to refrain from running across an expressway than is needed by a driver to avoid striking the pedestrian. Nor need the acceptability of such precedents be based on restitution since, as these precedents become known, their long-run effect is to deter accidents at least cost. If courts are to ignore wealth, religion, or family in deciding such conflicts, if persons before the courts are to be treated with regard only to the cause of action and available proof, then, as a normative proposition, it is difficult to suggest any criterion for deciding liability other than placing it on the party able to avoid the costly interaction most easily.

NOTES AND QUESTIONS

1. In order for incentives to be economically efficient, costs must be internalized "on the margin," and not just on average. In other words, a person considering a decision that imposes costs on another must not only pay the incremental or *marginal costs* that result, but must also perceive this payment as connected to the decision. This principle is illustrated by Calabresi's discussion of the incentive effects of auto insurance on persons contemplating the purchase of a second car (supra at page 44). Note that in both of Calabresi's hypothetical jurisdictions car owners ultimately pay for all the costs of all auto accidents. In Athens these costs are paid through the medium of auto insurance premiums, while in Sparta they are paid through the medium of general tax revenues. Can you explain why this makes a difference to the owners' incentives? Do the differing regimes have any effects on car owners' incentives aside from the decision to buy a second car? Could they influence decisions about how frequently or how safely to drive?

2. Collective or public goods such as streetlights or national defense, which can be enjoyed by many consumers simultaneously, are often classified as a distinct category of market failure, but they can also be seen as a type of externality. The conventional Pigouvian wisdom regarding public goods is that it is difficult to get individual consumers to contribute voluntarily toward their purchase. Because each consumer regards his contribution as a small fraction of the total funds available, each has the incentive to "free ride"—to stint on his own share while enjoying the benefits of others' contributions. As a result, the market will not provide an adequate supply, and it is necessary to have public provision, financed by involuntary taxation. Viewed in terms of externality, the underlying problem is that each individual contribution toward the public good confers an external benefit on all other consumers, but there is no way for the other consumers to pay for this benefit within the constraints of a private market. From the Coasian viewpoint, of course, the consumers' coordination problem is just another transaction cost; and market arrangements such as private clubs may in practice be just as effective at addressing this cost as state action would be, if not more so. Compare Paul Samuelson, The Pure Theory of Public Expenditure, 13 *Review of Economics and Statistics* 387 (1954) (Pigouvian approach), with Harold Demsetz, The Private Production of Public Goods, 13 *Journal of Law and Economics* 293 (1970) (Coasian approach).

3. Not all actions that affect the welfare of other individuals count as externalities. For instance, a person who buys a large quantity of goods in a thin but competitive market may drive up the price, to the detriment of others who also wish to buy, but there is no externality because the bulk buyer bears the full social cost of his actions in the form of the higher price. Because the bulk buyer is forced to bid against his rivals, the goods also wind up in the highest and best use, and the result is efficient (although it may be open to separate criticism on grounds of distributional equity.) Similarly, a merchant who bids customers away from another merchant through superior service or price injures the latter's welfare, but there is no externality. Some older writers use the term *pecuniary externality* to refer to this intramarket effect and the term *technological externality* for what we have simply been calling an externality, but this usage is today generally regarded as obsolete. Since Coase, the term *externality* has come to be reserved for extramarket phenomena. See generally E. J. Mishan, The Postwar Literature on Externalities: An Interpretive Essay, 9 *Journal of Economic Literature* 1 (1971).

4. The standard approach to externality assumes that the externality under consideration is the only inefficiency in the market. If there are other sources of market failure present as well, however, then internalizing the externality may not be the best policy. An example would be a monopolist that restricts supply below the competitive level but that also produces a small amount of air pollution. The negative externality from the pollution would ordinarily mean that the firm is overproducing, but this is more than made up for by the inefficient monopoly quantity restriction. Adding a governmental Pigouvian tax in this case would exacerbate the inefficiency. This illustrates what economists call the *problem of the second best:* in the best of all possible worlds we would wish to remove both the monopoly and the externality, but if we cannot remove the monopoly, it may be second best to tolerate the externality. Calabresi's discussions of monopoly power and of the relationship between liability and industrial development illustrate the problems raised by the theory of second best; the doctrine of charitable immunity provides another example. See generally R. G. Lipsey and K. Lancaster, The General Theory of Second Best, 24 *Review of Economic Studies* 11 (1956–57); Richard Markovits, A Basic Structure for Microeconomic Policy Analysis in Our Worse-than-Second-Best World: A Proposal and Critique of the Chicago Approach to the Study of Law and Economics, 1975 *Wisconsin Law Review* 950 (1975).

5. It has been argued both that the Coase theorem is a tautology (because transaction costs can be defined as anything that stands in the way of an efficient bargain) and that it is false (because transaction costs in the real world are not zero). A perhaps more telling criticism, first put forward by Donald Regan, The Problem of Social Cost Revisited, 15 *Journal of Law and Economics* 427 (1972), is that the Coase theorem is too vague to be a true theorem. Regan's objection is that Coase never explained why it was individually rational for his negotiators to bargain their way to an efficient result, rather than holding out for a larger share of the surplus. Regan argues that whether it is rational to bargain efficiently depends on institutional arrangements. In fairness to Coase, it is worth noting that he never referred to his argument as a theorem or offered it as an affirmative claim about the real world. Rather, his critique was that economists had not paid enough attention to institutional structures. For Coase's comments on the Coase theorem literature, see Ronald Coase, *The Firm, the Market, and the Law* (Chicago: University of Chicago Press, 1988), chap. 6 ("Notes on the Problem of Social Cost"). Many of Coase's students and colleagues, however, did take the theorem as an affirmative claim, and it is worth evaluating it as such. In this regard, do you find the Coase theorem persuasive in the farmer-rancher setting that Coase used to motivate it? Why or why not? What about in the case of *Sturges v. Bridgman,* which Coase discusses supra at pages 68–70?

6. One should distinguish between two versions of the Coase theorem: a strong version, which claims that in the absence of transaction costs parties will bargain their way to the *same allocation of resources* regardless of the legal regime, and a weaker version, which claims that

they need only bargain their way to *an efficient allocation of resources,* albeit possibly a different one. It is generally agreed that the strong version is invalid in the presence of wealth effects. Because changes in legal rules affect the distribution of wealth, they can change the parties' willingness to pay for legal entitlements, possibly altering the particular allocation they reach. Under both versions of the theorem, however, the legal regime still matters for purposes of distributional equity, if not for efficiency. Indeed, to the extent that Coase's theorem deprives opponents of wealth redistribution of the efficiency argument, it can offer substantial support to advocates of leveling of income and wealth.

7. Many have taken Coase's main lesson to be the importance of *reciprocal cause*—that since both victim and injurer contribute to the risk of an accident, both need to be given incentives to take precautions. Accordingly, it is not necessarily efficient for the state to intervene on behalf of the victim. This interpretation is misleading, since, as Cooter's article illustrates, reciprocal cause is equally important from the Pigouvian perspective. Coase's emphasis on reciprocal cause, however, is rhetorically connected to his critique of the model of market failure. His implication is that because the interaction between the parties' activities is more complex than may initially be apparent, private ordering is likely to do a better job of coordinating the parties' activities than public regulation.

8. The model of market failure is premised on the assumption that those competing for resource use are not linked up by any market—that is, that they are strangers. The model of cooperation is premised on the assumption that the parties can bargain with each other. Accordingly, the former might alternatively be called a model of tort, and the latter a model of contract—the tension between them mirroring the tension between the principles of these two fields of law. In this regard, it is useful to ask whether legal interactions conventionally seen as being on the borderline between tort and contract are better described by a Coasian or a Pigouvian model. Consider, for example, consumer products liability and workplace injuries. In both these areas of law, victims and injurers are often nominally in a market relationship. Is there then any externality for public regulators to worry about?

9. It is useful to consider how the Coase theorem and other propositions of law and economics might be tested empirically. There is a burgeoning experimental literature on the Coase theorem, much of which is worth consulting; see, for example, Elizabeth Hoffman and Matthew Spitzer, The Coase Theorem: Some Experimental Tests, 25 *Journal of Law and Economics* 73 (1982) [finding that experimental subjects tended to bargain to an efficient outcome but that their degree of self-interestedness in bargaining depended on the assignment of entitlements]; Elizabeth Hoffman and Matthew Spitzer, Experimental Law and Economics: An Introduction, 85 *Columbia Law Review* 991–1036 (1985) [general survey]. Another approach is to look for natural experiments arising from sudden and clear changes in the legal regime. For example, in a section of his article not reproduced here, Demsetz discusses the economic consequences of the then-controversial reserve clause in major league baseball, which prohibited an individual player from negotiating to play for any ball club other than the one that currently employed him. This clause, which gave clubs legal control over the movement of players, was alleged by its defenders to prevent wealthy clubs from acquiring too large a share of the best players; it was replaced by a system of free agency several years later as the result of an unanticipated labor arbitration ruling. How could you design a test of Demsetz's prediction, based on his reading of the Coase theorem, that the switch from the reserve clause to free agency would have no allocative effects on the assignment of players to teams? See Mark Kelman, Consumption Theory, Production Theory, and Ideology in the Coase Theorem, 52 *Southern California Law Review* 669 (1979) [arguing that the substantial movement of players after the reserve clause was lifted disproves the Coase theorem]; Matthew Spitzer and Elizabeth Hoffman, Reply to Kelman, 53 *Southern California Law Review* 1187 (1980); Mark Kelman, Rejoinder, 53 *Southern California Law Review* 1215 (1980).

10. For more thorough analysis of the model of market failure, see Francis Bator, The Anatomy of Market Failure, 72 *Quarterly Journal of Economics* 351 (1958); Kenneth Arrow, The Organization of Economic Activity: Issues Pertinent to the Choice of Market versus Nonmarket Allocation, in *Public Expenditure and Policy Analysis*, ed. Robert Haveman and Julius Margolis (Boston: Houghton Mifflin, 1983). For more thorough expositions of the model of cooperation, with emphasis on the concept of transaction costs, see Oliver E. Williamson, Transaction-cost Economics: The Governance of Contractual Relations, 22 *Journal of Law and Economics* 233 (1979); Neil Komesar, *Imperfect Alternatives: Choosing Institutions in Law, Economics, and Public Policy* (Chicago: University of Chicago Press, 1994) [offering an extended comparative analysis of legal institutions]; and Howard Shelanski and Peter Klein, Empirical Research in Transaction Cost Economics: A Review and Assessment, 11 *Journal of Law, Economics, & Organization* 335 (1995).

3

A Survey of Basic Applications

This chapter offers a survey of basic applications of the economic approach to law, drawn from the standard first-year legal curriculum. These applications are meant to illustrate more concretely the basic issues introduced in the preceding chapters: rationality, incentives, efficiency, market failure, and transaction costs. While each individual application is intended to be of independent interest, the collection as a whole is presented with a common goal: to train readers to recognize fundamental economic issues as they arise in legal contexts, and to compare functionally similar problems across doctrinal fields.

The essays by Harold Demsetz and by Guido Calabresi and A. Douglas Melamed examine the basic economics of property rights. Demsetz's essay outlines the relationship between property and economic incentives. In his view, one of the primary functions of the institution of property is to encourage the efficient use of resources by internalizing externalities; without property rights, individuals lack the proper incentives to invest in resource development and conservation. Taking as his primary illustration the development of hunting rights in land among American Indians, he argues that the choice among private, communal, and state systems of ownership is primarily determined by pressures to conserve on transaction costs. Specifically, he claims that in most circumstances private property rights reduce the cost of negotiating over externalities; accordingly, societies will tend to institute such rights in areas where the economic value of resource development is high.

Calabresi and Melamed extend Demsetz's analysis by distinguishing between two methods of protecting entitlements to property: *property rules* and *liability rules*. Under the former method, entitlements are protected by punishing anyone who uses or damages property without its owner's consent; under the latter, entitlements are protected by requiring infringers to pay damages in compensation. The authors argue that the choice between these two alternatives (as well as a third—making the entitlement inalienable) will influence both economic efficiency and the distribution of wealth. Liability rules are more efficient at internalizing externalities, they claim, when the transaction costs of market exchange is high; conversely, property rules are to be preferred when it is costly to determine the value of resources in administrative proceedings.

The essays by Richard Posner and by Steven Shavell explore the model of market failure in the context of tort law, focusing on that field's central doctrinal controversy—the standard of liability. Posner argues that efficient precaution can be promoted either by strict liability or by a negligence standard, so long as the definition of negligence comes from comparing costs and benefits, in the manner of the well-known *Learned Hand formula*. In Posner's view, if potential injurers understand that they will be held liable for negligence when they fail to take cost-justified precautions, they will have an incentive to behave efficiently. A negligence rule may also be desirable on grounds of administrative cost, since it results in fewer lawsuits than does a system of strict liability; it also helps to give precaution incentives to potential victims of accidents. Shavell's article focuses on this last point, showing how negligence helps to achieve what Cooter has called double responsibility on the margin. Under an idealized negligence system, injurers are led to take efficient precautions and thus are not subject to liability. This leaves the residual costs of accidents to be paid by victims, who then have efficient incentives to take care on their part. Because courts face transaction costs in making negligence determinations, however, the Hand formula works imperfectly and cannot encourage potential injurers to act efficiently along all margins of decision. In particular, Shavell argues that a negligence rule cannot provide incentives to engage in an efficient level of the underlying risky activity. Accordingly, the choice between negligence and strict liability turns on whether it is more important to provide the parties with incentives in those categories of precaution that courts can supervise at low cost, or in categories that courts cannot supervise except at high cost.

Thomas Ulen investigates analogous issues in the subject of contracts, focusing on the central doctrinal controversy in that field: remedy for breach. Ulen compares the standard remedial measures of the common law—restitution, reliance, and expectation—with the equitable remedy of specific performance. While he concludes that the expectation measure is the only one that, in theory, provides promisors with efficient incentives with regard to the decision to breach, he views specific performance, modified by whatever stipulated remedies the parties choose to provide on their own, as more efficient than money damages. The reason comes down to comparative transaction costs; in Ulen's view, the private negotiations fostered by specific performance do a better job than courts do at protecting subjective values attached to performance. Furthermore, Ulen argues that specific performance, by placing the costs of post-breach adjustments on the contracting parties rather than on public officials, encourages the parties to economize on those costs both when entering contracts initially and after a dispute has arisen.

Gary Becker's article on the economics of crime and punishment sheds light on deterrence arguments in criminal law. Becker presents crime as a type of economic behavior and the decision whether to become a criminal as equivalent to the problem of occupational choice. Just as law-abiding citizens choose among possible occupations based on the relative monetary and nonmonetary returns and on their individual talents and preferences, so does the criminal choose a life of crime. Furthermore, just as ordinary persons decide how hard to work based on the relative benefits and costs of labor and leisure, criminals have an incentive to increase the amount and severity of their lawbreaking up to the point where their private benefits equal their private costs on the margin. This analysis yields a positive economic theory of deterrence: by raising the expected punishment, society can reduce the relative returns to crime, and hence crime itself. More controversially, Becker also offers a normative theory of the criminal law, showing how the system of law enforcement and criminal punishment might be designed to achieve the efficient amount of crime—that is, the amount that minimizes the total social costs of criminal behavior, including costs of enforcement, prosecution, and punishment.

Finally, Richard Posner and A. Mitchell Polinsky apply economics to the design of legal procedures. Posner offers a general analysis of the economics of procedure, observing that all procedural systems produce two kinds of social costs—direct expenditures on litigation and dispute resolution, and the less tangible but nonetheless real costs of mistaken substantive outcomes. He proposes that, just as the law of torts can be viewed as a regulatory system for controlling the total costs of accidents, we can and should evaluate legal procedures according to how well they minimize procedural costs. Such an approach, however, is complicated by two factors: first, there may be a tradeoff between direct costs and the costs of error; and second, the decision to spend resources on civil litigation is controlled by private parties. As a result, reforms intended to lower the operating costs of the legal system may, through their effects on private decisions to bring lawsuits and to carry them forward to trial, ultimately increase social costs on balance.

Polinsky applies this general approach to the specific policy debate over treble damages in antitrust, focusing on the question of how best to encourage the efficient number of lawsuits. As he explains, individual incentives to bring and pursue lawsuits do not generally conform to the social optimum; there may be either too many civil actions or too few. This is because a typical lawsuit has both positive and negative externalities: it imposes costs on opposing litigants and on the public court system, while conferring the benefits of enhanced deterrence on the population at large.

Treble damages increase potential defendants' incentives to comply with substantive law, but they also increase plaintiffs' incentives to sue, perhaps beyond the optimal level. Just as in the substantive area, the problem is providing double responsibility on the margin. Polinsky argues that in order to provide proper incentives to both plaintiffs and defendants, it is necessary in general to decouple liability—that is, to separate the amounts paid by losing defendants from those received by successful plaintiffs.

3.1 PROPERTY

Toward a Theory of Property Rights
HAROLD DEMSETZ

The Concept and Role of Property Rights

In the world of Robinson Crusoe property rights play no role. Property rights are an instrument of society and derive their significance from the fact that they help a man form those expectations which he can reasonably hold in his dealings with others. These expectations find expression in the laws, customs, and mores of a society. An owner of property rights possesses the consent of fellowmen to allow him to act in particular ways. An owner expects the community to prevent others from interfering with his actions, provided that these actions are not prohibited in the specifications of his rights.

It is important to note that property rights convey the right to benefit or harm oneself or others. Harming a competitor by producing superior products may be permitted, while shooting him may not. A man may be permitted to benefit himself by shooting an intruder but be prohibited from selling below a price floor. It is clear, then, that property rights specify how persons may be benefited and harmed, and, therefore, who must pay whom to modify the actions taken by persons. The recognition of this leads easily to the close relationship between property rights and externalities.

. . . No harmful or beneficial effect is external to the world. Some person or persons always suffer or enjoy these effects. What converts a harmful or beneficial effect into an externality is that the cost of bringing the effect to bear on the decisions of one or more of the interacting persons is too high to make it worthwhile, and this is what the term shall mean here. "Internalizing" such effects refers to a process, usually a change in property rights, that enables these effects to bear (in greater degree) on all interacting persons.

A primary function of property rights is that of guiding incentives to achieve a greater internalization of externalities. Every cost and benefit associated with social interdependencies is a potential externality. One condition is necessary to make costs and benefits externalities. The cost of a transaction in the rights between the parties (internalization) must exceed the gains from internalization. In general, transacting cost can be large relative to gains because of "natural" difficulties in trading or they can be large because of legal reasons. In a lawful society the prohibition of voluntary negotiations makes the cost of transacting infinite. Some costs and benefits are not taken into account by users of resources whenever externalities exist, but allowing transactions increases the degree to which internalization takes place. For example, it might be thought that a firm which uses slave labor will not recognize all the costs of its activities, since it can have its slave labor by paying subsistence wages only. This will not be true if negotiations are permitted, for the slaves can offer to the firm a payment for their freedom based on the expected return to them of being free men. The cost of slavery can thus be internalized in the calculations of the firm. The transition from serf to free man in feudal Europe is an example of this process. . . .

The Emergence of Property Rights

If the main allocative function of property rights is the internalization of beneficial and harmful effects, then the emergence of property rights can be understood best by their association with the emergence of new or different beneficial and harmful effects.

Changes in knowledge result in changes in production functions, market values, and aspirations. New techniques, new ways of doing the same things, and doing new things—all invoke harmful and beneficial effects to which society has not been accustomed. It is my thesis in this part of the paper that the emergence of new property rights takes place in response to the desires of the interacting persons for adjustment to new benefit-cost possibilities.

The thesis can be restated in a slightly different fashion: property rights develop to internalize externalities when the gains of internalization become larger than the cost of internalization. Increased internalization, in the main, results from changes in economic values, changes which stem from the development of new technology and the opening of new markets, changes to which old property rights are poorly attuned. A proper interpretation of this assertion requires that account be taken of a community's preferences for private ownership. Some communities will have less well-developed private ownership systems and more highly developed state ownership systems. But, given a community's tastes in this regard, the emergence of new private or state owned property rights will be in response to changes in technology and relative prices.

I do not mean to assert or to deny that the adjustments in property rights which take place need be the result of a conscious endeavor to cope with new externality problems. These adjustments have arisen in Western societies largely as a result of gradual changes in social mores and in common law precedents. At each step of this adjustment process, it is unlikely that externalities per se were consciously related to the issue being resolved. These legal and moral experiments may be hit-and-miss procedures to some extent but in a society that weights the achievement of efficiency heavily, their viability in the long run will depend on how well they modify behavior to accommodate to the externalities associated with important changes in technology or market values.

A rigorous test of this assertion will require extensive and detailed empirical work. A broad range of examples can be cited that are consistent with it: the development of air rights, renters' rights, rules for liability in automobile accidents, etc. In this part of the discussion, I shall present one group of such examples in some detail. They deal with the development of private property rights in land among American Indians. These examples are broad ranging and come fairly close to what can be called convincing evidence in the field of anthropology.

The question of private ownership of land among aboriginals has held a fascination for anthropologists. It has been one of the intellectual battlegrounds in the attempt to assess the "true nature" of man unconstrained by the "artificialities" of civilization. In the process of carrying on this debate, information has been uncovered that bears directly on the thesis with which we are now concerned. What appears to be accepted as a classic treatment and a high point of this debate is Eleanor Leacock's memoir on *The Montagnes "Hunting Territory" and the Fur Trade.* Leacock's research followed that of Frank G. Speck who had discovered that the Indians of the Labrador Peninsula had a long-established tradition of property in land. This finding was at odds with what was known about the Indians of the American Southwest and it prompted Leacock's study of the Montagnes who inhabited large regions around Quebec.

Leacock clearly established the fact that a close relationship existed, both historically and geographically, between the development of private rights in land and the development of the commercial fur trade. The factual basis of this correlation has gone unchallenged. However, to my knowledge, no theory relating privacy of land to the fur trade has yet been articulated. The factual material uncovered by Speck and Leacock fits the thesis of this paper well, and in doing so, it reveals clearly the role played by property right adjustments in taking account of what economists have often cited as an example of an externality—the overhunting of game.

Because of the lack of control over hunting by others, it is in no person's interest to invest in increasing or maintaining the stock of game. Overly intensive hunting takes place. Thus a successful hunt is viewed as imposing external costs on subsequent hunters—costs that are not taken into account fully in the determination of the extent of hunting and of animal husbandry.

Before the fur trade became established, hunting was carried on primarily for purposes of food and the relatively few furs that were required for the hunter's family. The externality was clearly present. Hunting could be practiced freely and was carried on without assessing its impact on other hunters. But these external effects were of such small significance that it did not pay for anyone to take them into account. There did not exist anything resembling private ownership in land. And in the *Jesuit Relations,* particularly Le Jeune's record of the winter he spent with the Montagnes in 1633–34 and in the brief account given by Father Druilletes in 1647–48, Leacock finds no evidence of private land holdings. Both accounts indicate a socioeconomic organization in which private rights to land are not well developed.

We may safely surmise that the advent of the fur trade had two immediate consequences. First, the value of furs to the Indians was increased considerably. Second, and as a result, the scale of hunting activity rose sharply. Both consequences must have increased considerably the importance of the externalities associated with free hunting. The property right system began to change, and it changed specifically in the direction required to take account of the economic effects made important by the fur trade. The geographical or distributional evidence collected by Leacock indicates an unmistakable correlation between early centers of fur trade and the oldest and most complete development of the private hunting territory.

> By the beginning of the eighteenth century, we begin to have clear evidence that territorial hunting and trapping arrangements by individual families were developing in the area around Quebec. . . . The earliest references to such arrangements in this region indicates a purely temporary allotment of hunting territories. They [Algonkians and Iroquois] divide themselves into several bands in order to hunt more efficiently. It was their custom . . . to appropriate pieces of land about two leagues square for each group to hunt exclusively. Ownership of beaver houses, however had already become established, and when discovered, they were marked. A starving Indian could kill and eat another's beaver if he left the fur and the tail.

The next step toward the hunting territory was probably a seasonal allotment system. An anonymous account written in 1723 states that the "principle of the Indians is to mark off the hunting ground selected by them by blazing the trees with their crests so that they may never encroach on each other. . . . By the middle of the century these allotted territories were relatively stabilized." . . .

Two factors suggest that the thesis is consistent with the absence of similar rights among the Indians of the southwestern plains. The first of these is that there were no plains animals of commercial importance comparable to the fur-bearing animals of the forest, at least not until cattle arrived with Europeans. The second factor is that animals of the plains are primarily grazing species whose habit is to wander over wide tracts of land. The value of establishing boundaries to private hunting territories is thus reduced by the relatively high cost of preventing the animals from moving to adjacent parcels. Hence both the value and cost of establishing private hunting lands in the Southwest are such that we would expect little development along these lines. The externality was just not worth taking into account.

The lands of the Labrador Peninsula shelter forest animals whose habits are considerably different from those of the plains. Forest animals confine their territories to relatively small areas, so that the cost of internalizing the effects of husbanding these animals is considerably reduced. This reduced cost, together with the higher commercial value of fur-bearing forest animals, made it productive to establish private hunting lands. Frank G. Speck finds that family proprietorship among the Indians of the Peninsula included retaliation against trespass. Animal resources were husbanded. Sometimes conservation practices were carried on extensively. Family hunting territories were divided into quarters. Each year the family hunted in a different quarter in rotation, leaving a tract in the center as a sort of bank, not to be hunted over unless forced to do so by a shortage in the regular tract.

To conclude our excursion into the phenomenon of private rights in land among the American Indians, we note one further piece of corroborating evidence. Among the Indians of the Northwest, highly developed private family rights to hunting lands had also emerged—rights which went so far as to include inheritance. Here again we find that forest animals predominate and that the West Coast was frequently visited by sailing schooners whose primary purpose was trading in furs.

The Coalescence and Ownership of Property Rights

I have argued that property rights arise when it becomes economic for those affected by externalities to internalize benefits and costs. But I have not yet examined the forces which will govern the particular form of right ownership. Several idealized forms of ownership must be distinguished at the outset. These are communal ownership, private ownership, and state ownership.

By communal ownership, I shall mean a right which can be exercised by all members of the community. Frequently the rights to till and to hunt the land have been communally owned. The right to walk a city sidewalk is communally owned. Communal ownership means that the community denies to the state or to individual citizens the right to interfere with any person's exercise of communally-owned rights. Private ownership implies that the community recognizes the right of the owner to exclude others from exercising the owner's private rights. State ownership implies that the state may exclude anyone from the use of a right as long as the state follows accepted political procedures for determining who may not use state-owned property. I shall not examine in detail the alternative of state ownership. The object of the analysis which follows is to discern some broad principles governing the development of property rights in communities oriented to private property.

It will be best to begin by considering a particularly useful example that focuses our attention on the problem of land ownership. Suppose that land is communally owned. Every person has the right to hunt, till, or mine the land. This form of ownership fails to concentrate the cost associated with any person's exercise of his communal right on that person. If a person seeks to maximize the value of his communal rights, he will tend to overhunt and overwork the land because some of the costs of his doing so are borne by others. The stock of game and the richness of the soil will be diminished too quickly. It is conceivable that those who own these rights, i.e., every member of the community, can agree to curtail the rate at which they work the lands if negotiating and policing costs are zero. Each can agree to abridge his rights. It is obvious that the costs of reaching such an agreement will not be zero. What is not obvious is just how large these costs may be.

Negotiating costs will be large because it is difficult for many persons to reach a mutually satisfactory agreement, especially when each hold-out has the right to work the land as fast as he pleases. But, even if an agreement among all can be reached, we must yet take account of the costs of policing the agreement, and these may be large, also. After such an agreement is reached, no one will privately own the right to work the land; all can work the land but at an agreed upon shorter workweek. Negotiating costs are increased even further because it is not possible under this system to bring the full expected benefits and expected costs of future generations to bear on current users.

If a single person owns land, he will attempt to maximize its present value by taking into account alternative future time streams of benefits and costs and selecting that one which he believes will maximize the present value of his privately-owned land rights. We all know that this means that he will attempt to take into account the supply and demand conditions that he thinks will exist after his death. It is very difficult to see how the existing communal owners can reach an agreement that takes account of these costs.

In effect, an owner of a private right to use land acts as a broker whose wealth depends on how well he takes into account the competing claims of the present and the future. But with communal rights there is no broker, and the claims of the present generation will be given an uneconomically large weight in determining the intensity with which the land is worked. Future generations might desire to pay present generations enough to change the present intensity of land usage. But they have no living agent to place their claims on the market. Under a communal property system, should a living person pay others to reduce the rate at which they work the land, he would not gain anything of value for his efforts. Communal property means that future generations must speak for themselves. No one has yet estimated the costs of carrying on such a conversation.

The land ownership example confronts us immediately with a great disadvantage of communal property. The effects of a person's activities on his neighbors and on subsequent generations will not be taken into account fully. Communal property results in great externalities. The full costs of the activities of an owner of a communal property right are not borne directly by him, nor can they be called to his attention easily by the willingness of others to pay him an appropriate sum. Communal property rules out a "pay-to-use-the-property" system and high negotiation and policing costs make ineffective a "pay-him-not-to-use-the-property" system.

The state, the courts, or the leaders of the community could attempt to internalize the external costs resulting from communal property by allowing private parcels owned by small groups of persons with similar interests. The logical groups in terms of similar interests, are, of course, the family and the individual. Continuing with our use of the land ownership example, let us initially distribute private titles to land randomly among existing individuals and, further, let the extent of land included in each title be randomly determined.

The resulting private ownership of land will internalize many of the external costs associated with communal ownership, for now an owner, by virtue of his power to exclude others, can generally count on realizing the rewards associated with husbanding the game and increasing the fertility of his land. This concentration of benefits and costs on owners creates incentives to utilize resources more efficiently.

But we have yet to contend with externalities. Under the communal property system the maximization of the value of communal property rights will take place without regard to many costs, because the owner of a communal right cannot exclude others from enjoying the fruits of his efforts and because negotiation costs are too high for all to agree jointly on optimal behavior. The development of private rights permits the owner to economize on the use of those resources from which he has the right to exclude others. Much internalization is accomplished in this way. But the owner of private rights to one parcel does not himself own the rights to the parcel of another private sector. Since he cannot exclude others from their private rights to land, he has no direct incentive (in the absence of negotiations) to economize in the use of his land in a way that takes into account the effects he produces on the land rights of others. If he constructs a dam on his land, he has no direct incentive to take into account the lower water levels produced on his neighbor's land.

This is exactly the same kind of externality that we encountered with communal property rights, but it is present to a lesser degree. Whereas no one had an incentive to store water on any land under the communal system, private owners now can take into account directly those benefits and costs to their land that accompany water storage. But the effects on the land of others will not be taken into account directly.

The partial concentration of benefits and costs that accompany private ownership is only part of the advantage this system offers. The other part, and perhaps the most important, has escaped our notice. The cost of negotiating over the remaining externalities will be reduced greatly. Communal property rights allow anyone to use the land. Under this system it becomes necessary for all to reach an agreement on land use. But the externalities that accompany private ownership of property do not affect all owners, and, generally speaking, it will be necessary for only a few to reach an agreement that takes these effects into account. The cost of negotiating an internalization of these effects is thereby reduced considerably. The point is important enough to elucidate.

Suppose an owner of a communal land right, in the process of plowing a parcel of land, observes a second communal owner constructing a dam on adjacent land. The farmer prefers to have the stream as it is, and so he asks the engineer to stop his construction. The engineer says, "Pay me to stop." The farmer replies, "I will be happy to pay you, but what can you guarantee in return?" The engineer answers, "I can guarantee you that I will not continue constructing the dam, but I cannot guarantee that another engineer will not take up the task because this is communal property; I have no right to exclude him." What would be a simple negotiation between two persons under a private property arrangement turns out to be a rather complex negotiation between the farmer and everyone else. This is the basic explanation, I believe, for the preponderance of single rather than multiple owners of property. Indeed, an increase in the number of owners is an increase in the communality of property and leads, generally, to an increase in the cost of internalizing.

The reduction in negotiating cost that accompanies the private right to exclude others allows most externalities to be internalized at rather low cost. Those that are not are associated with activities that generate external effects impinging upon many people. The soot from smoke affects many homeowners, none of whom is willing to pay enough to the factory to get its owner to reduce smoke output. All homeowners together might be willing to pay enough, but the cost of their getting together may be enough to discourage effective market bargaining. The negotiating problem is compounded even more if the smoke comes not from a single smoke stack but from an industrial district. In such cases, it may be too costly to internalize effects through the marketplace. . . .

The greater are diseconomies of scale to land ownership the more will contractual arrangement be used by the interacting neighbors to settle these differences. Negotiating and policing costs will be compared to costs that depend on the scale of ownership, and parcels of land will tend to be owned in sizes which minimize the sum of these costs.

The interplay of scale economies, negotiating cost, externalities, and the modification of property rights can be seen in the most notable "exception" to the assertion that ownership tends to be an individual affair: the publicly-held corporation. I assume that significant economies of scale in the operation of large corporations is a fact and, also, that large requirements for equity capital can be satisfied more cheaply by acquiring the capital from many purchasers of equity shares. While economies of scale in operating these enterprises exist, economies of scale in the provision of capital do not. Hence, it becomes desirable for many "owners" to form a joint-stock company.

But if all owners participate in each decision that needs to be made by such a company, the scale economies of operating the company will be overcome quickly by high negotiating cost. Hence a delegation of authority for most decisions takes place and, for most of these, a small management group becomes the de facto owners. Effective ownership, i.e., effective control of property, is thus legally concentrated in management's hands. This is the first legal modification, and it takes place in recognition of the high negotiating costs that would otherwise obtain.

The structure of ownership, however, creates some externality difficulties under the law of partnership. If the corporation should fail, partnership law commits each shareholder to meet the debts of the corporation up to the limits of his financial ability. Thus, managerial de facto ownership can have considerable external effects on shareholders. Should property rights remain unmodified, this externality would make it exceedingly difficult for entrepreneurs to acquire equity capital from wealthy individuals. (Although these individuals have recourse to reimbursements from other shareholders, litigation costs will be high.) A second legal modification, limited liability, has taken place to reduce the effect of this externality. De facto management ownership and limited liability combine to minimize the overall cost of operating large enterprises. Shareholders are essentially lenders of equity capital and not owners, although they do participate in such infrequent decisions as those involving mergers. What shareholders really own are their shares and not the corporation. Ownership in the sense of control again becomes a largely individual affair. The shareholders own their shares, and the president of the corporation and possibly a few other top executives control the corporation.

To further ease the impact of management decisions on shareholders, that is, to minimize the impact of externalities under this ownership form, a further legal modification of rights is required. Unlike partnership law, a shareholder may sell his interest without first obtaining the permission of fellow shareholders or without dissolving the corporation. It thus becomes easy for him to get out if his preferences and those of the management are no longer in harmony. This "escape hatch" is extremely important and has given rise to the organized trading of securities. The increase in harmony between managers and shareholders brought about by exchange and by competing managerial groups helps to minimize the external effects associated with the corporate ownership structure. Finally, limited liability considerably reduces the cost of exchanging shares by making it unnecessary for a purchaser of shares to examine in great detail the liabilities of the corporation and the assets of other shareholders; these liabilities can adversely affect a purchaser only up to the extent of the price per share.

The dual tendencies for ownership to rest with individuals and for the extent of an individual's ownership to accord with the minimization of all costs is clear in the land ownership paradigm. The applicability of this paradigm has been extended to the corporation. But it may not be clear yet how widely applicable this paradigm is. Consider the problems of copyright and patents. If a new idea is freely appropriable by all, if there exist communal rights to new ideas, incentives for developing such ideas will be lacking. The benefits derivable from these ideas will not be concentrated on their originators. If we extend some degree of private rights to the originators, these ideas will come forth at a more rapid pace. But the existence of the private rights does not mean that their effects on the property of others will be directly taken into account. A new idea makes an old one obsolete and another old one more valuable. These effects will not be directly taken into account, but they can be called to the attention of the originator of the new idea through market negotiations. All problems of externalities are closely analogous to those which arise in the land ownership example. The relevant variables are identical.

What I have suggested in this paper is an approach to problems in property rights. But it is more than that. It is also a different way of viewing traditional problems. An elaboration of this approach will, I hope, illuminate a great number of social-economic problems.

Property Rules, Liability Rules, and Inalienability: One View of the Cathedral

GUIDO CALABRESI AND A. DOUGLAS MELAMED

. . . The first issue which must be faced by any legal system is one we call the problem of "entitlement." Whenever a state is presented with the conflicting interests of two or more people, or two or more groups of people, it must decide which side to favor. Absent such a decision, access to goods, services, and life itself will be decided on the basis of "might makes right"—whoever is stronger or shrewder will win. Hence the fundamental thing that law does is to decide which of the conflicting parties will be entitled to prevail. The entitlement to make noise versus the entitlement to have silence, the entitlement to pollute versus the entitlement to breathe clean air, the entitlement to have children versus the entitlement to forbid them—these are the first order of legal decisions.

Having made its initial choice, society must enforce that choice. Simply setting the entitlement does not avoid the problem of "might makes right"; a minimum of state intervention is always necessary. Our conventional notions make this easy to comprehend with respect to private property. If Taney owns a cabbage patch and Marshall, who is bigger, wants a cabbage, he will get it unless the state intervenes. But it is not so obvious that the state must also intervene if it chooses the opposite entitlement, communal property. If large Marshall has grown some communal cabbages and chooses to deny them to small Taney, it will take state action to enforce Taney's entitlement to the communal cabbages. . . .

The state not only has to decide whom to entitle, but it must also simultaneously make a series of equally difficult second-order decisions. These decisions go to the manner in which entitlements are protected and to whether an individual is allowed to sell or trade the entitlement. In any given dispute, for example, the state must decide not only which side wins but also the kind of protection to grant. It is with the latter decisions, decisions which shape the subsequent relationship between the winner and the loser, that this article is primarily concerned. We shall consider three types of entitlements—entitlements protected by property rules, entitlements protected by liability rules, and inalienable entitlements. The categories are not, of course, absolutely distinct; but the categorization is useful since it reveals some of the reasons which lead us to protect certain entitlements in certain ways.

An entitlement is protected by a *property rule* to the extent that someone who wishes to remove the entitlement from its holder must buy it from him in a voluntary transaction in which the value of the entitlement is agreed upon by the seller. It is the form of entitlement which gives rise to the least amount of state intervention: once the original entitlement is decided upon, the state does not try to decide its value. It lets each of the parties say how much the entitlement is worth to him, and gives the seller a veto if the buyer does not offer enough. Property rules involve a collective decision as to who is to be given an initial entitlement but not as to the value of the entitlement.

Whenever someone may destroy the initial entitlement if he is willing to pay an objectively determined value for it, an entitlement is protected by a *liability rule*. This value may be what it is thought the original holder of the entitlement would have sold it for. But the holder's complaint that he would have demanded more will not avail him once the objectively determined value is set. Obviously, liability rules involve an additional stage of state intervention: not only are entitlements protected, but their transfer or destruction is allowed on the basis of a value determined by some organ of the state rather than by the parties themselves.

An entitlement is inalienable to the extent that its transfer is not permitted between a willing buyer and a willing seller. The state intervenes not only to determine who is initially entitled and to determine the compensation that must be paid if the entitlement is taken or destroyed, but also to forbid its sale under some or all circumstances. *Inalienability rules* are thus quite different from property and liability rules. Unlike those rules, rules of inalienability not only "protect" the entitlement; they may also be viewed as limiting or regulating the grant of the entitlement itself.

It should be clear that most entitlements to most goods are mixed. Taney's house may be protected by a property rule in situations where Marshall wishes to purchase it, by a liability rule where the government decides to take it by eminent domain, and by a rule of inalienability in situations where Taney is drunk or incompetent. . . .

The Setting of Entitlements

What are the reasons for deciding to entitle people to pollute or to entitle people to forbid pollution, to have children freely or to limit procreation, to own property or to share property? They can be grouped under three headings: economic efficiency, distributional preferences, and other justice considerations.

Economic Efficiency

Perhaps the simplest reason for a particular entitlement is to minimize the administrative costs of enforcement. This was the reason Holmes gave for letting the costs lie where they fall in accidents unless some clear societal benefit is achieved by shifting them. By itself this reason will never justify any result except that of letting the stronger win, for obviously that result minimizes enforcement costs. Nevertheless, administrative efficiency may be relevant to choosing entitlements when other reasons are taken into account. This may occur when the reasons accepted are indifferent between conflicting entitlements and one entitlement is cheaper to enforce than the others. It may also occur when the reasons are not indifferent but lead us only slightly to prefer one over another and the first is considerably more expensive to enforce than the second.

But administrative efficiency is just one aspect of the broader concept of economic efficiency. Economic efficiency asks that we choose the set of entitlements which would lead to that allocation of resources which could not be improved in the sense that a further change would not so improve the condition of those who gained by it that they could compensate those who lost from it and still be better off than before. This is often called Pareto optimality. To give two examples, economic efficiency asks for that combination of entitlements to engage in risky activities and to be free from harm from risky activities which will most likely lead to the lowest sum of accident costs and of costs of avoiding accidents. It asks for that form of property, private or communal, which leads to the highest product for the effort of producing.

Recently it has been argued that on certain assumptions, usually termed the absence of transaction costs, Pareto optimality or economic efficiency will occur regardless of the initial entitlement. . . . But no one makes an assumption of no transaction costs in practice. Like the physicist's assumption of no friction or Say's law in macroeconomics, the assumption of no transaction costs may be a useful starting point, a device which helps us see how, as different elements which may be termed transaction costs become important, the goal of economic efficiency starts to prefer one allocation of entitlements over another. . . .

Distributional Goals

There are, we would suggest, at least two types of distributional concerns which may affect the choice of entitlements. These involve distribution of wealth itself and distribution of certain specific goods, which have sometimes been called merit goods.

All societies have wealth distribution preferences. They are, nonetheless, harder to talk about than are efficiency goals. For efficiency goals can be discussed in terms of a general concept like Pareto optimality to which exceptions—like paternalism—can be noted. Distributional preferences, on the other hand, cannot usefully be discussed in a single conceptual framework. There are some fairly broadly accepted preferences—caste preferences in one society, more rather than less equality in another society. There are also preferences which are linked to dynamic efficiency concepts—producers ought to be rewarded since they will cause everyone to be better off in the end. Finally, there are a myriad of highly individualized preferences as to who should be richer and who poorer which need not have anything to do with either equality or efficiency—silence lovers should be richer than noise lovers because they are worthier.

Difficult as wealth distribution preferences are to analyze, it should be obvious that they play a crucial role in the setting of entitlements. For the placement of entitlements has a fundamental effect on a society's distribution of wealth. It is not enough, if a society wishes absolute equality, to start everyone off with the same amount of money. A financially egalitarian society which gives individuals the right to make noise immediately makes the would-be noisemaker richer than the silence-loving hermit. Similarly, a society which entitles the person with brains to keep what his shrewdness gains him implies a different distribution of wealth from a society which demands from each according to his relative ability but gives to each according to his relative desire. One can go further and consider that a beautiful woman or handsome man is better off in a society which entitles individuals to bodily integrity than in one which gives everybody use of all the beauty available. . . .

If the choice of entitlements affects wealth distribution generally, it also affects the chances that people will obtain what have sometimes been called merit goods. Whenever a society wishes to maximize the chances that individuals will have at least a minimum endowment of certain particular goods—education, clothes, bodily integrity—the society is likely to begin by giving the individuals an entitlement to them. If the society deems such an endowment to be essential regardless of individual desires, it will, of course, make the entitlement inalienable. Why, however, would a society entitle individuals to specific goods rather than to money with which they can buy what they wish, unless it deems that it can decide better than the individuals what benefits them and society; unless, in other words, it wishes to make the entitlement inalienable?

We have seen that an entitlement to a good or to its converse is essentially inevitable. We either are entitled to have silence or entitled to make noise in a given set of circumstances. We either have the right to our own property or body or the right to share others' property or bodies. We may buy or sell ourselves into the opposite positions but we must start somewhere. Under these circumstances, a society which prefers people to have silence, or own property, or have bodily integrity, but which does not hold the grounds for its preference to be sufficiently strong to justify overriding contrary preferences by individuals, will give such entitlements according to the collective preference, even though it will allow them to be sold thereafter.

Whenever transactions to sell or buy entitlements are very expensive, such an initial entitlement decision will be nearly as effective in assuring that individuals will have the merit good as would be making the entitlement inalienable. Since coercion is inherent because of the fact that a good cannot practically be bought or sold, a society can choose only whether to make an individual have the good, by giving it to him, or to prevent him from getting it by giving him money instead. In such circumstances society will pick the entitlement it deems favorable to the general welfare and not worry about coercion or alienability; it has increased the chances that individuals will have a particular good without increasing the degree of coercion imposed on individuals. A common example of this may occur where the good involved is the present certainty of being able to buy a future benefit and where a futures market in that good is too expensive to be feasible.

Other Justice Reasons

The final reasons for a society's choice of initial entitlements we termed other justice reasons, and we may as well admit that it is hard to know what content can be poured into that term, at least given the very broad definitions of economic efficiency and distributional goals that we have used. Is there, in other words, a reason which would influence a society's choice of initial entitlements that cannot be comprehended in terms of efficiency and distribution? A couple of examples will indicate the problem.

Taney likes noise; Marshall likes silence. They are, let us assume, inevitably neighbors. Let us also assume there are no transaction costs which may impede negotiations between them. Let us assume finally that we do not know Taney's and Marshall's wealth or, indeed, anything else about them. Under these circumstances we know that Pareto optimality—economic efficiency—will be reached whether we choose an entitlement to make noise or to have silence. We also are indifferent, from a general wealth distribution point of view, as to what the initial entitlement is because we do not know whether it will lead to greater equality or inequality. This leaves us with only two reasons on which to base our choice of entitlement. The first is the relative worthiness of silence lovers and noise lovers. The second is the consistency of the choice, or its apparent consistency, with other entitlements in the society.

The first sounds appealing, and it sounds like justice. But it is hard to deal with. Why, unless our choice affects other people, should we prefer one to another? To say that we wish, for instance, to make the silence lover relatively wealthier because we prefer silence is no answer, for that is simply a restatement of the question. Of course, if the choice does affect people other than Marshall and Taney, then we have a valid basis for decision. But the fact that such external effects are extremely common and greatly influence our choices does not help us much. It does suggest that the reaching of Pareto optimality is, in practice, a very complex matter precisely because of the existence of many external effects which markets find hard to deal with. And it also suggests that there often are general distributional considerations between Taney-Marshall and the rest of the world which affect the choice of entitlement. It in no way suggests, however, that there is more to the choice between Taney-Marshall than Pareto optimality and distributional concerns. . . . [W]hat sounds like a justice standard is simply a handy way of importing efficiency and distributional notions too diverse and general in their effect to be analyzed fully in the decision of a specific case.

The second sounds appealing in a different way since it sounds like "treating like cases alike." If the entitlement to make noise in other people's ears for one's pleasure is viewed by society as closely akin to the entitlement to beat up people for one's pleasure, and if good efficiency and distributional reasons exist for not allowing people to beat up others for sheer pleasure, then there may be a good reason for preferring an entitlement to silence rather than noise in the Taney-Marshall case. Because the two entitlements are apparently consistent, the entitlement to silence strengthens the entitlement to be free from gratuitous beatings which we assumed was based on good efficiency and distributional reasons. It does so by lowering the enforcement costs of the entitlement to be free from gratuitous beatings; the entitlement to silence reiterates and reinforces the values protected by the entitlement to be free from gratuitous beatings and reduces the number of discriminations people must make between one activity and another, thus simplifying the task of obedience.

The problem with this rationale for the choice is that it too comes down to efficiency and distributional reasons. We prefer the silence maker because that entitlement, even though it does not of itself affect the desired wealth distribution or lead us away from efficiency in the Taney-Marshall case, helps us to reach those goals in other situations where there are transaction costs or where we do have distributional preferences. It does this because people do not realize that the consistency is only apparent. If we could explain to them, both rationally and emotionally, the efficiency and distributional reasons why gratuitous beating up of people was inefficient or led to undesirable wealth distribution, and if we could also explain to them why an entitlement to noise rather than silence in the Taney-Marshall case would not lead to either inefficiency or maldistribution, then the secondary undermining of the entitlement to bodily integrity would not occur. It is only because it is expensive, even if feasible, to point out the difference between the two situations that the apparent similarity between them remains. And avoiding this kind of needless expense, while a very good reason for making choices, is clearly no more than a part of the economic efficiency goal. . . .

Rules for Protecting and Regulating Entitlements

. . .

Property and Liability Rules

Why cannot a society simply decide on the basis of the already mentioned criteria who should receive any given entitlement, and then let its transfer occur only through a voluntary negotiation? Why, in other words, cannot society limit itself to the property rule? To do this it would need only to protect and enforce the initial entitlements from all attacks, perhaps through criminal sanctions, and to enforce voluntary contracts for their transfer. Why do we need liability rules at all?

In terms of economic efficiency the reason is easy enough to see. Often the cost of establishing the value of an initial entitlement by negotiation is so great that even though a transfer of the entitlement would benefit all concerned, such a transfer will not occur. If a collective determination of the value were available instead, the beneficial transfer would quickly come about.

Eminent domain is a good example. A park where Guidacres, a tract of land owned by 1,000 owners in 1,000 parcels, now sits would, let us assume, benefit a neighboring town enough so that the 100,000 citizens of the town would each be willing to pay an average of $100 to have it. The park is Pareto desirable if the owners of the tracts of land in Guidacres actually value their entitlements at less than $10,000,000 or an average of $10,000 a tract. Let us assume that in fact the parcels are all the same and all the owners value them at $8,000. On this assumption, the park is, in economic efficiency terms, desirable—in values forgone it costs $8,000,000 and is worth $10,000,000 to the buyers. And yet it may well not be established. If enough of the owners hold-out for more than $10,000 in order to get a share of the $2,000,000 that they guess the buyers are willing to pay over the value which the sellers in actuality attach, the price demanded will be more than $10,000,000 and no park will result. The sellers have an incentive to hide their true valuation and the market will not succeed in establishing it.

An equally valid example could be made on the buying side. Suppose the sellers of Guidacres have agreed to a sales price of $8,000,000 (they are all relatives and at a family banquet decided that trying to hold-out would leave them all losers). It does not follow that the buyers can raise that much even though each of 100,000 citizens in fact values the park at $100. Some citizens may try to free-load and say the park is only worth $50 or even nothing to them, hoping that enough others will admit to a higher desire and make up the $8,000,000 price. Again there is no reason to believe that a market, a decentralized system of valuing, will cause people to express their true valuations and hence yield results which all would in fact agree are desirable.

Whenever this is the case an argument can readily be made for moving from a property rule to a liability rule. If society can remove from the market the valuation of each tract of land, decide the value collectively, and impose it, then the holdout problem is gone. Similarly, if society can value collectively each individual citizen's desire to have a park and charge him a "benefits" tax based upon it, the freeloader problem is gone. If the sum of the taxes is greater than the sum of the compensation awards, the park will result.

Of course, one can conceive of situations where it might be cheap to exclude all the freeloaders from the park, or to ration the park's use in accordance with original willingness to pay. In such cases the incentive to free-load might be eliminated. But such exclusions, even if possible, are usually not cheap. And the same may be the case for market methods which might avoid the holdout problem on the seller side.

Moreover, even if holdout and freeloader problems can be met feasibly by the market, an argument may remain for employing a liability rule. Assume that in our hypothetical, freeloaders can be excluded at the cost of $1,000,000 and that all owners of tracts in Guidacres can be convinced, by the use of $500,000 worth of advertising and cocktail parties, that a sale will only occur if they reveal their true land valuations. Since $8,000,000 plus $1,500,000 is less than $10,000,000, the park will be established. But if collective valuation of the tracts and of the benefits of the prospective park would have cost less than $1,500,000, it would have been inefficient to establish the park through the market—a market which was not worth having would have been paid for.

Of course, the problems with liability rules are equally real. We cannot be at all sure that landowner Taney is lying or holding out when he says his land is worth $12,000 to him. The fact that several neighbors sold identical tracts for $10,000 does not help us very much. Taney may be sentimentally attached to his land. As a result, eminent domain may grossly undervalue what Taney would actually sell for, even if it sought to give him his true valuation of his tract. In practice, it is so hard to determine Taney's true valuation that eminent domain simply gives him what the land is worth "objectively," in the full knowledge that this may result in over or under compensation. The same is true on the buyer side. "Benefits" taxes rarely attempt, let alone succeed, in gauging the individual citizen's relative desire for the alleged benefit. They are justified because, even if they do not accurately measure each individual's desire for the benefit, the market alternative seems worse. For example, fifty different households may place different values on a new sidewalk that is to abut all the properties. Nevertheless, because it is too difficult, even if possible, to gauge each household's valuation, we usually tax each household an equal amount.

The example of eminent domain is simply one of numerous instances in which society uses liability rules. Accidents is another. If we were to give victims a property entitlement not to be accidentally injured we would have to require all who engage in activities that may injure individuals to negotiate with them before an accident, and to buy the right to knock off an arm or a leg. Such pre-accident negotiations would be extremely expensive, often prohibitively so. To require them would thus preclude many activities that might, in fact, be worth having. And, after an accident, the loser of the arm or leg can always very plausibly deny that he would have sold it at the price the buyer would have offered. Indeed, where negotiations after an accident do occur—for instance, pretrial settlements—it is largely because the alternative is the collective valuation of the damages. . . .

Inalienable Entitlements

Thus far we have focused on the questions of when society should protect an entitlement by property or liability rules. However, there remain many entitlements which involve a still greater degree of societal intervention: the law not only decides who is to own something and what price is to be paid for it if it is taken or destroyed, but also regulates its sale—by, for example, prescribing preconditions for a valid sale or forbidding a sale altogether. Although these rules of inalienability are substantially different from the property and liability rules, their use can be analyzed in terms of the same efficiency and distributional goals that underlie the use of the other two rules.

While at first glance efficiency objectives may seem undermined by limitations on the ability to engage in transactions, closer analysis suggests that there are instances, perhaps many, in which economic efficiency is more closely approximated by such limitations. This might occur when a transaction would create significant externalities—costs to third parties.

For instance, if Taney were allowed to sell his land to Chase, a polluter, he would injure his neighbor Marshall by lowering the value of Marshall's land. Conceivably, Marshall could pay Taney not to sell his land; but, because there are many injured Marshalls, freeloader and information costs make such transactions practically impossible. The state could protect the Marshalls and yet facilitate the sale of the land by giving the Marshalls an entitlement to prevent Taney's sale to Chase but only protecting the entitlement by a liability rule. It might, for instance, charge an excise tax on all sales of land to polluters equal to its estimate of the external cost to the

Marshalls of the sale. But where there are so many injured Marshalls that the price required under the liability rule is likely to be high enough so that no one would be willing to pay it, then setting up the machinery for collective valuation will be wasteful. Barring the sale to polluters will be the most efficient result because it is clear that avoiding pollution is cheaper than paying its costs—including its costs to the Marshalls.

Another instance in which external costs may justify inalienability occurs when external costs do not lend themselves to collective measurement which is acceptably objective and nonarbitrary. This nonmonetizability is characteristic of one category of external costs which, as a practical matter, seems frequently to lead us to rules of inalienability. Such external costs are often called moralisms.

If Taney is allowed to sell himself into slavery, or to take undue risks of becoming penniless, or to sell a kidney, Marshall may be harmed, simply because Marshall is a sensitive man who is made unhappy by seeing slaves, paupers, or persons who die because they have sold a kidney. Again Marshall could pay Taney not to sell his freedom to Chase the slaveowner; but again, because Marshall is not one but many individuals, freeloader and information costs make such transactions practically impossible. Again, it might seem that the state could intervene by objectively valuing the external cost to Marshall and requiring Chase to pay that cost. But since the external cost to Marshall does not lend itself to an acceptable objective measurement, such liability rules are not appropriate.

In the case of Taney selling land to Chase, the polluter, they were inappropriate because we knew that the costs to Taney and the Marshalls exceeded the benefits to Chase. Here, though we are not certain of how a cost-benefit analysis would come out, liability rules are inappropriate because any monetization is, by hypothesis, out of the question. The state must, therefore, either ignore the external costs to Marshall, or if it judges them great enough, forbid the transaction that gave rise to them by making Taney's freedom inalienable.

Obviously we will not always value the external harm of a moralism enough to prohibit the sale. And obviously also, external costs other than moralisms may be sufficiently hard to value to make rules of inalienability appropriate in certain circumstances; this reason for rules of inalienability, however, does seem most often germane in situations where moralisms are involved.

There are two other efficiency reasons for forbidding the sale of entitlements under certain circumstances: self paternalism and true paternalism. Examples of the first are Ulysses tying himself to the mast or individuals passing a bill of rights so that they will be prevented from yielding to momentary temptations which they deem harmful to themselves. This type of limitation is not in any real sense paternalism. It is fully consistent with Pareto efficiency criteria, based on the notion that over the mass of cases no one knows better than the individual what is best for him or her. It merely allows the individual to choose what is best in the long run rather than in the short run, even though that choice entails giving up some short run freedom of choice. Self paternalism may cause us to require certain conditions to exist before we allow a sale of an entitlement; and it may help explain many situations of inalienability, like the invalidity of contracts entered into when drunk, or under undue influence or coercion. But it probably does not fully explain even these.

True paternalism brings us a step further toward explaining such prohibitions and those of broader kinds—for example the prohibitions on a whole range of activities by minors. Paternalism is based on the notion that at least in some situations the Marshalls know better than Taney what will make Taney better off. Here we are not talking about the offense to Marshall from Taney's choosing to read pornography, or selling himself into slavery, but rather the judgment that Taney was not in the position to choose best for himself when he made the choice for erotica or servitude. . . .

Finally, just as efficiency goals sometimes dictate the use of rules of inalienability, so, of course, do distributional goals. Whether an entitlement may be sold or not often affects directly who is richer and who is poorer. Prohibiting the sale of babies makes poorer those who can cheaply produce babies and richer those who through some nonmarket device get free an "unwanted" baby. Prohibiting exculpatory clauses in product sales makes richer those who were injured by a product defect and poorer those who were not injured and who paid more for the product because the exculpatory clause was forbidden. Favoring the specific group that has benefited may or may not have been the reason for the prohibition on bargaining. What is important is that, regardless of the reason for barring a contract, a group did gain from the prohibition.

This should suffice to put us on guard, for it suggests that direct distributional motives may lie behind asserted nondistributional grounds for inalienability, whether they be paternalism, self paternalism, or externalities. This does not mean that giving weight to distributional goals is undesirable. It clearly is desirable where on efficiency grounds society is indifferent between an alienable and an inalienable entitlement and distributional goals favor one approach or the other. It may well be desirable even when distributional goals are achieved at some efficiency costs. The danger may be, however, that what is justified on, for example, paternalism grounds is really a hidden way of accruing distributional benefits for a group whom we would not otherwise wish to benefit. For example, we may use certain types of zoning to preserve open spaces on the grounds that the poor will be happier, though they do not know it now. And open spaces may indeed make the poor happier in the long run. But the zoning that preserves open space also makes housing in the suburbs more expensive and it may be that the whole plan is aimed at securing distributional benefits to the suburban dweller regardless of the poor's happiness. . . .

The Framework and Criminal Sanctions

Obviously we cannot canvass the relevance of our approach through many areas of the law. But we do think it beneficial to examine one further area, that of crimes against property and bodily integrity. The application of the framework to the use of criminal sanctions in cases of theft or violations of bodily integrity is useful in that it may aid in understanding the previous material, especially as it helps us to distinguish different kinds of legal problems and to identify the different modes of resolving those problems.

Beginning students, when first acquainted with economic efficiency notions, sometimes ask why ought not a robber be simply charged with the value of the thing robbed. And the same question is sometimes posed by legal philosophers. If it is worth more to the robber than to the owner, is not economic efficiency served by such a penalty? Our answers to such a question tend to move quickly into very high sounding and undoubtedly relevant moral considerations. But these considerations are often not very helpful to the questioner because they depend on the existence of obligations on individuals not to rob for a fixed price and the original question was why we should impose such obligations at all.

One simple answer to the question would be that thieves do not get caught every time they rob and therefore the costs to the thief must at least take the unlikelihood of capture into account. But that would not fully answer the problem, for even if thieves were caught every time, the penalty we would wish to impose would be greater than the objective damages to the person robbed.

A possible broader explanation lies in a consideration of the difference between property entitlements and liability entitlements. For us to charge the thief with a penalty equal to an objectively determined value of the property stolen would be to convert all property rule entitlements into liability rule entitlements.

The question remains, however, why not convert all property rules into liability rules? The answer is, of course, obvious. Liability rules represent only an approximation of the value of the object to its original owner and willingness to pay such an approximate value is not indication that it is worth more to the thief than to the owner. In other words, quite apart from the expense of arriving collectively at such an objective valuation, it is no guarantee of the economic efficiency of the transfer. If this is so with property, it is all the more so with bodily integrity, and we would not presume collectively and objectively to value the cost of a rape to the victim against the benefit to the rapist even if economic efficiency is our sole motive. Indeed when we approach bodily integrity we are getting close to areas where we do not let the entitlement be sold at all and where economic efficiency enters in, if at all, in a more complex way. But even where the items taken or destroyed are things we do allow to be sold, we will not without special reasons impose an objective selling price on the vendors.

Once we reach the conclusion that we will not simply have liability rules, but that often, even just on economic efficiency grounds, property rules are desirable, an answer to the beginning student's question becomes clear. The thief not only harms the victim, he undermines rules and distinctions of significance beyond the specific case. Thus even if in a given case we can be sure that the value of the item stolen was no more than X dollars, and even if the thief has been caught and is prepared to compensate, we would not be content simply to charge the thief X dollars. Since in the majority of cases we cannot be sure of the economic efficiency of the transfer by theft, we must add to each case an undefinable kicker which represents society's need to keep all property rules from being changed at will into liability rules. In other words, we impose criminal sanctions as a means of deterring future attempts to convert property rules into liability rules. . . .

3.2 TORT

The Learned Hand Formula for Determining Liability

RICHARD POSNER

In *United States v. Carroll Towing Co.*, 159 F.2d 169 (2d Cir. 1947), the question was presented whether it was negligent for the Conners Company, the owner of a barge, to leave it unattended for several hours in a busy harbor. While unattended, the barge broke away from its moorings and collided with another ship. Judge Learned Hand stated for the court (at p. 173):

> There is no general rule to determine when the absence of a bargee or other attendant will make the owner of the barge liable for injuries to other vessels if she breaks away from her moorings. . . . It becomes apparent why there can be no such general rule, when we consider the grounds for such a liability. Since there are occasions when every vessel will break from her moorings, and since, if she does, she becomes a menace to those about her, the owner's duty, as in other similar situations, to provide against resulting injuries is a function of three variables: (1) The probability that she will break away; (2) the gravity of the resulting injury, if she does; (3) the burden of adequate precautions. Possibly it serves to bring this notion into relief to state it in algebraic terms: if the probability be called P; the injury, L; and the burden, B; liability depends upon whether B is less than L multiplied by P: i.e., whether $B < PL$. . . . In the case at bar the bargee left at five o'clock in the afternoon of January 3rd, and the flotilla broke away at about two o'clock in the afternoon of the following day, twenty-one hours afterwards. The bargee had been away all the time, and we hold that his fabricated story was affirmative evidence that he had no excuse for his absence. At the locus in quo—especially during the short January days and in the full tide of war activity—barges were being constantly "drilled" in and out. Certainly it was not beyond reasonable expectation that, with the inevitable haste and bustle, the work might not be done with adequate care. In such circumstances we hold—and it is all that we do hold—that it was a fair requirement that the Conners Company should have a bargee aboard (unless he had some excuse for his absence), during the working hours of daylight.

By redefinition of two terms in the Hand formula it is easy to bring out its economic character. B, the burden of precautions, is the cost of avoiding the accident, while L, the loss if the accident occurs, is the cost of the accident itself. P times L ($P \times L$)—the cost of the accident if it occurs, multiplied (or as is sometimes said, "discounted") by the probability that the accident will occur, is what an economist would call the "expected cost" of the accident. Expected cost is most easily understood as the average cost that will be incurred over a period of time long enough for the predicted number of accidents to be the actual number. For example, if the probability that a certain type of accident will occur is .001 (one in a thousand) and the accident cost if it does occur is $10,000, the expected accident cost is $10 ($10,000 × .001); and this is equivalent to saying that if we observe the activity that gives rise to this type of accident for a long enough period of time we will observe an average accident cost of $10. Suppose the activity in question is automobile trips from point A to point B. If there are 100,000 trips, there will be 100 accidents, assuming that our probability of .001 was correct. The total cost of the 100 accidents will be $1 million ($10,000 × 100). The average cost, which is simply the total cost ($1 million) divided by the total number of trips (100,000), will be $10. This is the same as the expected cost.

Another name for expected accident costs—for $P \times L$, the right-hand side of the Hand formula—is the benefits from accident avoidance. If one incurs B, the burden of precautions or cost of accident avoidance, one produces a benefit—namely, avoidance of the expected accident costs. The Hand formula is simply an application to accidents of the principle of cost-benefit analysis. Negligence means failing to avoid an accident where the benefits of accident avoidance exceed the costs.

The Hand formula shows that it is possible to think about tort law in economic terms—that, in fact, a famous judge thought about it so. . . .

Let us imagine the world divided strictly into injurers and victims— those who inflict injury (though they are themselves unharmed) and those who are injured. . . . Under strict liability, and assuming that the court assesses damages equal to the loss suffered by the victim, the right-hand side of the Hand formula becomes an expected-damages figure. L is replaced by D (standing for damages) because every time a loss is suffered by an accident victim the injurer must pay for it. $P \times D$ is then the expected damages of the injurer if he does not avoid the accident.

In a system of strict liability the injurer must pay the injured victim regardless of the left-hand side of the Hand formula—regardless of whether B is greater than or less than $P \times L$. That is what strict liability means. It does not follow, however, that the Hand formula becomes irrelevant to the level of safety. Although the court will not apply the Hand formula—it will inquire simply whether the accident occurred and not whether the accident could have been prevented at a lower cost than the expected accident cost—rational potential injurers will apply the formula (i.e., use cost-benefit analysis) in order to determine whether to take steps to avoid the accident. And if the cost of avoidance turns out to be greater than the expected accident cost, the rational profit-maximizing (or utility-maximizing) potential injurer will not incur the cost of avoidance. He will prefer a lower expected damages cost to a higher accident-avoidance cost.

A negligence standard differs from strict liability in that under negligence the injurer is liable only for those accidents that he could have avoided at a lower cost than the expected accident cost. This is what Judge Hand suggests in the quotation at the beginning of this chapter. Notice that, at least as a first approximation, the number of accidents is the same under strict liability and negligence. This is because, as just mentioned, a potential injurer in a system of strict liability will not invest in safety beyond the point at which the costs of safety are equal to the benefits in accident avoidance. The difference between strict and negligence liability is that under the former system injurers must pay the losses of victims of accidents that are not worth preventing, whereas under the latter system injurers have no duty to pay victims of such accidents. . . .

The fact that negligence and strict liability lead, at least as a rough first approximation, to the same level of accidents may seem to provide an economic argument in favor of negligence. The argument is this: A system of strict liability provides no more (or less) safety than a negligence system. Thus the allocation of resources to safety is unchanged. But strict liability involves more claims than negligence, since under strict liability victims of unavoidable as well as avoidable accidents have claims against their injurers. The additional claims are costly to process, and the added costs confer no benefit in preventing additional accidents: no additional accidents are prevented.

In fact, the economic comparison of negligence and strict liability is more complicated than this. Most of the complications will be postponed but one needs to be considered here. It is as follows. There are two ways of avoiding an accident. One is to conduct the activity giving rise to the accident more carefully; the other is to reduce the amount, or change the nature, of the activity. . . . [N]egligence usually (though not always) connotes failure to use the right amount of care rather than failure to reduce the amount of activity to the correct level or change the activity (e.g., from railroading to trucking, or from growing wheat to burning trash). If so, negligence and strict liability may result in a different number of accidents after all. Strict liability will deter certain accidents where the cost of avoiding the accident by reducing the amount of activity is less than the expected accident cost; negligence will not deter such accidents.

However, before concluding that strict liability yields a more efficient allocation of resources to accident avoidance than negligence does, we must consider accident avoidance by victims. We assumed that victims were helpless, but this is not true in general. Suppose that a particular accident, having an expected cost of $100, could be avoided either by the injurer at a cost of $50 or by the victim at a cost of $10. Clearly, it would be more efficient for the victim to avoid than for the injurer to do so. This is a case where we want the victim to bear the full costs of the accident so that he will be induced to avoid it. We want a principle of contributory negligence whereby if the victim could have avoided the accident at a cost lower than the expected accident cost and also lower than the injurer's cost of avoidance, the injurer is not liable for the victim's loss.

It might appear that the easiest way to implement this principle would be simply to apply the Hand formula to victims as well as injurers—to say that if the cost of avoidance to the victim is less than the expected accident cost, he is guilty of contributory negligence (because B is smaller than $P \times L$) and therefore "liable" (i.e., not entitled to recover his damage from the injurer). This approach will yield the economically correct result in many cases, but not in all. The victim might be negligent under the Hand formula because his accident-avoidance cost was lower than the expected accident cost, but at the same time the injurer's accident-avoidance cost might be lower still. In that event we would want the injurer to be liable after all, even though the victim was contributorily negligent. . . .

In comparing strict liability and negligence, we have emphasized possible differences in the allocation of resources to safety. This is one important dimension of efficiency, but another, which should always be kept in mind as well, is administrative cost. Strict liability increases the number of claims but reduces the cost of processing each one by eliminating the issue of negligence; whether the total costs of administering a strict liability system are lower or higher than those of administering a negligence system are unclear. In any event, the optimum liability rule is the one that minimizes the sum of allocative (expected-accident plus accident-avoidance) and administrative costs.

The Hand formula was announced in a negligence case but, it should be clear by now, can be applied to other issues in tort law besides the determination of negligence. It can be used, as we have seen, to answer the question whether a victim is contributorily negligent, or to justify a rule of strict liability in a class of accidents where it is clear that the cost of accident avoidance to injurers is lower than the accident-avoidance cost to victims. The Hand formula can also be used where, unlike the typical accident case, the probability of harm is one. Many pollution cases are of this type, and . . . they provide particularly clear examples of cost-benefit analysis in tort law since they do not involve the added complication that the accident costs are uncertain.

The Hand formula can be applied even to cases of deliberate injury. Because the injurer in such a case is trying to hurt the victim, rather than just hurting him as a by-product of other activity, P tends to be very high in such cases. At the same time, B will often be very low, and even negative, especially if it is interpreted, as it should be, to refer to social as distinct from purely private costs. What is the social cost of refraining from spitting in someone's face? It is very low—zero or even negative (because it takes an effort to spit, not to avoid spitting). Although the injury cost is also small, it is higher than the social cost of avoidance and is inflicted with a probability close to one. Thus, application of the Hand formula to cases of deliberate harm will typically result in a decision to condemn the defendant's conduct. Nor would a different result be reached by considering the victim's conduct; in most cases of a deliberate or malicious infliction of harm, the injurer is clearly the cheaper harm-avoider than the victim.

Granted that it is possible to derive a system of liability rules from cost-benefit principles as capsulized in the Hand formula, the question remains why common-law judges might want to do so. One reason is that they might want to increase the wealth of society. . . . What should be noted here is that for tort law to increase the wealth of society, liability rules must influence people's behavior, potential injurers' and potential victims' alike. If the threat of liability (or of not being compensated, if a victim) does not deter accidents but only shifts wealth about after the accident occurs, tort law will not increase the wealth of society but will serve only to redistribute the diminished wealth that remains after the accident has occurred; it will not increase efficiency as that term is being used here.

The proposition that liability rules deter is a controversial one. It seems to presuppose not only that people behave rationally with regard to extreme situations of personal danger but that they have sufficient information regarding the principles of tort law and the probabilities of what are often quite remote contingencies (e.g., that a cigar will explode in one's mouth if certain safety precautions in the manufacture of the cigar are not taken), and also that damages are assessed under correct economic principles. The deterrent efficacy of tort law is a matter of fair debate. . . . An alternative formulation of the economic theory of torts, however, holds that tort principles express our moral indignation at certain types of invasion of bodily integrity or property rights—an indignation that is rooted, or at least can be expressed, in economic terms. In this view, we are indignant at a negligent injury because our moral natures are offended by economic waste, illustrated by an accident avoidable at a lower cost than the expected accident cost. Thus, even if tort rules operate to provide an outlet for strong emotions rather than to regulate conduct (except perhaps insofar as having such an outlet reduces the incidence of violent retaliation), the rules might still have an implicit economic logic. . . .

Strict Liability versus Negligence

STEVEN SHAVELL

The aim of this article is to compare strict liability and negligence rules on the basis of the incentives they provide to "appropriately" reduce accident losses. It will therefore be both convenient and clarifying to abstract from other issues in respect to which the rules could be evaluated. In particular, there will be no concern with the bearing of risk—for parties will be presumed risk neutral—nor with the size of "administrative costs"—for the legal system will be assumed to operate free of such costs—nor with distributional equity—for the welfare criterion will be taken to be the following aggregate: the benefits derived by parties from engaging in activities less total accident losses less total accident prevention costs. . . .

Accidents will be conceived of as involving two types of parties, "injurers" and "victims," only the latter of which are assumed to suffer direct losses. The category of accidents that will be examined initially are *unilateral* in nature, by which is meant that the actions of injurers but not of victims are assumed to affect the probability or severity of losses. The unilateral case is studied for two reasons. First, it is descriptive of situations in which whatever changes in the behavior of victims that could reasonably be expected to result from changes in liability rules would have only a small influence on accident losses. The second reason is pedagogical; it is easier to understand the general *bilateral* case after having studied the unilateral case.

Unilateral Case

This case (as well as the bilateral case) will be considered in each of several situations distinguished by the nature of the relationship between injurers and victims.

Accidents between Strangers

In this subcase it is supposed that injurers and victims are strangers, that neither are sellers of a product, and that injurers may choose to engage in an activity which puts victims at risk.

By definition, under the negligence rule all that an injurer needs to do to avoid the possibility of liability is to make sure to exercise due care if he engages in his activity. Consequently *he will not be motivated to consider the effect on accident losses of his choice of whether to engage in his activity or, more generally, of the level at which to engage in his activity;* he will choose his level of activity in accordance only with the personal benefits so derived. But surely any increase in his level of activity will typically raise expected accident losses (holding constant the level of care). Thus he will be led to choose too high a level of activity; the negligence rule is not "efficient."

Consider by way of illustration the problem of pedestrian-automobile accidents (and, as we are now discussing the unilateral case, let us imagine the behavior of pedestrians to be fixed). Suppose that drivers of automobiles find it in their interest to adhere to the standard of due care but that the possibility of accidents is not thereby eliminated. Then, in deciding how much to drive, they will contemplate only the enjoyment they get from doing so. Because (as they exercise due care) they will not be liable for harm suffered by pedestrians, drivers will not take into account that going more miles will mean a higher expected number of accidents. Hence, there will be too much driving; an individual will, for example, decide to go for a drive on a mere whim despite the imposition of a positive expected cost to pedestrians.

However, under a rule of strict liability the situation is different. Because an injurer must pay for losses whenever he is involved in an accident, he will be induced to consider the effect on accident losses of both his level of care and his level of activity. His decisions will therefore be efficient. Because drivers will be liable for losses sustained by pedestrians, they will decide not only to exercise due care in driving but also to drive only when the utility gained from it outweighs expected liability payments to pedestrians.

Accidents between Sellers and Strangers

In this subcase it is assumed that injurers are sellers of a product or service and that they conduct their business in a competitive market. (The assumption of competition allows us to ignore monopoly power, which is for the purposes of this article a logically tangential issue.) Moreover, it is assumed that victims are strangers; they have no market relationship with sellers either as their customers or as their employees.

Under the negligence rule the outcome is inefficient, but the reasoning is slightly different from that of the last subcase. While it is still true that all a seller must do to avoid liability is to take due care, why this results in too high a level of activity has to do with market forces. Because the seller will choose to avoid liability, the price of his product will not reflect the accident losses associated with production. This means that buyers of the product will face too low a price and will purchase too much, which is to say that the seller's level of activity will be too high. Imagine that the drivers are engaged in some business activity—let us say that they are taxi drivers. Then, given that they take due care, the taxi drivers will not have liability expenses, will set rates equal to "production" cost (competition among taxi drivers is assumed), will experience a greater demand than if rates were appropriately higher, and will therefore carry too many fares and cause too many accidents.

Under strict liability, the outcome is efficient, and again the reasoning is a little different from that in the last subcase. Since sellers have to pay for accident losses, they will be led to take the right level of care. And since the product price will reflect accident losses, customers will face the "socially correct" price for the product; purchases will therefore be appropriately lower than what they would be if the product price did not reflect accident losses. Taxi drivers will now increase rates by an amount equal to expected accident losses suffered by pedestrians, and the demand for rides in taxis will fall.

Accidents between Sellers and Customers—or Employees

It is presumed here that victims have a market relationship with sellers as either their customers or their employees; and since both situations are essentially the same, it will suffice to discuss only that when victims are customers. In order to understand the role (which is important) of customers' knowledge of risk, three alternative assumptions will be considered: customers know the risk presented by each seller; they do not know the risk presented by each seller but they do know the average seller's risk; they misperceive even this average risk.

Under the negligence rule, the outcome is efficient only if customers correctly perceive risks. As before, when the victims were strangers, sellers will take due care in order to avoid liability, so that the product price will not reflect accident losses. However, now the accident losses are borne by the customers. Thus, the "full" price in the eyes of customers is the market price plus imputed perceived accident losses. Therefore, if risks are correctly perceived, the full price equals the socially correct price, and the quantity purchased will be appropriate. But if risks are not correctly perceived, the quantity purchased will be inappropriate; if customers underestimate risks, what they regard as the full price is less than the true full price and they will buy too much of the product, and conversely if they overestimate risks.

Think, for example, of the risk of food poisoning from eating at restaurants. Under the negligence rule, restaurants will decide to avoid liability by taking appropriate precautions to prepare meals under sanitary conditions. Therefore, the price of meals will not reflect the expected losses due to the (remaining) risk of food poisoning. If customers know this risk, they will correctly consider it in their decisions over the purchase of meals. But if they underestimate the risk, they will purchase too many meals; and if they overestimate it, too few.

Under strict liability, the outcome is efficient regardless of whether customers misperceive risks. As in the last subcase, because sellers have to pay for accident losses, they will decide to take appropriate care and will sell the product at a price reflecting accident losses. Thus customers will face the socially correct price and will purchase the correct amount. Their perception of the risk is irrelevant since it will not influence their purchases; as they will be compensated under strict liability for any losses, the likelihood of losses will not matter to them. Restaurant-goers will face a price that reflects expected losses due to food poisoning when meals are prepared under sanitary conditions; they will buy the same—and appropriate—number of meals whether they think the probability of food poisoning is low or high, for they will be compensated for any losses suffered.

When sellers are simply not liable for accident losses, then the outcome is efficient only if customers know the risk presented by each seller. For, given this assumption, because customers will seek to buy products with the lowest full price (market price plus expected accident losses), sellers will be induced to take appropriate care (since this will lower the accident-loss component of the full price). While it is true that if a restaurant took inadequate precautions to prevent food poisoning, it could offer lower-priced meals, it is also true that customers would respond not just to the market price of meals but also to the likelihood of food poisoning—which they are presumed to know. Therefore customers would decide against giving the restaurant their business. Consequently, restaurants will be led to take adequate precautions and to charge accordingly. Moreover, because customers will base purchases on the correctly perceived full price, they will buy the correct amount.

If, however, customers do not know the risk presented by individual sellers, there are two sources of inefficiency when sellers are not liable. The first is that, given the risk of loss, the quantity purchased by customers may not be correct; of course, this will be true if customers misperceive the risk. The second source of inefficiency is that sellers will not be motivated by market forces to appropriately reduce risks. To understand why, consider the situation when customers do correctly perceive the average risk (when they do not correctly perceive this risk, an explanation similar to the one given here could be supplied). That is, assume that customers know the risk presented by sellers as a group but do not have the ability to "observe" the risk presented by sellers on an individual basis. Then sellers would have no inducement to exercise adequate care. Suppose that restaurant-goers know the risk of food poisoning at restaurants in general and it is, say, inappropriately high. Then if a particular restaurant were to take sufficient precautions to lower the risk, customers would not recognize this (except insofar as it eventually affected the average risk—but under the assumption that there are many competing restaurants, this effect would be negligible). Thus the restaurant could not charge a higher price for its meals—customers would have no reason not to go to the cheaper restaurants. In consequence, a situation in which sellers take inadequate care to reduce risks would persist; and similar reasoning shows that a situation in which they take adequate care would not persist. (Notice, however, that since customers are assumed to correctly perceive the average risk, at least they will purchase the correct number of meals—correct, given the high risk.). . . .

Bilateral Case

In this case, account is taken of the possibility that potential victims as well as injurers may influence the probability or magnitude of accident losses by their choices of both level of care and of level of activity.

Accidents between Strangers

Under the negligence rule, the outcome is not efficient. As was true in the unilateral case, since all that an injurer needs to do to avoid liability is to exercise due care, he will choose too high a level of activity. In regard to victims, however, the situation is different. Since a victim bears his accident losses, he will choose an appropriate level of care and an appropriate level of his activity, given the (inefficient) behavior of injurers. The drivers will exercise due care but will go too many miles. And the pedestrians, knowing that they must bear accident losses, will exercise due care (in crossing streets and so forth) and they will also reduce the number of miles they walk in accordance with expected accident losses per mile.

Under strict liability with a defense of contributory negligence, the outcome is symmetrical to the last—and again inefficient. Because all that a victim needs to do to avoid bearing accident losses is to take due care, he will have no motive to appropriately reduce his level of activity; this is the inefficiency. However, because injurers now bear accident losses, they will take the appropriate amount of care and choose the right level of activity, given the inefficient behavior of victims. Drivers will exercise due care and go the correct number of miles. Pedestrians will also exercise due care but will walk too many miles.

From this discussion it is apparent that the choice between strict liability with a defense of contributory negligence and the negligence rule is a choice between the lesser of two evils. Strict liability with the defense will be superior to the negligence rule when it is more important that injurers be given an incentive through a liability rule to reduce their activity level than that victims be given a similar incentive; that is to say, when it is more important that drivers go fewer miles than that pedestrians walk fewer miles.

Because neither of the familiar liability rules induces efficient behavior, the question arises, "Is there any conceivable liability rule depending on parties' levels of care and harm done that induces efficient behavior?" . . . [T]he answer is "No." The problem in essence is that for injurers to be induced to choose the correct level of activity, they must bear all accident losses; and for victims to choose the correct level of their activity, they also must bear all accident losses. Yet it is in the nature of a liability rule that both conditions cannot hold simultaneously; clearly, injurers and victims cannot each bear all accident losses.

Accidents between Sellers and Strangers

Because the reader will be able to appeal to arguments analogous to those already made and in order to avoid tedious repetition, explanation of the results stated in this and the next subcase will be abbreviated or will be omitted.

Under both the negligence rule and strict liability with a defense of contributory negligence, the outcome is inefficient, as was true in the last subcase. Under the negligence rule, sellers will take appropriate care, but since the product price will not reflect accident losses, too much will be purchased by customers. Also, since victims bear accident losses, they will take appropriate care and choose the right level of activity. Under strict liability with the defense, sellers will take appropriate care and the product price will reflect accident losses, so the right amount will be purchased. Victims will exercise due care but will choose too high a level of activity. In addition, as in the last subcase, there does not exist any liability rule that induces efficient behavior.

Accidents between Sellers and Customers/Employees

As before it will be enough to discuss here only the situation when victims are customers. If customers have perfect knowledge of the risk presented by each seller, then the outcome is efficient under strict liability with a defense of contributory negligence or the negligence rule or if sellers are not subject to liability at all. For instance, in the latter situation, since customers wish to buy at the lowest full price, sellers will be led to take appropriate care; and since customers will make their purchases with the full price in mind, the quantity they buy will be correct; and since they bear their losses, they will take appropriate care.

There is, however, a qualification that needs to be made concerning the way in which it is imagined that customers influence accident losses. If one assumes that customers influence losses only by their choice of level of care and of the amount purchased, then what was stated in the previous paragraph is correct; and in regard to services and nondurables (such as meals at restaurants) this assumption seems entirely natural. But in regard to durable goods, it might well be thought that customers influence accident losses not only by their choice of level of care and of purchases, but also by their decision as to frequency of use per unit purchased. The expected number of accidents that a man will have when using a power lawn mower would seem to be influenced not only by whether he in fact purchases one (rather than, say, a hand mower) and by how carefully he mows his lawn with it, but also by how frequently he chooses to mow his lawn. In order for customers to be led to efficiently decide the frequency of use, they must bear their own accident losses. Thus, in regard to durables, the outcome is efficient under the negligence rule or if sellers are not liable, but the outcome is inefficient under strict liability with a defense of contributory negligence; for then if the man buys a power lawn mower, he will have no motive to appropriately reduce the number of times he mows his lawn.

Now suppose that customers correctly perceive only average risks. Then, subject again to a qualification concerning durables, the results are as follows. The outcome is efficient under strict liability with a defense of contributory negligence or under the negligence rule, but the outcome is not efficient if sellers are not liable, for then they will not take sufficient care. The qualification is that if the sellers produce durables, strict liability with the defense is inefficient, leaving the negligence rule as the only efficient rule.

Last, suppose that customers misperceive risks. Then the outcome is efficient only under strict liability with a defense of contributory negligence; and the qualification to this is that, if sellers produce durables, even strict liability with the defense is inefficient, so that there does not exist a liability rule which is efficient. . . .

Concluding Comments

1. A question which is in a sense logically prior to the analysis of this article must be mentioned, namely, "Why isn't the level of activity usually considered in the formulation of a due care standard?" After all, the inefficiencies discussed here were viewed in the main as deriving from the fact that in order to avoid being found negligent (or contributorily negligent), parties are not motivated to alter their level of activity. The answer to the question appears to be that the courts would run into difficulty in trying to employ a standard of due care expanded in scope to include the level of activity. In formulating such a broadened due care standard, courts would, by definition, have to decide on the appropriate level of activity, and their competence to do this is problematic. How would courts decide the number of miles an individual ought to drive or how far or how often a pedestrian ought to walk? How would courts decide the level of output an industry—much less a firm within an industry—ought to produce? To decide such matters, courts would likely have to know much more than would normally have to be known to decide whether care, conventionally interpreted, was adequate.

2. From the logic of the arguments presented here, it can be seen that what is important about the variable "level of activity" is only that it is not included in the due care standard. Any other variable omitted from the standard would also be inappropriately chosen in many of the circumstances in which we said the same of the level of activity. For example, in regard to accidents involving firms and strangers it has been noted that, if the scale of a firm's research in safety technology is not comprehended by the standard of due care, then under the negligence rule the firm would not be expected to invest sufficiently in such research.

3. . . . The analysis presented here does appear to help to explain certain features of tort law. A notable example is provided by the so-called pockets of strict liability: for ultrahazardous activities, ownership of wild animals, and so forth. These areas of strict liability seem to have two characteristics. First, they are such that injurer activity has a distinctive aspect (which makes the activity easy for the law to single out) and imposes nonnegligible risks on victims (which make the activity worthwhile controlling). And, second, they are such that victim activity is usually not at all special—on the contrary, it is typically entirely routine in nature, part of what it is to carry on a normal life—and is therefore activity that cannot and ought not be controlled. Consequently, it is appealing to explain the pockets of strict liability by the idea that strict liability is preferable if it is more desirable to control injurers' activity than victims'.

However, there are many features of tort law which the analysis by itself does not seem to satisfactorily explain. And this is not unexpected, for it is in the nature of the formal approach to isolate selected factors of interest by ignoring others; the formal approach aims for a particular kind of insight, not for true balance or comprehensiveness. Two examples will illustrate various limitations in our ability to employ in a direct way the results of this article. The first concerns the trend in decisions in product-liability cases toward expansion of manufacturer's liability. If this trend can be likened to one toward holding manufacturers strictly liable, we may be tempted to explain it as broadly rational given some of our results. . . . However, realism requires us to look at other, complementary explanations of the trend, such as that strict liability may provide a better means of risk sharing than the negligence rule, or that strict liability may be easier to apply than the negligence rule. Moreover, realism requires us to ask whether there even is an explanation of the trend based on its social rationality—whether in fact the trend might be socially undesirable, say on the ground that the expansion in the scope of liability has led to an excessively costly volume of disputes. Similar questions may be asked in respect to the second example, which concerns the fact that the negligence rule is the dominant form of tort liability in Anglo-American and in Western European legal systems today. Our analysis certainly does not suggest why this should be so, since, at least as often as otherwise, strict liability (with a defense of contributory negligence) is superior to the negligence rule. We are therefore led to ask again about such matters as risk sharing, administrative simplicity, and (especially) the social costs of expansion of the scope of liability. . . .

3.3   CONTRACT

The Efficiency of Specific Performance: Toward a Unified Theory of Contract Remedies

THOMAS ULEN

. . .The purpose of this essay is to begin the development of an integrated theory of contract remedies by delineating the circumstances under which courts should simply enforce a stipulated remedy clause or grant relief to the innocent party in the form of damages or specific performance. The conclusion, in brief, is that in the absence of stipulated remedies in the contract that survive scrutiny on the usual formation defenses, specific performance is more likely than any form of money damages to achieve efficiency in the exchange and breach of reciprocal promises. If specific performance is the routine remedy for breach, there are strong reasons for believing, first, that more mutually beneficial exchanges of promises will be concluded in the future and that they will be exchanged at a lower cost than under any other contract remedy, and, second, that under specific performance postbreach adjustments to all contracts will be resolved in a manner most likely to lead to the promise being concluded in favor of the party who puts the highest value on the completed performance and at a lower cost than under any alternative.

The argument proceeds by examining the relationship between different contract

Thomas Ulen, the Efficiency of Specific Performance: Toward a Unified Theory of Contract Remedies, 83 *Michigan Law Review* (1984), pp. 341–403. Reprinted with permission.

remedies and the costs imposed on contracting parties and on society at the time that promises are exchanged and during negotiations, if any, after the breach. A central tenet of the argument is that the transaction costs facing parties who have already concluded a contract are less, even if there has been a breach, than the costs of a court's resolving the dispute. . . .

Efficient Breach

There are circumstances in which performance of an otherwise legitimate contractual promise would be inefficient. Suppose, for example, that A promises to sell B a house for $100,000. Let us assume that B values the house at $115,000. Thus, at A's asking price, B realizes a consumer surplus of $15,000. Before the sale is completed, C offers A $125,000 for the same house. Should the law compel A to deliver on his promise to B, or should it allow, indeed encourage, him to breach his promise to B in order to sell to C?

From an economic standpoint the answer is clear. Economic efficiency will be served if resources are allocated to their highest-valued uses while minimizing the cost of reallocation. Thus, if, as previously assumed, efficiency is our goal, contract law should specify a remedy for breach that will lead to ownership of the house by the person who values it the most, and should attempt to reach this result at the lowest possible resource cost. In this case, the house apparently has the greatest value to C: we know that he places a value of at least $125,000 on the house; B, by assumption, values it at $115,000; and A values it at something less than $100,000. . . .

Legal Remedies for Breach of Contract

Restitution

Legal remedies in the common law of contract remedies seek to protect three interests of the innocent party in a breach of contract. First is the restitution interest. The goal of awarding a restitution interest in damages is to return the innocent party, insofar as possible, to his position before the contract was formed. Thus, any benefits, in the form of money or property, that the innocent party transferred to the breacher between the formation of the contract and the breach are to be returned. This situation typically arises where a contract is only partially fulfilled. The law holds that restitution is to be made to the innocent party even if performance might have resulted in loss to the innocent party.

It is not entirely clear when or why courts prefer restitution to other damage measures. A frequently cited reason for ordering a restoration of values on both sides is to prevent unjust enrichment from breach of contract. Little attention has been given to the possibility of an efficiency reason for preferring restitution to other contract remedies. One possible efficiency explanation is that, in general, it is extremely inexpensive to measure damages in terms of benefit conferred, especially in comparison to the other damage measures available to the courts.

However, despite the inexpensive nature of this remedy, restitution as a measure of damages will not necessarily lead to economically efficient breach of contract. Consider, again, the example in which A has agreed to sell a house to B for $100,000 when B values the house at $115,000. If A knows that upon breach he will have to pay B restitution damages, A may breach when economic efficiency would urge him to perform. Suppose that B gives A a good faith deposit of $5,000 when the contract is formed and that before A conveys to B, C offers A $110,000 for the house. If A anticipates that the court will award restitution damages to B for breach of contract, A will breach, pay $5,000 to B, and sell the house to C. Because C values the house less than does B but more than the contractual price, the breach does not lead to a Pareto-efficient reallocation.

Nor is it the case that B can protect his subjective valuation arising from completion of the contract by leaving a good faith deposit with A that is just equal to the difference between the price he is willing to pay for the house and the contractual price. In this instance, suppose that B attempts to protect his subjective valuation by leaving $15,000 with A. The fact of the matter is that A will still breach if he receives any offer above B's $100,000. All A is required to do under a restitution formula is to return B's $15,000 in order to place B in as good a position as he was prior to the formation of the contract. He is not required to place B in as good a position as B would have been upon completion of the contract. Thus, the size of the good faith deposit that B leaves with A has no bearing on A's decision to breach. It is in A's interest to breach, under the restitution formula, whenever he is offered a higher price than the contractual price agreed upon with B. It is efficient, however, for him to breach only if he receives an offer that is higher than B's subjective valuation of the house. Since restitution offers no way to induce only this breach, it must be rejected, on efficiency grounds, as the routine contract remedy.

Reliance

A second interest of the innocent party that damage measures seek to protect is the reliance interest. Expenses incurred by the innocent party in reliance on the performance of the other party's promise or in preparing to accept the fruits of the contract are recoverable as reliance loss when the contract is breached. The purpose of this measure of damages is to prevent punishing the innocent party for relying on the contract.

In the simple transaction we have been examining between A and B for the sale of a house for $100,000, suppose that B, in anticipation of A's performing the contract, hires a mover, an interior decorator, a painter, and so on. B's reliance on A's promise to convey the house at $100,000 has induced him to spend, say, $8,000. B continues to value the house at $115,000. Now comes C to bid $110,000 to A for his house. If A must pay reliance loss to B as damages for breaching his contract, then A will breach, pay B $8,000 in damages, and sell to C for $110,000. A has increased his profit by $2,000, and in the eyes of the court B is no worse off than if he had never entered the contract in the first place. Yet, as with restitution as the damage measure, the breach, where reliance loss is the damage measure, is not necessarily Pareto efficient: the house has not passed to the party who values it most. . . .

It might be objected that those with a high subjective valuation on performance might be induced to time their expenses inefficiently under the reliance measure. If it were well known that the routine remedy for breach was the payment of any expenses incurred in reasonable reliance on the performance of the contract, then B could fully protect himself against breach by incurring reliance expenses up to the difference between the contractual price and his reservation price for performance. In my example, B could guarantee that a breach by A would be efficient by incurring $15,000 in anticipation of moving into the house.

There is an efficiency problem in that, if reliance is the routine remedy, there seems to be little reason for B to stop at $15,000 in his expenses before A has performed. If he knows that the court will hold A liable for his reliance expenses, why should he not make $20,000 in pre-completion commitments? He may well have intended to make that commitment anyway in order, say, to have the house remodeled. The cost to him of re-timing that expenditure is small: the additional interest on the loan to effect the improvements. The result, however, is that the house may not now be efficiently allocated. A now faces a damage payment of $20,000 and may, therefore, perform when breach is more efficient. If C were willing to pay $118,000 for the house, then it should pass to him, since he values it more than does either A or B. But it will not if A is responsible to B for $20,000 in reliance damages.

An answer to this objection to reliance damages as the routine remedy is that the law already discourages B's over-reliance by protecting only reasonable reliance. That is, the law will not allow B to pile on expenses willy-nilly so as to bind A to perform. There is no doubt from an efficiency standpoint that this is the proper stance for the law, but there are other reasons to believe that allowing only reasonable reliance will still not guarantee only efficient breach. The most important of these is that determining which reliance expenses were reasonable and which were not is likely to be expensive. This is all the more true because the breachee has a strong incentive to demonstrate large reliance interest at the same time that the breacher has the contrary incentive to minimize that interest.

Expectation

The third interest of the innocent party that the law of contract damages protects is the expectation interest. Whereas with restitution and reliance the goal of the remedy is to place the innocent party in the position he was in before formation of the contract, with expectation damages the breachee is to be put in the position he expected to assume had the contract been performed. . . . In addition to being the most widely used contract remedy, expectation has attracted the favorable attention of economists because it is the only measure of contract damages that induces breach only where breach is more efficient than is performance.

Consider again the contract between A and B for delivery of a house at a price of $100,000, a house that B values at $115,000. If it can be easily determined that B's position under the completed contract will be a net gain of $15,000, then the expectation loss that B suffers from A's nonperformance is precisely $15,000. If A is aware that that measure is the one that will be used if he breaches, then A will breach only when it is economically efficient to breach. Since A will be responsible to B for

$15,000, he will breach only when C offers him more than $115,000 for his house. If C does make such an offer, then it may be concluded that the house is worth at least that much to C and that, therefore, he places a higher value on the house than does anyone else involved. Since our presumed goal for contract remedies is to move the house to its highest-valued use at the lowest cost, expectation damages seem to be the routine remedy for which we have been looking. Still, the law found it difficult until recently to justify expectancy as the routine contract remedy.

While the theoretical attractiveness of expectation loss as a damage measure is straightforward, practical application of the measure is not. The crucial problem is to determine, after the breach has occurred, what the innocent party's expected profit was at the time that the contract was formed. Clearly, the promisee has a powerful urge, after a breach, to exaggerate the gain he had expected from performance while the promisor has equally strong feelings that the gain could never have been so large. Recall that this is precisely the problem noted above with reliance damages. . . .

The conclusion of the literature on the efficiency goal in damage measures for breach of contract may be briefly stated: the expectation interest is the only measure of damages that will lead to efficient breach. There are, nonetheless, some widely recognized problems that arise in the measurement of the breachee's expectation interest. In the case of buyer's breach, the seller's reasonably foreseeable profits from the completed contract should be taken as the measure of damages if the goal is to encourage only efficient breach. Yet there are well known pitfalls in computing lost profits. In the case of seller's breach, the buyer's expectation interest is his consumer surplus or subjective valuation of the completed contract. There are serious evidentiary problems in determining this amount. It is probably the case that stipulated damages, if enforced even with a seemingly punitive element, are a less costly way to protect subjective valuation than are expectation damages. It is as yet an open question whether the inefficiencies of the alternative legal remedies, restitution and reliance, are severe enough to offset the far greater ease and precision of measuring those alternatives to expectation. We have yet to discuss whether expectation, with its high costs of measurement, is a superior guarantor of only efficient breach than is the equitable remedy, specific performance. I turn to that discussion in the next section.

The Efficiency of Specific Performance

Damage payments are the legal remedy for contract breach; specific performance is the equitable remedy. Specific performance is a judicial order requiring the promisor to perform his contractual promise or forbidding him from performing the promise with any other party. If, for example, A has promised to sell a house to B for $100,000 but breaches in order to sell to C for $125,000, B might seek relief in the form of a court order requiring A to sell to B. Alternatively, B might ask the court for an injunction forbidding A to sell to anyone but B. As a general rule, court invokes equitable remedies only when it thinks that legal remedies are likely to offer inadequate relief, that is, to be under-compensatory. The granting of equitable relief is at the discretion of the court upon a demonstration by the plaintiff that damages will not adequately compensate him. The typical cases in which this under-compensation is said to arise are in the sale of "unique goods," the sale of land (considered by the law, largely for historical reasons, to be a unique good), and long-term input contracts. When an otherwise innocent party asks for specific performance, the breacher is permitted to mount defenses that are not usually available against a damage award: insufficient certainty of terms, inadequate security for the innocent party's performance, the breacher's unilateral mistake, and the high level of supervisory costs that the court might incur in enforcing performance.

The economic efficiency of this state of affairs is open to question. In particular, it is not obvious that the efficient exchange of reciprocal promises or the enforcement of valid contractual promises is best served where specific performance is reserved for the circumstances noted above. . . . In this section, I shall attempt to show that specific performance should be, on efficiency grounds, the routine contract remedy. The reasons for this conclusion may be briefly summarized here. First, if contractual parties are on notice that valid promises will be specifically enforced, they will more efficiently exchange reciprocal promises at formation time. In particular, they will have a stronger incentive than currently exists under the dominant legal remedy to allocate efficiently the risks of loss from breach rather than leaving that task, in whole or in part, to the court or to post-breach negotiations conducted under the threat of a potentially inefficient legal remedy. Second, and perhaps most importantly, specific performance offers the most efficient mechanism for protecting subjective values attached to performance. Thus, it promotes contract breach only if it is efficient, that is, if someone will be better off and no one will be worse off because of the breach. In this regard, specific performance and an expansive enforcement of stipulated remedies constitute integral and inseparable parts of a unified theory of efficient contract remedies. Third, if specific performance were the routine remedy, the post-breach costs of adjusting a contract in order to move the promise to the highest-valuing user would be lower than under the most efficient legal remedy. The central reason for this is that under specific performance the costs of determining various parties' valuation of performance are borne by those parties in voluntary negotiations. This means that the costs of determining willingness-to-pay are borne by those most efficiently placed to determine that amount. Finally, because the costs of ascertaining any subjective values of the innocent party through evidence presented to a court are so high and because, therefore, the possibility of undercompensating the innocent party through a damage remedy is high, specific performance is far less likely to be undercompensatory and far more likely to protect the breachee's subjective valuation than is any other judicially imposed contract remedy. . . .

The Effect of Specific Performance on Formation
(Pre-Breach) Negotiation Costs

. . .

In an important article Professor Kronman has suggested that, judged on an efficiency standard, specific performance is currently being correctly invoked. The heart of the argument is that, in contract for the sale of a unique good, promisor and promisee would agree, if the court would recognize their agreement, that the promisor should perform specifically. By the same token, in a contract for the sale of a fungible good, promisor and promisee would agree that money damages would be paid to the promisee in the event of the promisor's breach. . . .

A crucial tenet of this argument is that money damages in the case of unique goods are likely to be under-compensatory and will, therefore, lead to inefficient breach. Thus, it is important to be clear about the sources of this under-compensation. The costs incurred in pre-contract search—those of locating sellers, obtaining information, comparing quality, and so on—are not generally recoverable if the contract is breached. In this respect what distinguishes fungible from unique goods is that, despite the breach, pre-contract search costs will bear fruit in the case of fungible goods but not in the case of unique goods. Therefore, money damages are likely to be under-compensatory in the case of unique goods.

[Furthermore,] Professor Kronman argues that promisors of unique goods, facing a thin market, will be more willing to accept a provision for specific performance than would be the case with fungible goods because the chances of his receiving a more attractive third-party offer are not very high. For nonunique goods and services, the promisor is more likely to receive alternative offers before he has fulfilled his contractual promise and is, therefore, more adamant about retaining the flexibility that comes with a money damages rule. . . .

It is worth noting that in the civil law countries specific performance is the routine contract remedy. This is a difficult situation to understand if there is really something to Professor Kronman's contention that confining equitable relief to the case of unique goods corresponds to what freely contracting parties would prefer. Perhaps the tastes of contracting parties in Western Europe are vastly different from those in the common law countries, but this is very doubtful. More likely, there is no necessary connection between specific performance and uniqueness. . . .

[T]his having been said, it is nonetheless the case that there is an important germ of truth in Professor Kronman's hypothesis. I believe that his contention would be completely accurate if it were revised to read as follows: If contracting parties were free to specify any remedy that was mutually agreeable, they would be likely to opt for specific performance rather than damages where the promisee attached some particular subjective valuation to the promisor's performance. The key difference here is the insertion that it is subjective valuation rather than uniqueness that makes specific performance attractive. Clearly, there is a relationship between uniqueness and subjective valuation: someone is more likely to attach a value greater than market value to a rare, one-of-a-kind item than to a highly fungible item. However, the class of things to which someone attaches a subjective valuation is greater than the class of unique items. Once his category is expanded to include all promises to which there is a subjective valuation, then the rest of Professor Kronman's analysis stands. . . .

Defenses

. . .

High Supervisory Costs. One of the most troubling issues in making specific performance the routine remedy for breach of contract is that there may be circumstances in which the costs to the court of supervising the performance of the breacher may be inefficiently high. There may be a very high probability of noncompliance, owing, perhaps, to the breacher's having forcefully and convincingly indicated his refusal to comply. The court might find that, under those conditions, it will need to incur extraordinarily large expenses in order that the breacher be held to his promise. The prestige of the court is in some jeopardy and may be damaged in those circumstances; this fact should be taken into account in figuring the costs of specific performance.

Alternatively, the performance contemplated may be so complex as to defy effective supervision. Suppose that A contracts with B to play Hamlet at B's theater and subsequently A refuses to perform. A court will not give B a decree requiring A to act. It is no doubt true that the costs to the court of judging whether, after B has received a specific performance decree, A had discharged his contractual obligation, including the quality of his performance, are extraordinarily high. For example, how should the court assure the quality of A's performance as the Prince of Denmark? Perhaps because A is in such a pique about his dispute with B he might, without stringent supervision by the court, seek to embarrass B by the shoddiness of his Hamlet. But how far should the court go? Should it specify gestures, grimaces, smiles, tones of declamation? The problem is a real one that the design of efficient remedies must seriously confront.

The contention is, in part, that high supervision costs will increase the costs of using specific performance as the routine remedy to the point that money damages are more attractive. Consider, for example, that if defendants know that they may be relieved of specific performance when they are able to demonstrate high supervision costs, then they have an incentive to raise that defense in cases where it may be inappropriate. Since this would, in general, raise the costs of litigation, it might make the otherwise more efficient remedy of specific performance more expensive and, therefore, less clearly efficient than money damages. A further contention is that since a far wider set of contracts than is commonly assumed involve high supervision costs, specific performance is inappropriate as the routine contract remedy.

While there is something creditable to this criticism, it must be carefully considered. For this criticism of specific performance to hold true, high supervision costs must attend a large number of contracts, and money damages must be the most efficient means of resolving the breach of those contracts. Neither of these matters can be demonstrated. . . .

[E]ven where supervisory costs are likely to be high, the inefficiencies that would follow from granting specific performance are exaggerated. There are two reasons for this: the contract may never be performed, and if it is, the promisor's regard for his professional reputation and future employability will temper the incentive to misperform the contract.

The presumption that an award of specific performance will necessarily result in performance of the contract is incorrect. We have seen above that specific performance, like injunctive relief, should be understood as an instruction to the litigants to use the market, rather than the court, to solve their dispute. There is every reason to believe that if B is awarded a decree of specific performance against A to play Hamlet, the two will begin negotiations to resolve the dispute, with A presumably willing to pay B not to exercise his right to the contractual promise. B, for his part, may be willing to exchange that right rather than run the risk of incurring large expenses in policing A's portrayal of Hamlet. That is, it may be mutually beneficial to promisor and promisee to bargain out of performance. Although it is difficult to know a priori when this will happen, the possibility that there will be no performance in circumstances in which supervision costs of the performance would be high should lessen the concern about the inefficiencies that might result. Indeed, it may be that the proper way to consider the problem of high supervision costs is not that it puts extraordinary burdens on the legal system but rather that it merely gives the breacher a much better bargaining position in the post-breach negotiations than would be the case under a contract in which the quality of the breacher's performance was not solely in the breacher's hands. If that is the proper economic analysis of the matter of high supervision costs, then it may well be that specific performance is the preferred remedy there, too. Assuming that the promisor makes a credible threat that supervising the quality of his performance will be high, then the worst that can happen to the promisee is that he accepts, in return for not enforcing his right to specific performance, a price that reflects the contract price less the anticipated supervision costs. Such a conclusion would serve as an inducement for future contracts regarding personal services, or other high supervisory cost activities, to include liquidation clauses specifying responsibility for the costs of monitoring performance. Alternatively, promisees in situations of high supervision costs will discount the contract price they are willing to give a promisor by the probability of breach and by the level of anticipated supervision costs.

Even if there is no exchange of the right to performance results, the force of competition may temper the defendant's urge to misperform in some manner. A, for example, must take care for his future employability on the stage—with other promoters, if not necessarily B—and this fact may spur him to produce as creditable a Hamlet as if he were acting for the sheer love of it, rather than under threat of contempt of court.

Thirdly, and lastly, the high supervisory costs of equitable relief are objectionable in large part because they are incurred at public expense. This objection would be mitigated if the costs were borne by the litigants, not by taxpayers in general. The use of court-appointed special masters to oversee equitable decrees is one means of achieving this privatization, but one that, despite its attraction to economists, does not find much favor in the legal fraternity.

Summary and Conclusion

The contention of this paper is that remedies for breach of contract are not now entirely consistent with the goal of economic efficiency. The routine remedy is the awarding of money damages, whereas economic efficiency considerations urge specific performance as the routine remedy. Following Calabresi and Melamed's analysis of legal and equitable remedies in nuisance law, I propose that the efficient exchange of mutually beneficial promises would be better served by using the level of transaction costs as the guide for choosing a contract remedy: if transaction costs are low between the defaulter and the innocent party, then an award of specific performance will encourage the parties to exchange the right to performance voluntarily and efficiently; if, however, those costs are so high that no voluntary exchange can take place, then the court should intervene and compel an exchange at a collectively determined price; that is, the court should award money damages. Since it is most likely to be the case that parties to a contract have low transaction costs in that they, unlike, say, tortfeasors and their victims, have already established a relationship, courts should presume that specific performance is to be awarded, with money damages being the exceptional award. This is precisely the opposite of current practice. . . .

3.4 CRIMINAL LAW

Crime and Punishment: An Economic Approach

GARY BECKER

Since the turn of the century, legislation in Western countries has expanded rapidly to reverse the brief dominance of laissez-faire during the nineteenth century. The state no longer merely protects against violations of person and property through murder, rape, or burglary but also restricts "discrimination" against certain minorities, collusive business arrangements, "jaywalking," travel, the materials used in construction, and thousands of other activities. The activities restricted not only are numerous but also range widely, affecting persons in very different pursuits and of diverse social backgrounds, education levels, ages, races, etc. Moreover, the likelihood that an offender will be discovered and convicted and the nature and extent of punishments differ greatly from person to person and activity to activity. Yet, in spite of such diversity, some common properties are shared by practically all legislation, and these properties form the subject matter of this essay. . . .

Basic Analysis

The Cost of Crime

Although the word "crime" is used in the title to minimize terminological innovations, the analysis is intended to be sufficiently general to cover all violations, not just felonies—like murder, robbery, and assault, which receive so much newspaper coverage—but also tax evasion, the so-called white-collar crimes, and traffic and other violations. Looked at this broadly, "crime" is an economically important activity or "industry," notwithstanding the almost total neglect by economists. Some relevant evi-

Gary Becker, Crime and Punishment: An Economic Approach, 76 *Journal of Political Economy* (1968), pp. 169–217. Copyright © 1958 University of Chicago Press. Reprinted with permission.

dence recently put together by the President's Commission on Law Enforcement and Administration of Justice (the "Crime Commission") [reveals that p]ublic expenditures in 1965 at the federal, state, and local levels on police, criminal courts and counsel, and "corrections" amounted to over $4 billion, while private outlays on burglar alarms, guards, counsel, and some other forms of protection were about $2 billion. Unquestionably, public and especially private expenditures are significantly understated, since expenditures by many public agencies in the course of enforcing particular pieces of legislation, such as state fair-employment laws, are not included, and a myriad of private precautions against crime, ranging from suburban living to taxis, are also excluded.

[Additionally, the Crime Commission estimates that] gross income from expenditures on various kinds of illegal consumption, including narcotics, prostitution, and mainly gambling, amounted to over $8 billion. The value of crimes against property, including fraud, vandalism, and theft, amounted to almost $4 billion, while about $3 billion worth resulted from the loss of earnings due to homicide, assault, or other crimes. All [these] costs . . . total about $21 billion, which is almost 4 percent of reported national income in 1965. If the sizeable omissions were included, the percentage might be considerably higher. . . .

The Model

It is useful in determining how to combat crime in an optimal fashion to develop a model to incorporate the behavioral relations behind the costs listed [above]. These can be divided into five categories: the relations between (1) the number of crimes, called "offenses" in this essay, and the cost of offenses, (2) the number of offenses and the punishments meted out, (3) the number of offenses, arrests, and convictions and the public expenditures on police and courts, (4) the number of convictions and the costs of imprisonments or other kinds of punishments, and (5) the number of offenses and the private expenditures on protection and apprehension. . . .

Damages. Usually a belief that other members of society are harmed is the motivation behind outlawing or otherwise restricting an activity. The amount of harm would tend to increase with the activity level, as in the relation

$$H_i = H_i(O_i),$$

$$\text{with } H_i' = \frac{dH_i}{dO_i} > 0, \tag{1}$$

where H_i is the harm from the ith activity and O_i is the activity level.* The concept of harm and the function relating its amount to the activity level are familiar to econo-

---

Editor's note on mathematical notation: The expression $H_i = H_i(O_i)$ is an abbreviation indicating that the level of harm (H_i) depends on the level of activity (O_i). It should be read as "H_i is a function of O_i." Similarly, the expression dH_i/dO_i denotes the rate at which changes in O_i lead to changes in H_i. It should be read as "the derivative of H_i with respect to O_i." The expression H_i' is shorthand for "the derivative of H_i." All the symbols in these and subsequent equations should be understood to represent actual empirical quantities such as the values estimated by the Crime Commission.

mists from their many discussions of activities causing external diseconomies. From this perspective, criminal activities are an important subset of the class of activities that cause diseconomies, with the level of criminal activities measured by the number of offenses.

The social value of the gain *[G]* to offenders presumably also tends to increase with the number of offenses, as in

$$G = G(O),$$

$$\text{with } G' = \frac{dG}{dO} > 0. \tag{2}$$

The net cost or damage *[D]* to society is simply the difference between the harm and gain and can be written as

$$D(O) = H(O) - G(O). \tag{3}$$

...These values are important components of, but are not identical to, the net damages to society. For example, the cost of murder is measured by the loss in earnings of victims and excludes, among other things, the value placed by society on life itself; the cost of gambling excludes both the utility to those gambling and the "external" disutility to some clergy and others; the cost of "transfers" like burglary and embezzlement excludes social attitudes toward forced wealth redistribution and also the effects on capital accumulation of the possibility of theft. Consequently, the [Commission's] $15 billion estimate for the cost of crime . . . may be a significant understatement of the net damages to society, not only because the costs of many white-collar crimes are omitted, but also because much of the damage is omitted even for the crimes covered.

The Cost of Apprehension and Conviction. The more that is spent on policemen, court personnel, and specialized equipment, the easier it is to discover offenses and convict offenders. One can postulate a relation between the output of police and court "activity" and various inputs of manpower, materials, and capital, as in $A = f(m,r,c)$, where f is a production function summarizing the "state of the arts." Given f and input prices, increased "activity" would be more costly, as summarized by the relation

$$C = C(A)$$

$$\text{with } C' = \frac{dC}{dA} > 0. \tag{4}$$

It would be cheaper to achieve any given level of activity the cheaper were policemen, judges, counsel, and juries and the more highly developed the state of the arts, as determined by technologies like fingerprinting, wiretapping, computer control, and lie-detecting.

One approximation to an empirical measure of "activity" is the number of offenses cleared by conviction. . . . [The Crime Commission] indicates that in 1965 public expenditures in the United States on police and courts totaled more than $3 billion, by no means a minor item. Separate estimates were prepared for each of seven major felonies. Expenditures on them averaged about $500 per offense (reported) and about $2,000 per person arrested, with almost $1,000 being spent per murder. $500 is an estimate of the average cost of these felonies and would presumably be a larger figure if the number of either arrests or convictions were greater. . . .

The Supply of Offenses. Theories about the determinants of the number of offenses differ greatly, from emphasis on skull types and biological inheritance to family upbringing and disenchantment with society. . . . The approach taken here follows the economists' usual analysis of choice and assumes that a person commits an offense if the expected utility to him exceeds the utility he could get by using his time and other resources at other activities. Some persons become "criminals," therefore, not because their basic motivation differs from that of other persons, but because their benefits and costs differ. I cannot pause to discuss the many general implications of this approach, except to remark that criminal behavior becomes part of a much more general theory and does not require ad hoc concepts of differential association, anomies and the like, nor does it assume perfect knowledge, lightning-fast calculation, or any of the other caricatures of economic theory.

This approach implies that there is a function relating the number of offenses by any person to his probability of conviction, to his punishment if convicted, and to other variables, such as the income available to him in legal and other illegal activities, the frequency of nuisance arrests, and his willingness to commit an illegal act. This can be represented as

$$O_j = O_j(p_j, f_j, u_j), \tag{5}$$

where O_j is the number of offenses he would commit during a particular period, p_j his probability of conviction per offense, f_j his punishment per offense, and u_j a portmanteau variable representing all these other influences. . . .

An increase in either p_j or f_j would reduce the utility expected from an offense and thus would tend to reduce the number of offenses because either the probability of "paying" the higher "price" or the "price" itself would increase. . . . The effect of changes in some components of u_j could also be anticipated. For example, a rise in the income available in legal activities or an increase in law-abidingness due, say, to "education" would reduce the incentive to enter illegal activities and thus would reduce the number of offenses. Or a shift in the form of the punishment, say, from a fine to imprisonment, would tend to reduce the number of offenses, at least temporarily, because they cannot be committed while in prison. . . .

The total number of offenses is the sum of all the O_j and would depend on the set of p_j, f_j, and u_j. Although these variables are likely to differ significantly between persons because of differences in intelligence, age, education, previous offense history, wealth, family upbringing, etc., for simplicity I now consider only their average values, p, f, and u, and write the market offense function as

$$O = O(p, f, u). \tag{6}$$

This function is assumed to have the same kinds of properties as the individual functions, in particular, to be negatively related to p and f and to be more responsive to the former than the latter if, and only if, offenders on balance have risk preference. . . .

Punishments. . . . The cost of different punishments to an offender can be made comparable by converting them into their monetary equivalent or worth, which, of course, is directly measured only for fines. For example, the cost of an imprisonment is the discounted sum of the earnings foregone and the value placed on the restrictions in consumption and freedom. Since the earnings foregone and the value placed on prison restrictions vary from person to person, the cost even of a prison sentence of given duration is not a unique quantity but is generally greater, for example, to offenders who could earn more outside of prison. The cost to each offender would be greater the longer the prison sentence, since both foregone earnings and foregone consumption are positively related to the length of sentences.

Punishments affect not only offenders but also other members of society. Aside from collection costs, fines paid by offenders are received as revenue by others. Most punishments, however, hurt other members as well as offenders: for example, imprisonment requires expenditures on guards, supervisory personnel, buildings, food, etc. Currently about $1 billion is being spent each year in the United States on probation, parole, and institutionalization alone, with the daily cost per case varying tremendously from a low of $0.38 for adults on probation to a high of $11.00 for juveniles in detention institutions.

The total social cost of punishments is the cost to offenders plus the cost or minus the gain to others. Fines produce a gain to the latter that equals the cost to offenders, aside from collection costs, and so the social cost of fines is about zero, as befits a transfer payment. The social cost of probation, imprisonment, and other punishments, however, generally exceeds that to offenders, because others are also hurt. . . . [The ratio b of social cost to offender cost] varies greatly between different kinds of punishments: $b \approx 0$ for fines, while $b > 1$ for torture, probation, parole, imprisonment, and most other punishments. It is especially large for juveniles in detention homes or for adults in prisons and is rather close to unity for torture or for adults on parole. . . .

Fines

. . .

The Case for Fines

Just as the probability of conviction and the severity of punishment are subject to control by society, so too is the form of punishment: legislation usually specifies whether an offense is punishable by fines, probation, institutionalization, or some combination. Is it merely an accident, or have optimality considerations determined that today, in most countries, fines are the predominant form of punishment, with institutionalization reserved for the more serious offenses? This section presents several arguments which imply that social welfare is increased if fines are used *whenever feasible.*

In the first place, probation and institutionalization use up social resources, and fines do not, since the latter are basically just transfer payments, while the former use resources in the form of guards, supervisory personnel, probation officers, and the offenders' own time. [The Crime Commission] indicates that the cost is not minor either: in the United States in 1965, about $1 billion was spent on "correction," and this estimate excludes, of course, the value of the loss in offenders' time.

Moreover, the determination of the optimal number of offenses and severity of punishments is somewhat simplified by the use of fines. A wise use of fines requires knowledge of marginal gains and harm and of marginal apprehension and conviction costs; admittedly, such knowledge is not easily acquired. A wise use of imprisonment and other punishments must know this too, however, and, in addition, must know about the elasticities of response of offenses to changes in punishments. As the bitter controversies over the abolition of capital punishment suggest, it has been difficult to learn about these elasticities. . . .

Fines provide compensation to victims, and optimal fines at the margin fully compensate victims and restore the status quo ante, so that they are no worse off than if offenses were not committed. Not only do other punishments fail to compensate, but they also require "victims" to spend additional resources in carrying out the punishment. It is not surprising, therefore, that the anger and fear felt toward ex-convicts who in fact have not "paid their debt to society" have resulted in additional punishments, including legal restrictions on their political and economic opportunities and informal restrictions on their social acceptance. . . .

One argument made against fines is that they are immoral because, in effect, they permit offenses to be bought for a price in the same way that bread or other goods are bought for a price. A fine can be considered the price of an offense, but so too can any other form of punishment; for example, the "price" of stealing a car might be six months in jail. The only difference is in the units of measurement: fines are prices measured in monetary units, imprisonments are prices measured in time units, etc. If anything, monetary units are to be preferred here as they are generally preferred in pricing and accounting. . . .

We might detour briefly to point out some interesting implications for the probability of conviction of the fact that the monetary value of a given fine is obviously the same for all offenders, while the monetary equivalent or "value" of a given prison sentence or probation period is generally positively related to an offender's income. The discussion [above] suggested that actual probabilities of conviction are not fixed to all offenders but usually vary with their age, sex, race, and, in particular, income. Offenders with higher earnings have an incentive to spend more on planning their offenses, on good lawyers, on legal appeals, and even on bribery to reduce the probability of apprehension and conviction for offenses punishable by, say, a given prison term, because the cost to them of conviction is relatively large compared to the cost of these expenditures.

Similarly, however, poorer offenders have an incentive to use more of their time in planning their offenses, in court appearances, and the like to reduce the probability of conviction for offenses punishable by a given fine, because the cost to them of conviction is relatively large compared to the value of their time. The implication is that the probability of conviction would be systematically related to the earnings of offenders: negatively for offenses punishable by imprisonment and positively for those punishable by fines. Although a negative relation for felonies and other offenses punishable by imprisonment has been frequently observed and deplored . . . , I do not know of any studies of the relation for fines or of any recognition that the observed negative relation may be more a consequence of the nature of the punishment than of the influence of wealth.

Another argument made against fines is that certain crimes, like murder or rape, are so heinous that no amount of money could compensate for the harm inflicted. This argument has obvious merit and is a special case of the more general principle that fines cannot be relied on exclusively whenever the harm exceeds the resources of offenders. For then victims could not be fully compensated by offenders, and fines would have to be supplemented with prison terms or other punishments in order to discourage offenses optimally. This explains why imprisonments, probation, and parole are major punishments for the more serious felonies; considerable harm is inflicted, and felonious offenders lack sufficient resources to compensate. Since fines are preferable, it also suggests the need for a flexible system of installment fines to enable offenders to pay fines more readily and thus avoid other punishments.

This analysis implies that if some offenders could pay the fine for a given offense and others could not, the former should be punished solely by fine and the latter partly by other methods. In essence, therefore, these methods become a vehicle for punishing "debtors" to society. . . .

Whether a punishment like imprisonment in lieu of a full fine for offenders lacking sufficient resources is "fair" depends, of course, on the length of the prison term compared to the fine. For example, a prison term of one week in lieu of a $10,000 fine would, if anything, be "unfair" to wealthy offenders paying the fine. Since imprisonment is a more costly punishment to society than fines, the loss from offenses would be reduced by a policy of leniency toward persons who are imprisoned because they cannot pay fines. Consequently, optimal prison terms for "debtors" would not be "unfair" to them in the sense that the monetary equivalent to them of the prison terms would be less than the value of optimal fines, which in turn would equal the harm caused or the "debt."

It appears, however, that "debtors" are often imprisoned at rates of exchange with fines that place a low value on time in prison. Although I have not seen systematic evidence on the different punishments actually offered convicted offenders, and the choices they made, many statutes in the United States do permit fines and imprisonment that place a low value on time in prison. For example, in New York State, Class A misdemeanors can be punished by a prison term as long as one year or a fine no larger than $1,000 and Class B misdemeanors, by a term as long as three months or a fine no larger than $500 (*Laws of New York,* 1965, chap. 1030, Arts. 70 and 80). According to my analysis, these statutes permit excessive prison sentences relative to the fines, which may explain why imprisonment in lieu of fines is considered unfair to poor offenders, who often must "choose" the prison alternative.

Compensation and the Criminal Law

. . . [I]f the case for fines were accepted, and punishment by optimal fines became the norm, the traditional approach to criminal law would have to be significantly modified.

First and foremost, the primary aim of all legal proceedings would become the same: not punishment or deterrence, but simply the assessment of the "harm" done by defendants. Much of traditional criminal law would become a branch of the law of torts, say "social torts," in which the public would collectively sue for "public" harm. A "criminal" action would be defined fundamentally not by the nature of the action but by the inability of a person to compensate for the "harm" that he caused. Thus an action would be "criminal" precisely because it results in uncompensated "harm" to others. Criminal law would cover all such actions, while tort law would cover all other (civil) actions.

As a practical example of the fundamental changes that would be wrought, consider the antitrust field. Inspired in part by the economist's classic demonstration that monopolies distort the allocation of resources and reduce economic welfare, the United States has outlawed conspiracies and other constraints of trade. In practice, defendants are often simply required to cease the objectionable activity, although sometimes they are also fined, become subject to damage suits, or are jailed.

If compensation were stressed, the main purpose of legal proceedings would be to levy fines equal to the harm inflicted on society by constraints of trade. There would be no point to cease and desist orders, imprisonment, ridicule, or dissolution of companies. If the economist's theory about monopoly is correct, and if optimal fines were levied, firms would automatically cease any constraints of trade, because the gain to them would be less than the harm they cause and thus less than the fines expected. On the other hand, if Schumpeter and other critics are correct, and certain constraints of trade raise the level of economic welfare, fines could fully compensate society for the harm done, and yet some constraints would not cease, because the gain to participants would exceed the harm to others.

One unexpected advantage, therefore, from stressing compensation and fines rather than punishment and deterrence is that the validity of the classical position need not be judged a priori. If valid, compensating fines would discourage all constraints of trade and would achieve the classical aims. If not, such fines would permit the socially desirable constraints to continue and, at the same time, would compensate society for the harm done. . . .

Summary and Concluding Remarks

. . .

The main contribution of this essay, as I see it, is to demonstrate that optimal policies to combat illegal behavior are part of an optimal allocation of resources. Since economics has been developed to handle resource allocation, an "economic" framework becomes applicable to, and helps enrich, the analysis of illegal behavior. At the same time, certain unique aspects of the latter enrich economic analysis: some punishments, such as imprisonments, are necessarily nonmonetary and are a cost to society as well as to offenders; the degree of uncertainty is a decision variable that enters both the revenue and cost functions, etc.

Lest the reader be repelled by the apparent novelty of an "economic" framework for illegal behavior, let him recall that two important contributors to criminology during the eighteenth and nineteenth centuries, Beccaria and Bentham, explicitly applied an economic calculus. Unfortunately, such an approach has lost favor during the last hundred years, and my efforts can be viewed as a resurrection, modernization, and thereby I hope improvement on these much earlier pioneering studies.

3.5 PROCEDURE

An Economic Approach to Legal Procedure and Judicial Administration

RICHARD POSNER

An important purpose of substantive legal rules (such as the rules of tort and criminal law) is to increase economic efficiency. It follows . . . that mistaken imposition of legal liability, or mistaken failure to impose liability, will reduce efficiency. Judicial error is therefore a source of social costs and the reduction of error is a goal of the procedural system. . . .

Even when the legal process works flawlessly, it involves costs—the time of lawyers, litigants, witnesses, jurors, judges, and other people, plus paper and ink, law office and court house maintenance, telephone service, etc. These costs are just as real as the costs resulting from error: in general we would not want to increase the direct costs of the legal process by one dollar in order to reduce error costs by 50 (or 99) cents. The economic goal is thus to minimize the sum of error and direct costs.

Despite its generality, this formulation provides a useful framework in which to analyze the problems and objectives of legal procedure. It is usable even when the purpose of the substantive law is to transfer wealth or to bring about some other noneconomic goal, rather than to improve efficiency. All that is necessary is that it be possible, in principle, to place a price tag on the consequences of failing to apply the substantive law in all cases in which it was intended to apply, so that our two variables, error cost and direct cost, remain commensurable. . . .

The Costs of Error in Civil Actions

Suppose a company inflicts occasional injuries on people with whom it cannot contract due to very high transaction costs. Victims of these injuries could prevent them only at prohibitive cost . . . , but the company can purchase various relatively inexpensive safety devices that would reduce the accident rate significantly. In the absence of legal sanctions it has no incentive to purchase such devices since, due to the costs of transacting, it cannot sell anyone the benefits of the devices in increasing safety. If the tort law makes it liable for the costs of these accidents, and is enforced flawlessly, the company will purchase the optimum quantity of safety devices. If the law is not enforced flawlessly, a suboptimum quantity of safety equipment will be procured.

The goal of a system of accident liability is to minimize the total costs of accidents and of accident avoidance. If we assume that the only feasible method of accident avoidance is the purchase of a particular type of safety equipment, then those to-

Richard Posner, An Economic Approach to Legal Procedure and Judicial Administration, 2 *Journal of Legal Studies* (1973), pp. 399–458. Copyright © 1973 The University of Chicago Law School. Reprinted by permission.

tal costs are minimized by purchasing the quantity of that equipment at which the marginal product of safety equipment in reducing accident costs is equal to the marginal cost of the equipment. This marginal product is the rate at which the number of accidents inflicted by the company declines as the quantity of safety equipment purchased increases, multiplied by the cost per accident. . . .

The company, however, is not interested in minimizing the social costs of accidents and accident avoidance; it is interested only in minimizing its private accident and accident-avoidance costs. The former are the social costs of the firm's accidents multiplied by the probability that the firm will actually be held liable—forced to pay—for those costs. Since legal error presumably causes erroneous impositions as well as erroneous denials of liability, we must add a third term to the firm's cost function: the amount of money that it is forced to pay out in groundless claims. That amount is a function of the legal error rate and disappears when that rate is zero. . . .

The company minimizes its private accident and accident-avoidance costs by equating the marginal product of safety equipment in reducing its accident liability to the marginal private cost of that equipment (which we assume is the same as the marginal social cost). This marginal private product is simply the marginal social product weighted by the probability of the firm's being held liable. If that probability is one, the marginal social and private products are the same. But when the probability is less than one—that is, when the legal-error rate is positive—they diverge, leading to a social loss. . . . The higher the error rate, the greater the reduction in the purchase of safety equipment and the greater the social loss.

The analysis is incomplete because we have ignored the possible effect of a positive error rate, operating through the third term in the company's cost function (liability resulting from groundless claims), on the firm's purchase of safety equipment. Suppose that the errors against the company took the form exclusively of accident victims' exaggerating the extent of their injuries. By increasing the company's private accident costs, these errors would increase the marginal private product of safety equipment. . . . In fact, however, although all errors in favor of the company operate to lower its marginal-product curve, only some errors against the company operate to raise it. The purchase of additional safety equipment will not prevent the erroneous imposition of liability in a case in which no accident would have occurred in any event—the victim fabricated it—or in which the accident was inflicted by someone else and could not have been prevented by the defendant. Such errors do not increase the value of safety equipment to the firm and hence the marginal private product of that equipment. But even here a qualification is necessary. Additional safety equipment might strengthen the company's defense against a suit arising out of an accident actually caused by someone else. The company might be able to argue that, in view of all of the safety precautions it had taken, it could not have caused the accident. . . .

Settlement Out of Court

. . .

When Are Cases Settled?

Since settlement costs are normally much lower than litigation costs, the fraction of cases settled is an important determinant of the total direct cost of legal dispute resolution. The necessary condition for settlement is that the plaintiff's minimum offer—the least amount he will take in settlement of his claim—be smaller than the defendant's maximum offer. This is not a sufficient condition: the parties may find it impossible to agree upon a mutually satisfactory settlement price. But we shall assume that settlement negotiations are rarely unsuccessful for this reason and therefore that litigation occurs only when the plaintiff's minimum offer is greater than the defendant's maximum offer. The plaintiff's minimum offer is the expected value of the litigation to him plus his settlement costs, the expected value of the litigation being the present value of the judgment if he wins, multiplied by the probability (as he estimates it) of his winning, minus the present value of his litigation expenses. The defendant's maximum offer is the expected cost of the litigation to him and consists of his litigation expenses, plus the cost of an adverse judgment multiplied by the probability as he estimates it of the plaintiff's winning (which is equal to one minus the probability of his winning), minus his settlement costs. Anything that reduces the plaintiff's minimum offer or increases the defendant's maximum offer, such as an increase in the parties' litigation expenditures relative to their settlement costs, will reduce the likelihood of litigation. Hence measures to reduce litigation costs might actually increase the total costs of legal dispute resolution, by making trials, which are usually costlier than settlements, more attractive than before the measures were introduced.

Anything that increases the plaintiff's minimum settlement offer or reduces the defendant's maximum offer will increase the likelihood of litigation. An increase in the plaintiff's subjective probability of prevailing or in his stakes will do this, but so will an increase in the defendant's subjective probability of prevailing since it will induce him to reduce his maximum settlement offer. An increase in the defendant's stakes in the case will reduce the likelihood of litigation by leading him to increase his maximum settlement offer. In the important special case where the stakes to the parties are the same, it can be shown that an increase in those stakes will increase the likelihood of litigation. In that case, litigation cannot possibly occur unless the plaintiff's subjective probability of prevailing is greater than one minus the defendant's subjective probability, for otherwise the plaintiff's minimum settlement offer will be equal to or smaller than the defendant's maximum offer. Assuming that this minimum condition for litigation is satisfied, any increase in the stakes must increase the likelihood of litigation by making the plaintiff's minimum settlement offer grow faster than the defendant's maximum settlement offer. . . .

Expenditures on Litigation

Determinants of Parties' Expenditures on Litigation

. . .

At the point where settlement attempts fail and the parties decide to litigate, it is tempting to switch from the cooperative model of legal dispute resolution (in which the parties are viewed as attempting to work out a mutually advantageous contract) of the previous part to a competitive model in which each party is viewed as expending resources on litigation in much the same way as a seller expends resources on advertising—in order to persuade the "customer" (the tribunal) of the superior merits of his "product" (case). So abrupt a shift of emphasis would be difficult to justify, however. One reason why a competitive model is appropriate in the case of advertising expenditures is that competing sellers are forbidden to agree to limit those expenditures. Sellers would often be better off if they were permitted to negotiate mutual reductions in advertising and such agreements might be commonplace were it not for antitrust policy. The case of litigation expenditures is similar but in this case agreements to limit competition are condoned, indeed encouraged, by public policy. They are in fact common: it is the rare case where there is no cooperation between the litigants' attorneys to reduce the expense of litigation. To be sure, agreements to limit litigation expenditures as such would be costly to enforce, and are rare. But the purpose of such agreements can be accomplished indirectly (and much more cheaply) by agreements to dispense with proof of particular facts, to limit the number of witnesses, etc., and such agreements are common. . . .

Despite the prevalence of "collusion" in the process by which parties to a lawsuit decide how much to spend, negotiation will sometimes fail—we have indicated a possible source of high transaction costs in the difficulty of policing an agreement to limit expenditures—so the addition of a competitive model seems indicated. . . .

Presumably each party chooses the level of investment in the litigation that maximizes the expected value of the litigation to him, which is equal to the stakes in the case multiplied by the probability of prevailing, minus the costs of litigation. The probability of prevailing is a function of what the party spends, what his opponent spends, and various exogenous factors (such as the state of the precedents and the availability of evidence) that weight the effect of expenditures by either party on the probability of a particular outcome. To determine each party's optimum expenditure would require that we specify the precise relationships among the relevant variables, which we will not attempt to do. . . . For our purposes it is sufficient to note that each party will seek to equate the marginal product of the resources he invests in the litigation in enhancing the expected value of the litigation to their marginal cost (which we can assume to be constant) and that this marginal product will be greater, and hence the party's expenditures on litigation greater, the larger the party's stakes in the case and the more favorable the law or the evidence is to him; either circumstance will tend to make a dollar of additional expenditures on the litigation more productive. The effect on his optimum expenditure of the level of expenditures chosen by his opponent is less clear-cut. An increase in the opponent's expenditures may induce him to increase his own to overcome their effect or it may so reduce the value of his own expenditures as to induce him to reduce them. Which effect dominates depends on the precise form of the model and the specific values of its parameters. . . .

[This approach] implies, quite reasonably, that an increase either in the plaintiff's stakes or in the effectiveness of his litigation expenditures, or a decrease either in defendant's stakes or in the effectiveness of his litigation expenditures, will induce the plaintiff to spend at a higher rate than the defendant, and vice versa. When the stakes to the parties are the same, the ratio of their litigation expenditures will be positively correlated with the ratio of their subjective probabilities of prevailing if they made the same expenditures. This helps to explain why, as mentioned earlier, we can expect most cases to be decided correctly even if the plaintiff need establish his case only by a bare preponderance of the evidence. If the allegations essential to one party's claim are in fact true, ordinarily it will be easier for him to prove them than for his opponent to disprove them, assuming they spend the same amount of money on the trial. Stated another way, the effectiveness of the expenditures of the party with the meritorious claim will be high relative to the effectiveness of his opponent's expenditures. . . .

The Role of Procedure in Optimizing Litigation Expenditures

Economizing Procedures and Their Effects. Many familiar procedural devices appear to be designed, in part at least, to reduce the expense of litigation. Some examples are summary judgment, judicial notice, presumptions, collateral estoppel, requests for admissions by the adverse party, allocation of the burden of pleading and of production of evidence, exclusion of evidence that is merely cumulative, and perhaps the hearsay rule. A particularly clear example is provided by the rules governing venue which are designed to place the trial in the cheapest location for the parties. But whether such devices actually reduce the amount of money spent on litigation is not obvious. If the judge, by taking judicial notice that January 11, 1973, was a Thursday, saves a party the expense of hiring a witness to testify to the fact, it does not follow that the party's litigation expenditures will be lower. The effect of judicial notice is to enable the party to develop the same amount of evidence favorable to his contentions at lower cost. . . . The party's demand for evidence is equal to the marginal product of evidence in enhancing the expected value of litigation to him. . . . A reduction in the supply price induces the party to purchase a greater quantity of evidence, q_1, for which he pays a lower price, p_1. Whether pq is larger than $p_1 q_1$ depends on the elasticity of demand.* . . . If it is greater than one, $p_1 q_1$ (the party's litigation expenditures after the increase in productivity) will be larger than before; if it is one they will be the same; if it is less than one they will be smaller.

Whether demand is likely to be inelastic in the relevant region depends once again on our assumptions about the parties' reaction patterns. If a measure that makes it cheaper for one party to establish facts favorable to his position also makes it cheaper for the other party to establish facts favorable to his position, the perceived marginal product of evidence may be slight, since the effect of additional evidence in enhancing the expected value of litigation will be expected to be offset by the opponent's matching purchase of additional evidence. . . . But regardless of their impact on expenditures, such devices can probably be justified as reducing the error costs discussed [above]. The use of witness time to establish a fact that is clear beyond doubt does not advance the search for truth. The elimination of such a method of presenting evidence encourages the parties to increase the purchase of evidence that does dispel genuine factual uncertainties.

Similarly, the principal significance of the liberal pleading and discovery provisions that are the most distinctive features of the Federal Rules may be that they re-

---

Editor's note: The *elasticity of demand* (or *price elasticity of demand*) measures the sensitivity of demand for a good to changes in its price. It is defined as the percentage decrease in quantity demanded that accompanies a 1 percent increase in price. If quantity changes more than proportionately with price, demand is referred to as *elastic;* if it changes less than proportionately, demand is *inelastic.*

duce error costs, not that they reduce aggregate expenditures on litigation or increase the settlement rate. The abrogation of the traditional strict pleading requirements has probably reduced the number of meritorious cases dismissed because of a lawyer's oversight. Discovery enables each party to obtain the facts bearing on the merits of his contentions and the deficiencies of the adversary's. . . .

The Interaction between Error Costs and Direct Costs

The relationship between error costs and direct costs can be summarized in a loss function having three terms. The first term is error cost. This is a function of the probability of error, which in turn is a function of the fraction of cases litigated, the amount of private expenditures on litigation, and the amount of public expenditures. The second term is the sum of the private and public expenditures in cases that are litigated, and is equal to the total of those expenditures in all cases multiplied by the fraction of cases litigated. The third term is the total expenditures (all private) on cases that are settled, and is equal to the total private expenditures in all cases multiplied by the fraction of cases settled multiplied by the fractional cost of settling rather than litigating. An increase in the fraction of cases litigated, or in the public or private expenditures on litigation, will reduce the probability and hence cost of an erroneous judicial determination. An increase in public expenditures on litigation will reduce the relative cost advantage of settling rather than litigating (the government's subsidy of litigation has increased), and an increase in the relative cost advantages of settling will reduce the fraction of cases tried.

These relationships make clear why it is difficult to predict a priori the effect on overall efficiency of changes in the relevant variables. For example, an increase in the fraction of cases litigated will increase the social costs of legal dispute resolution only if the difference between the total costs of litigating cases and the total costs of settling them is greater than the reduction in error costs brought about by increasing the fraction of litigated cases. Otherwise it will reduce the total costs of legal dispute resolution. An increase in public expenditures will reduce error costs both directly and by inducing a larger fraction of cases to be tried, but it will increase the total direct costs of legal dispute resolution both directly and by making litigation relatively more attractive than settlement. Thus there can be no presumption that increasing the public expenditures on the court system will increase social welfare. . . . Finally, an increase in the fractional cost of settlement versus litigation, by lowering the cost of settlement relative to that of litigation, will reduce the direct costs of legal dispute resolution but indirectly increase the error costs. Thus, as argued earlier, measures that increase the attractiveness of settlement in comparison to litigation cannot be regarded as unequivocally desirable. . . .

Detrebling versus Decoupling Antitrust Damages: Lessons from the Theory of Enforcement

A. MITCHELL POLINSKY

. . .

The modern theory of enforcement began with a seminal paper by Gary Becker on crime and punishment. Becker's theory, which assumes that the government does the enforcing, is easily explained. . . . Suppose firms obtain some gain from engaging in an activity that imposes costs on others. Examples of such activities include polluting the air, evading taxes, and attempting to monopolize an industry. If it were costless for the public enforcement authority to catch or observe firms when they engage in a harm-creating activity, presumably every firm would be caught and fined an amount equal to the external cost of the activity. Firms then would engage in the activity only if their private benefits exceed the external cost. And, from society's perspective, such behavior would be efficient.

However, in most situations it is difficult or costly for the enforcement authority to catch firms that impose external costs. If, as a result, firms are not always caught, they would engage in the harm-creating activity too often unless they were made to pay more than the harm caused when they are caught. Then, according to Becker's theory, the fine could be raised to a level such that, as before, firms would engage in the activity only if their private gains exceed the external cost. Since this outcome can be achieved for any given probability of catching firms and since it is costlier to catch a larger fraction of those engaging in the activity, Becker argued that the enforcement authority should set the probability very low and the fine correspondingly high. This low probability/high fine combination characterizes the optimal system of public enforcement.

In a subsequent paper, . . . Landes and Posner claimed that private enforcement would lead to too much enforcement relative to optimal public enforcement. Their intuitive explanation was based on the following observations. Under public enforcement, if detection were certain, the fine should be set equal to the external damage caused by the activity. By raising the fine and lowering the probability of detection, the same level of deterrence can be achieved at less cost. Under private enforcement, however, they pointed out that raising the fine would lead to a higher probability of detection because self-interested private enforcers would be induced to invest more in enforcement. From this observation they concluded that a private system of enforcement would lead to "overenforcement."

A. Mitchell Polinsky, Detrebling versus Decoupling Antitrust Damages: Lessons from the Theory of Enforcement, 74 *Georgetown Law Journal* (1986), pp. 1231–1236. Copyright © 1986 Georgetown University. Reprinted with permission.

In a paper following this exchange, I showed that . . . private enforcement would lead in a wide range of circumstances to too little enforcement relative to optimal public enforcement. This result, which tends to occur when the external damage from the violation is large, can be explained as follows. Under private enforcement, individuals or firms are willing to invest in enforcement only if they at least break even—that is, only if their fine revenue is at least as great as their enforcement costs. Under public enforcement, however, the optimal probability/fine combination may result in fine revenue that is less than enforcement costs. This is particularly likely to occur when the damage from the violation is large since it is then optimal to deter most, if not all, potential violators. Because the fine that can be imposed is limited (by the net worth of the potential violators), optimal public enforcement may require a high probability of detection and correspondingly large enforcement costs. But if most potential violators are deterred and the fine that can be obtained from those who are not is limited, the fine revenue collected by the public enforcement authority may well be less than the cost of enforcement. If so, private enforcers would not be willing to invest enough in enforcement to achieve the same level of deterrence as under public enforcement since they would not be able to break even. In other words, a private system of enforcement could lead to "underenforcement." . . .

Lessons from the Theory

There are two lessons that I wish to draw from the theory of enforcement, one concerning the choice between detrebling and decoupling, and the other relating to the optimal design of a decoupling system.

Detrebling versus Decoupling

One of the principal conclusions of the theory was that if private enforcers receive the fine paid by the injurer, it is generally impossible to achieve the optimal combination of the probability of detection and the fine. If the same fine is used as under optimal public enforcement, the resulting probability of detection (generated by the self-interested choices of private enforcers) may be too high or too low. In other words, if the enforcing is done privately, there may be too much enforcement or too little enforcement. The same conclusion applies to private damage actions in antitrust law since, under the current system, the plaintiff generally receives what the defendant pays.

Advocates of detrebling [antitrust damages] presumably believe that awarding successful plaintiffs three times their damages induces them and their lawyers to invest too much in the detection and prosecution of antitrust violations. The only way to reduce enforcement under the present system is to reduce the damage multiplier. However, as the discussion of the theory of enforcement makes clear, this response may not be the cheapest way to attain the desired level of deterrence. It may be socially preferable to raise, not lower, the amount paid by the defendant, while at the same time reducing the incentives for plaintiffs and their lawyers to invest in enforcement. If antitrust damages are decoupled, the lower level of deterrence that is desired can be achieved more cheaply by awarding the plaintiff less than what the defendant pays.

It should be stressed, however, that the advantage of decoupling over detrebling does not depend on whether it is desirable to reduce the level of deterrence from that currently generated by treble damages. The reasoning behind this conclusion is essentially the same as that used in the previous paragraph—specifically, that the decoupling approach can attain the same level of deterrence as any damage multiplier, but with a lower probability of detection, and therefore with lower enforcement costs. The details of the argument follow.

First select the best possible damage multiplier in a system in which the plaintiff receives what the defendant pays. This multiplier could be less than or greater than three. Whatever the multiplier is, it will generate some probability of detection as a result of the investment incentives of private enforcers.

Now consider a system of decoupled damages. Under this system, raise the amount paid by the defendant from the level determined by the best damage multiplier. If the plaintiff still were to receive the same amount as under the damage multiplier approach, the level of deterrence would be higher in the decoupled system because the probability of detection would be the same but the defendant would be paying more. Therefore, without changing what the defendant pays (from the now higher level), reduce the amount awarded to the plaintiff until the resulting probability of detection falls to a level such that the defendant is deterred to the same degree under both systems.

It is now easy to see why the decoupling approach is superior to the damage multiplier approach. Each approach can achieve the desired level of deterrence of antitrust violations. But the decoupling approach can attain this level of deterrence with a lower probability of detection, and therefore with lower enforcement costs. Thus, regardless of whether it is desirable to lower, raise, or leave unchanged the present damage multiplier of three, the decoupling approach is preferable to the damage multiplier approach.

Optimal Decoupling

Because of the focus on reducing private antitrust enforcement by detrebling antitrust damages, it is not surprising that the few individuals who have considered decoupling have taken for granted that the optimal system of decoupling would award the plaintiff less than what the defendant pays. This presumption would be correct if private enforcement is excessive under the damage multiplier approach, since it would then be desirable to discourage plaintiffs and their lawyers from investing too much in detection and prosecution.

However, as noted earlier, private enforcement may lead to underenforcement rather than overenforcement. If private enforcement is inadequate, then the optimal system of decoupling would require that the plaintiff receive more than what the defendant pays (with the subsidy presumably coming from the government). For reasons explained earlier, this outcome is most likely to occur when the damage from the violation is large. Thus, the optimal system of decoupling could award the plaintiff more or less than what the defendant pays.

The implications of this conclusion for antitrust policy are straightforward. In those areas of antitrust law in which it is thought that overenforcement currently is a problem—for example, with respect to joint research ventures—the plaintiff could be given less than what the defendant pays. In areas in which underenforcement might otherwise occur—for example, with respect to horizontal price fixing—the plaintiff could be awarded more than what the defendant pays.

It should be pointed out, in passing, that the conclusion that the optimal system of decoupling could award the plaintiff more than what the defendant pays is not inconsistent with the argument used to show that decoupling is superior to detrebling. The earlier argument demonstrated that there always exists some system of decoupling (in the example used, the plaintiff receives less than what the defendant pays) that is preferable to the best damage multiplier. It did not purport to derive the best system of decoupling, as is done here.

Concluding Remarks

The concept of decoupling is not an abstract curiosity derived from the economic theory of enforcement. There are several instances in which damages already are decoupled, although not always for the reasons suggested in this comment. For example, given current tax laws, antitrust damages are in effect decoupled in all private antitrust actions that follow successful criminal prosecutions by the government. In these cases, all of the plaintiff's award may be treated as taxable income, while only one-third of the defendant's payment can be deducted. Thus, with the tax consequences taken into account, the plaintiff receives less than what the defendant pays (and the difference goes to government). Although the tax treatment of antitrust damages is not designed to promote optimal deterrence, this example shows that an explicit policy of decoupling antitrust damages would not be as radical a departure from current practice as might be thought.

Before decoupling can be recommended to policy makers, several additional issues need to be considered. Since these issues have not yet been analyzed in a systematic way, I will only list some of the questions that remain to be answered:

How should a system of decoupling deal with out-of-court settlements? For example, if at trial the plaintiff would receive less than what the defendant pays, should the settlement be "taxed" by the same amount? By the same percentage? What if the court is unable to monitor the settlement? Will out-of-court settlements tend to subvert or enhance the desirable effects on deterrence of the decoupling approach?

If at trial the plaintiff would receive more than what the defendant pays, why won't the parties "fabricate" an offense in order to obtain the implicit governmental subsidy? How should a system of decoupling respond to this possibility? Can fabricated offenses be adequately deterred simply by the threat of penalties for such behavior? . . .

If a system of decoupling can deal satisfactorily with the issues raised by these questions, decoupling may be superior to detrebling not only in theory, but also in practice.

NOTES AND QUESTIONS

1. Demsetz suggests that private property tends to encourage greater conservation of resources than does communal property. Do you think this generalization is accurate? Does it properly apply to his example of native American hunting rights? Does it apply to his example of property rights in large business organizations? See Benjamin Klein, Robert Crawford, and Armen Alchian, Vertical Integration, Appropriable Rents, and the Competitive Contracting Process, 21 *Journal of Law and Economics* 297 (1978); Oliver Williamson, Corporate Governance, 93 *Yale Law Journal* 1197 (1984); Oliver Williamson, The Organization of Work, 1 *Journal of Economic Behavior and Organization* 5 (1980).

2. As Calabresi and Melamed indicate, the choice between property and liability rules has distributional as well as efficiency consequences. The holder of an entitlement protected by a property rule will be in a strong bargaining position and may be able to get the bulk of the surplus from exchange. Under a liability rule, in contrast, the person who infringes and pays compensatory damages may get the surplus. Does this suggest to you any generalizations regarding when each type of rule is most appropriate? Can you think of instances in which distributional considerations argue in favor of one kind of rule and efficiency considerations in favor of the other? See A. Mitchell Polinsky, Resolving Nuisance Disputes: The Simple Economics of Injunctive and Damage Remedies, 32 *Stanford Law Review* 1075 (1980).

3. Although Calabresi and Melamed admit the possible existence of what they call "other justice considerations," they appear to suggest that such considerations ultimately reduce to economic efficiency and distributional equity. Are they correct in this suggestion? In particular, consider their claim that moral considerations, which they label "moralisms," can be understood as a sort of nonpecuniary externality (supra at page 102), as well as their explanation of why not all property rules should be converted into liability rules (supra at page 104).

4. Posner's economic argument in favor of the negligence standard is more fully expounded in Richard Posner, A Theory of Negligence, 1 *Journal of Legal Studies* 29 (1972). Shavell's comparative analysis of negligence and strict liability were anticipated, without mathematical formality, by Guido Calabresi and Jon Hirschoff, Toward a Test for Strict Liability in Torts, 81 *Yale Law Journal* 1055 (1972), Do you agree with Shavell that his analysis explains most of the pockets of strict liability in contemporary tort law? On the basis of his discussion, are there any specific doctrines that you can identify as clearly inefficient, or primarily motivated by goals other than efficiency? See generally Don Dewees, David Duff, and Michael Trebilcock, *Exploring the Domain of Accident Law* (New York: Oxford University Press, 1996) [surveying and critiquing empirical evidence on the efficiency and distributional effects of tort law across a variety of doctrinal areas]; Patricia Danzon, *Medical Malpractice: Theory, Evidence, and Public Policy* (Cambridge, Mass.: Harvard University Press, 1985); W. Kip Viscusi, *Reforming Products Liability* (Cambridge, Mass.: Harvard University Press, 1991).

5. Neither Posner nor Shavell discuss the tort liability regime that is most common in the United States today: comparative negligence. Can their analyses of strict liability and

negligence be extended to cover that regime's incentive properties? See Robert Cooter and Thomas Ulen, An Economic Case for Comparative Negligence, 61 *New York University Law Review* 1067 (1986) [suggesting that in the presence of legal error, comparative negligence may provide better incentives than either pure negligence or strict liability]. Similarly, how would you apply the general framework laid out by Calabresi, Posner, and Shavell to specific doctrines in the law of tort? See, for example, Alan Sykes, The Boundaries of Vicarious Liability: An Economic Analysis of the Scope of Employment Rule and Related Legal Doctrines, 101 *Harvard Law Review* 563 (1988) [suggesting that traditional scope of vicarious liability corresponds to situations in which employer is best able to reduce accident costs through increased precaution or reductions in activity]; Reinier Kraakman, Gatekeepers: The Anatomy of a Third-Party Enforcement Strategy, 2 *Journal of Law, Economics, & Organization* 53 (1986) [discussing when third parties in general are least-cost avoiders]; Steven Shavell, An Analysis of Causation and the Scope of Liability in the Law of Torts, 9 *Journal of Legal Studies* 463 (1980) [suggesting that doctrine of proximate cause promotes efficiency by limiting defendants' liability to the marginal social cost of their actions.]

6. A fuller analysis of the incentive properties of monetary damages for breach of contract can be found in Melvin Eisenberg and Robert Cooter, Damages for Breach of Contract, 73 *California Law Review* 1434 (1985); a more formal analysis can be found in Steven Shavell, The Design of Contracts and Remedies for Breach, 99 *Quarterly Journal of Economics* 1221 (1984). A related issue, raised in Cooter's article in the previous chapter but discussed only briefly by Ulen, is efficiency in reliance. Reliance is efficient when its benefits, measured by the resultant increase in the expected value of a completed exchange, outweighs its cost, measured by the risk that it will be wasted; and both liability and damage rules can affect incentives to choose reliance efficiently. On this issue, see, in addition to the sources already cited, William Rogerson, Efficient Reliance and Damage Measures for Breach of Contract, 15 *Rand Journal of Economics* 39 (1984). Other aspects of contracting for which efficient incentives are desirable include the decision whether to enter into a contract to begin with [see Richard Craswell, Offer, Acceptance, and Efficient Reliance, 48 *Stanford Law Review* 481 (1996)] and the decision whether to make an offer of contract initially [see Avery Katz, When Should an Offer Stick: The Economics of Promissory Estoppel in Preliminary Negotiations, 105 *Yale Law Journal* 1249 (1996).]

7. Despite the substantial literature on the economics of crime, it is fair to say that economic analysis has made less of a mark on criminal law than on torts, contracts, and property. This is in large part because traditional concepts of natural law and morality remain more influential in the criminal than the civil sphere, and these concepts are somewhat in tension with the principles of normative economics. Normative economics takes an ex ante perspective, stressing deterrence and incentives, while traditional morality often takes an ex post perspective, stressing retribution and desert. Thus, policies that promote efficient deterrence may be seen as unjust from a retributive viewpoint. See generally Mark Kelman, The Origins of Crime and Criminal Violence, in *The Politics of Law*, ed. David Kairys (New York: Pantheon, 1990); cf. Alon Harel, Efficiency and Fairness in Criminal Law: The Case for a Criminal Law Principle of Comparative Fault, 82 *California Law Review* 1181 (1994) [suggesting that traditional criminal law scholarship has ignored problem of providing victims with proper incentives to take precaution]. Consider in this regard two of Becker's recommendations: first, he proposes that the criminal law should impose fines instead of imprisonment whenever possible, in order to conserve on the costs of punishment. Second, in a portion of his article not reprinted here but referred to in the Polinsky excerpt, he proposes that society should impose high penalties with relatively low probability, in order to maintain a high level of deterrence while conserving on the costs of apprehending and prosecuting criminals. Is Becker

correct that these proposals are justifiable on efficiency grounds? If so, do considerations of retribution and desert outweigh any efficiency advantages they may have?

8. As Polinsky suggests, the economic analysis of crime has wider applications beyond the doctrinal boundaries of the criminal law, for its lessons are relevant in any setting where rules are being enforced. For example, the rule of treble damages in civil antitrust law can be understood as a way to lower the cost of achieving a given amount of deterrence; and punitive damages in tort may serve a similar function. Additionally, much of the theory of optimal punishment and procedure can be applied to the general problem of principal and agent. Stockholders wishing to motivate corporate managers and employers wishing to motivate employees may find Becker's, Posner's and Polinsky's insights useful in designing more cost-effective systems of organizational incentives. For discussions of this approach to employment law, see Stewart J. Schwab, Life-Cycle Justice: Accommodating Just Cause and Employment at Will, 92 *Michigan Law Review* 8 (1993). For a discussion of optimal monitoring of managers, see Michael C. Jensen and William H. Meckling, Theory of the Firm: Managerial Behavior, Agency Costs, and Ownership Structure, 3 *Journal of Financial Economics* 305 (1976).

9. The economics of procedure has attracted particular scholarly attention in recent years, in part as a response to increases in litigation expenditures in the United States. The seminal article on this topic is William Landes, An Economic Analysis of the Courts, 11 *Journal of Law and Economics* 61 (1971), and an excellent introductory survey is Robert Cooter and Daniel Rubinfeld, Economic Analysis of Legal Disputes and Their Resolution, 27 *Journal of Economic Literature* 1067 (1989). A good complement to the articles excerpted here is Steven Shavell, The Social versus the Private Incentive to Bring Suit in a Costly Legal System, 11 *Journal of Legal Studies* 333 (1982), which discusses externalities arising out of the litigation decision. There is also a growing literature on the allocation of litigation costs. Good starting points are Steven Shavell, Suit, Settlement and Trial: A Theoretical Analysis under Alternative Methods for the Allocation of Legal Costs, 11 *Journal of Legal Studies* 55 (1982) [suggesting that the English rule of "loser pays" encourages plaintiffs with relatively high probabilities of victory to bring suit and discourages those with relatively low probabilities of victory, but discourages settlement], and Avery Katz, Measuring the Demand for Litigation: Is the English Rule Really Cheaper?, 3 *Journal of Law, Economics, & Organization* 143 (1987) [suggesting that English rule increases incentives to spend resources in litigation.].

10. The readings in this and the previous chapter suggest that many traditional rules across a variety of common-law fields help to promote economic efficiency. A number of writers on law and economics, including most notably Richard Posner, have argued that this is no accident—that the common law tends generally toward efficiency, and that this offers a reason to prefer judicial lawmaking to either legislation or rulemaking by administrative agencies. These two claims have received much discussion in the literature. Do they seem plausible to you? Why might the common-law process tend toward efficiency, and why would it have a stronger tendency in this regard than other institutions? See Paul H. Rubin, Why is the Common Law Efficient?, 6 *Journal of Legal Studies* 51 (1977) and George L. Priest, The Common Law Process and the Selection of Efficient Rules, 6 *Journal of Legal Studies* 65 (1977) [both arguing that private parties will challenge inefficient precedents more frequently than efficient ones]; John Goodman, An Economic Theory of the Evolution of the Common Law, 7 *Journal of Legal Studies* 393 (1978) [private parties favoring efficient rules have greater stakes in litigation and hence greater incentive to spend resources in litigation]; Robert Cooter and Lewis Kornhauser, Can Litigation Improve the Law without the Help of Judges?, 9 *Journal of Legal Studies* 139 (1979) [arguing that such individual incentives cannot in the long run eliminate all inefficient rules from the common law].

Refining the Model I:
Strategic Behavior

With this chapter, we turn to a series of refinements of the basic economic model of rational choice. Our first refinement addresses the phenomenon of strategic behavior. Up till now, we have for the most part assumed that individual decisionmakers act atomistically—that is, that they regard themselves as too small to have any effect on the behavior of other individuals or on the legal system as a whole. This assumption considerably simplifies the problem of individual maximization, because it allows people to take others' decisions as exogenous constraints. In many settings, however, the assumption is unrealistic. Often people have an incentive to behave strategically—to do things that they would not otherwise choose to do, but for their effect on others' behavior. The classic example of this in microeconomics is monopoly. The theory of perfect competition holds that small firms selling homogenous products take market conditions as given. Accordingly, such firms have an incentive to increase their supply up till the point where the market price just covers their production costs. When the market is dominated by a monopolist, however, the outcome may be quite different. A rational monopolist recognizes that its actions affect the market price, and thus has an incentive to restrict supply in order to drive away bargain-seekers and earn extra profits from the high-price customers who remain. While competitive markets tend to lead as an invisible hand to the efficient equilibration of supply and demand, therefore, monopoly tends to lead to supracompetitive prices and inefficiency.

Many legal rules and institutions govern situations where there is substantial leeway for strategic behavior. For instance, a franchisee's investment incentives depend on how her franchisor and third-party adjudicators are likely to interpret the franchise contract in light of the investment; a tortfeasor's incentives to take precaution will depend on whether she thinks potential accident victims will take care before the fact or bring suit after the fact. What this observation implies for the design of legal policies, however, is open to dispute, as contemporary debates over competition policy illustrate. From a Pigouvian perspective, strategic behavior is a classic market failure, justifying state intervention in the form of antitrust or price regulation. From the Coasian perspective, in contrast, it is simply another transaction cost—one that arises in public and private settings alike. Because government institutions are just as subject to strategic manipulation as are private ones—for instance, through litigation, lobbying, control of legislative agendas, and regulatory capture—antitrust and price regulation may be no better at reaching efficient outcomes than private bargaining between the monopolist and its customers would be.

The first reading in this chapter, Robert Cooter's "The Cost of Coase," argues that strategic behavior is the central issue in evaluating the efficiency of private ordering. In Cooter's view, strategic behavior is the most important transaction cost private individuals face, because people engaged in bargaining have an incentive to maximize their individual gains from exchange rather than the total surplus available. This surplus can be divided among the parties in various ways; and if institutional arrangements allow it, the parties may invest resources in hopes of altering the division. A minor example of this is haggling, which takes up valuable time and delays the enjoyment of the bargain. Cooter concedes that if negotiating parties usually find it in their interest to cooperate, then the law should be structured to minimize impediments to private bargaining. On the other hand, if strategic behavior is more likely than cooperation, a policy of laissez-faire could lead to great harm. Cooter, recalling the unhappy account of self-interested behavior offered by Thomas Hobbes in his classic work *Leviathan,* suggests a "Hobbes theorem" as a counterweight to the Coase theorem: since human self-interest can lead to a war of all against all, in which life is nasty, brutish, and short, legal rules should be designed to minimize the costs of failed cooperation. He suggests that various legal doctrines, including the doctrine of duress in contract law, can be understood as serving such a purpose.

Determining whether a Coasian or a Hobbesian outcome is more likely, however, requires some account of what counts as rational behavior in negotiation. The second reading, by Cooter, Stephen Marks, and Robert Mnookin, presents such a model and applies it in the context of settlement bargaining in civil litigation. In the Cooter/Marks/Mnookin model, the negotiating parties are instructed simultaneously to submit a single sealed demand. If the parties' demands are compatible, each gets what she has demanded and any leftover surplus is split evenly; if the demands are incompatible, however, the bargaining ends and the parties get nothing. This situation, combined with the fact that neither party can predict her opponent's strategy with certainty, creates a potential for rational disagreement as the parties choose among relatively tough and relatively soft demands. A tough demand is more likely to turn out to be incompatible with the opponent's demand, and so risks the loss of the bargain entirely. A soft demand, conversely, lessens the chance of disagreement, but risks unnecessarily yielding surplus if the opponent has also made a soft demand. Accordingly, each party will balance on the margin the value of a more favorable bargain against the probability of reaching a bargain at all. Just where each individual party draws the balance will depend on the importance she attaches to these alternatives, but in general, unless the parties are infinitely averse to risk, each will prefer to live with at least some chance of disagreement. This implies that the outcome of bargaining cannot be fully Pareto efficient ex post, since occasionally two rationally tough bargainers will find themselves in the same negotiation.

The next two readings show how strategic behavior can distort incentives and lead to inefficiency within a state-imposed system of civil liability. Arthur Leff, focusing on the collection process in debtor-creditor law, shows how opportunistic debtors and creditors can exploit the costs of the legal system to avoid performing their legal obligations. This is because the civil damages that are in theory supposed to motivate the parties to behave efficiently are in practice often inadequate to cover the costs of litigation. As a result, promisors rationally anticipate that promisees will lack the incentive to enforce their formal contractual rights. In Leff's view, accordingly, incentives to perform contractual obligations and to take precautions against breach come less from governmentally imposed sanctions than from the private threats and promises made by the parties as they jockey for strategic advantage in the context of their relationship. This conclusion implies that legal rules will often fail to internalize externalities the way they might in the absence of such strategic competition.

Conversely, David Rosenberg and Steven Shavell show how strategic behavior, combined with the other transaction costs of litigation, can lead to opportunistic plaintiffs obtaining payments to which they are not entitled. This is the problem of frivolous or "strike" suits. Under the American rules for allocating legal costs, defendants find it cheaper to buy off frivolous suits than to go to trial, and therefore cannot credibly threaten to refuse to settle. As a result, frivolous plaintiffs cannot be deterred from bringing suit. Rosenberg and Shavell suggest that either a change in the allocation of legal costs or a change in the timing with which they are incurred could change the parties' incentives in this regard.

Finally, the last two excerpts in this chapter explore the consequences of strategic behavior for contract and property law, and suggest that conventional wisdom in these fields of law needs to be rethought to take fuller account of strategic incentives. The reading by Avery Katz discusses the law of contract formation. Katz argues that the rules of offer and acceptance fundamentally determine the outcome and efficiency of private exchange, because they form the institutional background to all private transactions. Such rules do not merely coordinate and facilitate exchange, as traditional wisdom would have it, but also regulate it, by attaching costs and benefits to various forms of strategic behavior. For this reason, he contends, contract formation should not be seen as a narrow and technical field but rather as logically prior to all other issues in the economics of contract law, and perhaps to law and economics as a whole.

Oliver Hart, in a critique of Coase's theory of the firm, makes an analogous argument about the law of property. Hart suggests that transaction costs within business organizations stem primarily from strategic interactions; in his view, property rights matter for incentive purposes not because they confer substantive economic entitlements in and of themselves, but because they determine the boundaries for subsequent bargaining. Ownership of a property right improves a party's negotiating position by conferring the power to make various threats and offers—including, most importantly, the threat to destroy the value of complementary assets nominally belonging to other parties. As a result, the allocation of rights will influence the expected division of spoils, determining indirectly the incentives to invest in such assets. This framework can be applied to various problems in business organization, including the optimal degree of vertical integration, the division of assets in bankruptcy, and the allocation of control rights between capital and labor.

4.1 THEORY

The Cost of Coase

ROBERT COOTER

Coase gave his name to a fundamental theorem on externalities and tort law, but he left to others the job of stating exactly what the theorem says. The basic idea of the theorem is that the structure of the law which assigns property rights and liability does not matter so long as transaction costs are nil; bargaining will result in an efficient outcome no matter who bears the burden of liability. The conclusion may be drawn that the structure of law should be chosen so that transaction costs are minimized, because this will conserve resources used up by the bargaining process and also promote efficient outcomes in the bargaining itself. . . .

We shall argue that the central version of the Coase Theorem cannot be deduced from economic assumptions. The widespread belief to the contrary is symptomatic of confusion about bargaining. This confusion results in blindness toward certain outcomes of policy. We shall try to restore accurate vision by explaining the relations among liability law, bargaining, and the economic assumptions of rational behavior.

Robert Cooter, The Cost of Coase, 11 *Journal of Legal Studies* (1982), pp. 1–29. Copyright © 1982 The University of Chicago Law School. Reprinted with permission.

[The] Bargaining Problem

The Coase Theorem identifies a set of conditions under which the legal assignment of liability makes no difference. The phrase "makes no difference" has two interpretations: (i) the allocation of resources is invariant, or (ii) the allocation of resources is efficient. There was a confused debate about invariance versus efficiency, but now there is agreement that the invariance version is untenable. It is the efficiency version which we shall consider. . . .

The mechanism for achieving efficiency in the absence of competitive markets is bargaining. For example, Calabresi formulated the Coase Theorem as follows: "If one assumes rationality, no transaction costs, and no legal impediments to bargaining, all misallocation of resources would be fully cured in the market by bargains." This formulation apparently presupposes a general proposition about bargaining, namely, "Bargaining games with zero transaction costs reach efficient solutions."

In order to evaluate this interpretation of Coase, we must explain the place of bargaining in game theory. A zero-sum game is a game in which total winnings minus total losses equals zero. Poker is an example. A zero-sum game is a game of pure redistribution, because nothing is created or destroyed. By contrast, a coordination game is a game in which the players have the same goal. For example, if a phone conversation is cut off, then the callers face a coordination problem. The connection cannot be restored unless someone dials, but the call will not go through if both dial at once. The players win or lose as a team, and winning is productive, so coordination games are games of pure production.

A bargaining game involves distribution and production. Typically, there is something to be divided called the stakes. For example, one person may have a car to sell and the other may have money to spend. The stakes are the money and the car. If the players can agree upon a price for the car, then both of them will benefit. The surplus is the joint benefits from cooperation, for example, consumer's surplus plus seller's surplus in our example of the car. If the players cannot agree upon how to divide the stakes, then the surplus will be lost. In brief, bargaining games are games in which production is contingent upon agreement about distribution.

The bargaining version of the Coase Theorem takes an optimistic attitude toward the ability of people to solve this problem of distribution. The obstacles to cooperation are portrayed as the cost of communicating, the time spent negotiating, the cost of enforcing agreements, etc. These obstacles can all be described as transaction costs of bargaining. Obviously, we can conceive of a bargaining game in which these costs are nil.

A pessimistic approach assumes that people cannot solve the distribution problem, even if there are no costs to bargaining. According to this view, there is no reason why rationally self-interested players should agree about how to divide the stakes. The distribution problem is unsolvable by rational players. To eliminate the possibility of noncooperation, we would have to eliminate the problem of distribution, that is, to convert the bargaining game into a coordination game. But it makes no sense to speak about a bargaining game without a problem of distribution.

Our example of selling a car illustrates the collision of these two viewpoints. The costs of communicating, writing a contract, and enforcing its terms are the transaction costs of buying or selling a car. These costs sometimes constitute an obstacle to exchange. However, there is another obstacle of an entirely different kind, namely the absence of a competitive price. The parties must haggle over the price until they can agree upon how to distribute the gains from trade. There is no guarantee that the rational pursuit of self-interest will permit agreement. If we interpret zero transaction costs to mean that there is no dispute over price, then we have dissolved the bargaining game.

The polar opposite of the optimistic bargaining theorem can be stated as follows: "Bargaining games have noncooperative outcomes even when the bargaining process is costless." This line of thought suggests the polar opposite of the Coase Theorem: "Private bargaining to redistribute external costs will not achieve efficiency unless there is an institutional mechanism to dictate the terms of the contract." We have already discussed one institutional mechanism to achieve efficiency, namely a competitive market, which eliminates the power of parties to threaten each other. Another such institution is compulsory arbitration.

The conception of law which is the polar opposite to Coase is articulated in Hobbes and is probably much older. It is based upon the belief that people will exercise their worst threats against each other unless there is a third party to coerce both of them. The third party for Hobbes is the prince or leviathan—we would say dictatorial government—who has unlimited power relative to the bargainers. Without his coercive threats, life would be "nasty, brutish, and short." We shall refer to the polar opposite of the Coase Theorem as the Hobbes Theorem.

The Coase Theorem identifies the problem of externalities with the cost of the bargaining process, whereas the Hobbes Theorem identifies the problem with the absence of an authoritative distribution of the stakes. We shall argue that both theorems are false. However, they are illuminating falsehoods because they offer a guide to structuring law in the interest of efficiency.

In real situations faced by policymakers, transaction costs are positive. The Coase Theorem suggests that the role of law is to assign entitlements to the party who values them the most, so that the costly process of exchanging the entitlement is unnecessary. There are many similar versions of this proposition, for example, liability for accidents should be assigned to the party who can prevent them at lowest cost, or the cost of breach of contract should be assigned to the party who is the best insurer against nonperformance. If the party who values the entitlement the most cannot be identified, then it should be assigned to the party who can initiate an exchange at the least cost.

The Hobbes Theorem suggests that the role of law is to minimize the inefficiency that results when bargaining fails, by restricting the threats which the parties can make against each other. In the jargon of game theory, law increases the value of the noncooperative solution by eliminating elements of the payoff matrix with low value. This function is obvious in criminal law, where threats of violence against property or persons are punished. This function of law is also apparent in regulation of collective bargaining and strike activity. We claim that the same principle is at work where the threat is, say, to pollute a stream, to not perform on a contract, or to not take precautions against accidents.

For example, suppose that efficiency requires both the injurer and the victim to take precautions against accidents. According to the Hobbes Theorem, liability should be assigned to the party whose lack of precaution is most destructive. Put technically, liability should be assigned to the party for whom the excess of joint benefits over the private costs of precaution is largest. Alternatively, consider the problem of nonperformance on a contract by the promisor. According to the Hobbes Theorem, liability should be assigned to the promisee if excessive reliance by the promisee results in more net damage than insufficient precaution by the promisor against the events causing nonperformance. . . .

The Coase Theorem and the Hobbes Theorem have contradictory implications for the size of government. We can see this point most clearly by considering the policy implications in the ideal world of zero transaction costs. According to the Coase Theorem, there is no continuing need for government under these conditions. Like the deist god, the government retires from the scene after creating some rights over externalities, and efficiency is achieved regardless of what rights were created. According to the Hobbes Theorem, the coercive threats of government or some similar institution are needed to achieve efficiency when externalities create bargaining situations, even though bargaining is costless. Like the theist god, the government continuously monitors private bargaining to insure its success.

The Coase Theorem represents extreme optimism about private cooperation and the Hobbes Theorem represents extreme pessimism. Perhaps the Coase Theorem is more accurate than the Hobbes Theorem in the sense that gains from trade in bargaining situations are more often realized than not, or perhaps the Hobbes Theorem is more accurate from the perspective of lawyers who must pick up the pieces when cooperation fails. We shall not attempt an allocation of truth. The strategic considerations are not normally insurmountable, as suggested by Hobbes, or inconsequential, as suggested by Coase. An informed policy choice must balance the Coase Theorem against the Hobbes Theorem in light of the ability of the parties to cooperate. . . .

Austin Instrument v. Loral

In order to give concrete meaning to our argument, we shall consider an example of an appellate decision which fits the Hobbes Theorem better than the Coase Theorem.

In 1971 the Court of Appeals of New York extended the concept of duress in deciding the case of *Austin Instrument, Inc. v. Loral Corporation*. Loral had been awarded a general contract from the Navy for the production of radar sets, and Austin was one of its subcontractors. A year later Loral was awarded a second contract from the Navy and solicited a bid from Austin to construct some of the component parts. This second bid was solicited at a time when Austin had not completed delivery on items prescribed in its first contract with Loral. Austin threatened nonperformance on the remainder of its first contract unless Loral awarded the second contract on terms that favored Austin, including a retroactive price increase on the remaining items from the first contract. Austin was not the lowest bidder on some items covered by the second contract. Loral needed the items from the first contract to meet its delivery deadline with the Navy and escape the penalty clauses for late delivery in its general contract. Loral could have challenged Austin and sued for breach of contract in the event of nonperformance. Instead, Loral capitulated to Austin's demands but made clear in a letter to Austin that it did so under duress. Deliveries proceeded on schedule, but after receiving all of the items specified in the two contracts with Austin, Loral withheld its last payment and both parties sued.

The New York Supreme Court decided the case in favor of Austin on the grounds that Loral had access to the traditional legal remedy for breach of contract. In that court's view Loral had not established that the normal legal remedy for breach would be inadequate or ineffectual. However, the Court of Appeals reversed the lower court's ruling and found in favor of Loral. The higher court held that duress was established by the fact that the threatened party could not obtain the goods from another source in the event of default.

We can understand the higher court's finding in terms of Hobbesian bargaining theory. If Loral had responded to Austin's initial threat by a counterthreat, namely the threat to sue for breach of contract, Austin might have capitulated and the outcome would have been efficient. But if Austin did not capitulate, then the outcome would have been costly to both parties. A judge following the Hobbes Theorem would want to reduce the destructiveness of noncooperation. This end was accomplished by allowing Loral the option of postponing its challenge until after Austin completed performance, as permitted by the rule of duress. The structure of the law was chosen to eliminate the worst consequences of nonperformance. . . .

. . . [T]he game of default has a higher cooperative and a lower noncooperative solution than the game of duress. A court concerned with efficiency would prefer the game of default if it were optimistic about cooperation in two-party bargaining, and it would prefer the game of duress if it were pessimistic. Apparently, the Court of Appeals was pessimistic and chose the game of duress, as commended by the Hobbes Theorem.

The Coase Theorem and the Hobbes Theorem are both too flexible and vague to permit construction of a decisive example in which they are contradictory. No doubt a true believer in the Coase Theorem could find a way to explain the higher court's decision. However, the Hobbes Theorem is the more immediate, less forced explanation. . . .

Bargaining in the Shadow of the Law: A Testable Model of Strategic Behavior

ROBERT COOTER, STEPHEN MARKS, AND ROBERT MNOOKIN

Pretrial bargaining may be described as a game played in the shadow of the law. There are two possible outcomes: settlement out of court through bargaining, and trial, which represents a bargaining breakdown. The courts encourage private bargaining but stand ready to step from the shadows and resolve the dispute by coercion if the parties cannot agree. Bargaining is successful from an economic viewpoint if an efficient solution to the dispute is found at little cost. . . .

The usual approach to bargaining in the legal setting assumes that trial is caused by excessive optimism on the part of plaintiff and defendant. If both parties are optimistic, then there is no way to split the stakes so that each receives as much as he or she expects to gain from trial. In these circumstances, trial is inevitable.

Our approach is different. In our model, excessive optimism is not the fundamental cause of trials. The fundamental cause is the problem of distribution faced by the players. The problem of distribution is to divide the stakes in dispute. A rational bargainer will make a demand such that the gain from settling on slightly more favorable terms is offset by the increased risk of a breakdown in negotiations. Thus, the optimal bargaining strategy of a litigant balances a larger share of the stakes against a higher probability of trial. There will be a positive risk of trial when strategies are optimal.

A player's strategy is optimal when he maximizes his expected utility. Our assumption that individuals maximize expected utility involves an innovation in game theory as applied to law. In the usual analysis, the players simultaneously maximize utility, and an equilibrium is achieved when everyone knows exactly what everyone else is doing (Nash equilibrium). Uncertainty is eliminated. In our analysis, the players simultaneously maximize expected utility, and an equilibrium is achieved when everyone knows the distribution of strategies pursued by others. . . . Uncertainty persists about individuals but not aggregates. The persistence of uncertainty allows some disputes to end in trial. . . .

Framework: Optimization and Equilibrium

. . .

Bargaining in the shadow of the law achieves a close fit to our abstract characterization of bargaining games. The players are usually well defined, consisting of a plaintiff and defendant in many cases. The stakes are also well defined, such as the cost of an accident, the damage from nonperformance on a contract, the property accumulated in a marriage, the estate of the deceased, the assets of a bankrupt company. In pretrial negotiations, everyone has an interest in avoiding a trial. The surplus from cooperation is usually obvious—for example, legal fees, cost of delaying resolution of the dispute, waste from a judicial outcome off the contract curve. The plaintiff and defendant have an incentive to avoid trial, but they have a disagreement over how to divide the stakes. There is a problem of efficiency and also one of distribution.

A legal dispute enters the public record when a complaint is filed. A trial date is often set after the complaint is filed, which puts a time limit on bargaining. The simplest characterization of the bargaining process is a sequence of offers and counteroffers for dividing the stakes. A settlement is reached if the plaintiff's demand in some round of negotiations does not exceed the defendant's offer. Bargaining terminates and a trial begins if the trial date is reached before a settlement occurs. . . . The outcome of a trial is the destruction of part of the stakes (the surplus) and distribution of the remainder.

Optimization

. . . Our next task is to characterize the behavior of the players. Each player must choose a bargaining strategy. We can reduce the choice of strategy to its simplest elements by imagining that the plaintiff writes down her final demand, seals it in an envelope, and mails it to the defendant. At the same time, the defendant writes down his final offer, seals it in an envelope, and mails it to the plaintiff. The envelopes are delivered just before the trial is scheduled to begin. If the offer is at least as great as the demand, a settlement occurs. If the demand is greater than the offer, then a trial occurs. The mailing of the envelopes is a device for portraying the uncertainty of each player concerning his opponent's strategy.

This example can be clarified by using some notation to describe how the plaintiff would compute her optimal final demand. Let the stakes be $100,000 and let x be the plaintiff's share of the stakes, where x is denominated in hundreds of thousands of dollars. Thus a settlement will occur if the plaintiff demands x for herself and the defendant demands no more than $1 - x$ for himself. Let $P(1 - x)$ be the probability that the defendant demands no more than $(1 - x)$. Thus the probability of settlement is P and the probability of trial is $1 - P$. Let the plaintiff's payoff from trial be T. The problem faced by the plaintiff can be written

$$\max_x P(1 - x)x + [1 - P(1 - x)]T.$$

This mathematical problem can be solved intuitively. Suppose that the plaintiff is contemplating whether x is her optimal demand. If she increases her demand by $1.00, then she stands to gain $1.00 with probability $P(1 - x)$. However, increasing her demand by $1.00 will increase the probability of a trial by the marginal value of $P(1 - x)$, denoted p. If a trial occurs, then she will lose the difference between her payoff from settlement x and her payoff from trial T. Thus the risk of loss from demanding more is $(x - T)p$. At the optimum, the probable gain is exactly offset by the risk of loss: $0 = -(x - T)p + P$. The plaintiff's optimal demand is the value of x which satisfies this equality. The same kind of computation is made to find the defendant's optimal demand.

In this example, each player makes one final offer. Real bargaining is a sequence of offers and counteroffers, not a single pair of offers. Consider a more complex example from family law. A divorcing couple without children disagree about how to divide the value of a house. If the spouses can agree on division of the stakes (the house), then they can settle the dispute without lawyers or a trial, and they can arrange for the house to go to the spouse who values it the most. In the pretrial period, the spouses exchange offers and counteroffers. As the trial date for the divorce approaches, both make concessions. Neither knows how much the other will concede. Each decides how much to concede by trading off a larger share of the value of the house against a higher probability of trial. If the concession rates are fast enough, then a settlement occurs, but otherwise there is a trial. . . .

Choosing the strategy of optimal hardness is almost identical mathematically with choosing the optimal final demand. If a player adopts a hard strategy, then he receives a larger share of the stakes in the event of settlement. But a harder strategy is less likely to result in settlement. If the pair of strategies chosen by the players is too hard, then the dispute will be resolved by trial. Thus a player finds his optimal strategy by trading off a larger share of the stakes against a higher probability of trial. There is no substantial change in the mathematical formulation of the choice problem, which we discussed above, except that the choice variable x is interpreted as an index of the hardness of the strategy. . . .

Expectations

We described how players find the strategy which maximizes expected payoffs, assuming that each player has expectations (denoted P) about what his opponent will do. Our next task is to explain how these expectations are formed. There are various economic models for the formation of expectations, one of which is called rational expectations. This phrase means that expectations contain no systematic bias, that is, the subjective expectations correspond to objective frequencies of the random event. We can characterize rational expectations in our model by explaining the source of uncertainty. Each player in our model has observable and unobservable traits. Players with the same observable traits, but different unobservable traits, will pursue different strategies. However, the expectations which a player has about his opponent are formed on the basis of the observable traits alone. Consequently, each player remains in doubt about what his opponent will do. Subjective expectations correspond to objective frequencies when a player expects his opponent to act in a way that corresponds to the actual randomness of the class of players who have the same observable traits as his opponent.

For example, there are many divorces each year which require dividing wealth and income. Each spouse will adopt a bargaining strategy which we assume can be characterized on a scale of hardness. The objective distribution is the actual frequency with which spouses with similar observable traits adopt strategies ranging over the scale of hardness. If the subjective expectations correspond to the objective frequency, then the spouses' expectations are rational.

What is to count as an observable trait? There is much latitude in the model for answering this question. Our ultimate interest is in predicting the way changes in variables observable to policymakers affect the ability of parties to settle out of court. From our perspective, the important observable variables are the ones which have explanatory power in an econometric model.

In this paper we shall stress the assumption of rational expectations, although many of our conclusions remain true if we introduce biased expectations. There is a sense in which rational expectations are more fundamental than biased expectations. If expectations of decision makers are biased, then they will be surprised by the consequences of their choices. If they are surprised, then they will revise their expectations. If their expectations are revised, then the system is not in equilibrium. The rational-expectations model is fundamental because it is intimately linked to the concept of equilibrium, as we shall now show in greater detail.

Equilibrium

Expectations will be rational if there is a learning process by which bias is corrected. There is a mechanism for learning in the legal setting. In our divorce example, the spouses might seek legal counsel. The lawyers would be experienced with such bargaining situations. It is easy to see how bias would be eliminated from a lawyer's expectations. A lawyer expects a particular strategy on the part of his client to result in noncooperation a certain proportion of the time. If his expectations are disappointed, then he will revise them repeatedly until they are accurate. For example, if the husband adopts the conciliatory strategy of demanding only 20 percent of the market value of the house, and the lawyer expects this strategy to produce cooperation in every case but the wife rejects the offer, then the lawyer will have different expectations the next time a similar case occurs.

The learning process which we have described will result in revision of the subjective probabilities until they correspond to objective frequencies. The process will cease when expectations are rational. Thus the equilibrium of the game is a situation in which subjective probabilities correspond to objective frequencies. . . .

We could modify the definition of equilibrium to allow for biased expectations. For instance, it is possible that litigants are biased toward optimism. Such bias might arise because of a natural impulse toward hope, or because lawyers earn more money by encouraging litigation. To incorporate bias into the model, we would define equilibrium as a condition in which objective frequencies differ from subjective expectations by the postulated amount of bias. A study of optimistic expectations can therefore flow from the rational-expectations model, which is more fundamental because it is the starting place from which to proceed to special cases of bias.

Let us summarize our bargaining model. Bargaining occurs in a context of uncertainty about how opponents will react to offers and counteroffers. The basic approach in economics to choice under uncertainty involves a two-step process: first, form your best expectations about the likelihood of each possible outcome from acting; and, second, calculate your optimal move. . . . The fundamental problem in game theory is that one player's expectations about another's move depend on the other player's expectations about the first player's move. How can the two-step process be applied simultaneously by everyone? The paradox is resolved by specifying expectations in the first step which are confirmed in the second step. Rational expectations have this characteristic because the subjective expectations correspond to objective frequencies. . . .

Legal Institutions

. . .

Externalities and Coase

The first application of our model is to harmful externalities. The inefficiencies caused by externalities can be avoided by private bargaining, as observed by Coase. One interpretation of the Coase theorem holds that efficiency will be achieved in the presence of harmful externalities, regardless of the structure of liability law, provided that the transaction costs of bargaining are nil. If there are no impediments to bargains, then bargaining will continue until the gains from trade are exhausted. The gains from trade are exhausted when it is impossible to make one person better off without making someone else worse off (Pareto efficiency). Thus efficiency will be achieved by contracting around inefficient laws. . . .

The term "transaction costs" has a curious history in economics and law. Most theoretical terms eventually acquire a precise mathematical meaning, such as competition, demand, utility, public good, etc. "Transaction costs" has never been pinned down. Most writers take this term to include such costs as communicating and policing agreements. If we define transaction costs to mean the cost of communicating and policing agreements, then it is not true in our model that bargaining games with zero transaction costs reach efficient solutions. We predict the opposite: settlement is more likely if a round of negotiations is more costly. . . .

Suppose there is a small laundry which suffers from smoke emitted by a large factory. At issue is whether the laundry should be able to enjoin the factory. Since there are only two parties to the dispute, the cost of communicating and policing an agreement is small. The optimistic view holds that the outcome will be efficient whether or not the laundry can enjoin the factory. The pessimistic view holds that bargaining fails in some cases even when communication is costless, so it would be dangerous to permit a small laundry to enjoin a large factory.

Divorce provides another example. Research in progress by the authors shows that 15–20 percent of divorces involving children are resolved by trial. Our research also shows that the outcome of trial is easy to predict. It seems difficult to explain why two-party disputes, in which the trial outcome is predictable, would not be settled if transaction costs and optimism are the causes of trial, rather than strategic bargaining. . . .

4.2 APPLICATIONS

Injury, Ignorance, and Spite: The Dynamics of Coercive Collection

ARTHUR LEFF

Collection in America: Transaction Costs in a Judicial-Coercion Game

Introduction: The Creditor's Dilemma

Under the American law of contracts, after the other party has fully performed his obligations it is absolutely irrational for you fully to perform yours. If you, D, refuse voluntarily to repay v in full, [the creditor] C would always be better off accepting from you some lesser performance, $v - x$, so long as $(v - x) > (v - t_C)$ [where t_C denotes C's transaction costs in collection]. Thus if v were 1,000, and if D could force C in recovering v to expend a t_C of 100 while suffering no t_D himself, it would be rational for both C and D not to allow the coercive collection process to go to completion, but instead to settle on a repayment of anything between 901 and 999. . . . If one views coercive collection as a two-person game, and one takes account of the costs of playing it, it is not a two-person zero-sum game, but a two-person minus-sum game, and the parties can both gain by avoiding the play altogether, agreeing instead on [a] settlement.

When D has no t_D, he is comparatively indifferent as to whether the game is played or not; playing cannot make him worse off than not playing can. But what if D were forced to suffer, as part of the same coercive transaction which generated the t_C of 100, a t_D, perhaps one in excess of 100? . . .

Again, playing the coercion game to completion would be minus-sum, and again both C and D would be well advised to find some point . . . at which to settle. But in this case D would no longer be indifferent as to whether the game were played or not, for a completed game, while hurting C, would hurt D more. Thus the ultimate settlement point . . . would tend to move . . . more to the taste of C. And if t_D could be made great enough relative to t_C, D might well decide to pay all of v in order to avoid the threatened playing of the game. Thus the actual posture of the parties will depend largely on the relative quantities of t_C and t_D which, in any particular context, coercive collection will predictably generate.

The Content of t_C

. . .

It is relatively rare in our society for one to be able to transfer things of value from another's possession to one's own solely by the use of one's own labor, without the other's cooperation, or exceedingly ostentatious assent. It is, in fact, at precisely the moment that one asserts in oneself a right to possession which is superior to another's actual possession that the requirement ordinarily attaches that one purchase a third-party source of information and force, the "law." But the "law," at least as embodied in legal process, is a very expensive mechanism for generating and channeling information and force.

One must to some extent be Hobbesian about it. If paid for and played from beginning to end, this expensive game does result in the acquisition of an immense bundle of power; behind every final judgment procured in any court in this country stands, ultimately, the United States Army, but even the intermediate mercenaries one buys—sheriffs, marshalls, judges—are usually sufficient unto the day. If someone has something that "belongs to" you and "the law" finally says so, so far as power can get it to you, that power will suffice. But the price one is supposed to pay for harnessing the Leviathan to one's cause is, essentially, the cost of moving it according to its own rather arcane principles, that is, the cost of due process.

The cost of due process is high for at least four reasons. First, due process demands that at the outset the court and its officers be wholly ignorant of what happened and it is expensive to educate them, at least using the pleading-and-playlet format of the common law. Second, the process of education cannot proceed on a generalized (mass-produced) basis; each case is theoretically hand-crafted. Third, save in a court of small claims it is usually specialists (e.g., lawyers) who do the crafting. Fourth, because the courts do not allocate docket space by competitive bidding between plaintiffs, the creditor with the largest claim at stake must take his place in a "queue" behind plaintiffs with smaller claims.

In reality, however, the practice is not quite as hard as all that. Fact finding and rule applying does tend to get stylized, and some of the specialists; through repetition, learn their jobs well. But there is another factor involved in using "the law" with respect to half-executed contracts which tends to increase the transaction costs of the party who has performed: the risk of wrongly losing, and of losing all. Assume again that C has done all that he promised D while D has done none. If nothing else were to happen at that point, C would be out v and D would be ahead v, for something "belonging to" C is at that point in D's hands. Now, assume further that D refuses to pay, and that C must go to law. Even if the law functions properly (that is, in accord with the assumed facts) the maximum that C can recover is $v - t_C$. But if the law miscarries and a "wrong" decision is made, C's post-litigation status could be as bad as $0 - t_C$. This, of course, is C's maximum exposure. Though the law does tend to formulate as many transactions as possible in yes-no terms, it is not always that rigid. While C might not get v, he might still get an amount greater than zero. But there are nonetheless innumerable accidents in legal proceedings which lead to status quo results, and for C any result which retains the status quo is more than a total loss. . . .

The Content of t_D

The judicial-coercive process is so designed that C may have to go through a number of steps, none cost free, even though D remains totally passive. Unless one counts the cost of tearing up and throwing away an occasional summons, the process can be free for D all the way from institution of the action to execution of the judgment. C's mercenaries will have little difficult or dangerous work to do if D chooses this course of inaction, and the resultant t_C will be low, but the choice is D's. In order to increase C's costs over the bare minimum for commencing a lawsuit and taking it through to judgment, D can expend some money and effort. But the critical point is that in most cases of breach, C will be forced to spend some t_C before D need even decide whether he will spend any t_D.

Were this where the story ended, the irrationality of paying one' debts would not be merely apparent. But we all know that not fully carrying out one's obligations can bring with it a retribution that is hardly limited to social disapprobation, a guilty conscience, or an unpleasant afterlife. Translated into our formal terms, there are components of t_D in addition to those which our analysis has heretofore taken into account. Most of these components do not take the form of a direct transfer payment from D to C, but a typical one does involve such a shift. Not only do the courts regularly enforce contractual clauses providing that the creditor who sues successfully also recover the expenses of collection, even in the absence of any specific provision certain costs of collection are allocated to the defaulting buyer. This particular transfer payment simultaneously decreases t_C and increases t_D. A complete transfer of t_C to D will not be possible because C still must bear the risk component of his t_C. . . . But this is only to say that these cost-shifting devices may not by themselves be sufficient to force D into a position where it would be wholly irrational to breach, where, that is, $t_C = 0$ and $t_D > 0$.

Beyond these procedures for shifting the cost of coercive collection to D, there are for C even more powerful and threatening devices which involve no transfer payment at all. They function not as payments to C, but as mere destruction of D's wealth.

But that too acts to coerce D. In a sense, of course, all t_D is like that; it hurts D without directly helping C. But some kinds of t_D, the costs of defense, for example, can be avoided by raising none. The components of t_D to be considered next cannot be avoided merely by remaining passive. Indeed such a response is likely to increase the injury.

When D has defaulted in the performance he owes to C, there are two courses open to C under modern sales law. He may attempt via self-help to get back what he gave D, or he may at law attempt to get what D promised to give him. In some situations he may reclaim goods delivered under a contract (so far as that is physically feasible) and then, to the extent that that does not get him to where he would have been had the transaction gone through as planned, he may attempt to get additional things of value from D to make up the deficiency.

The attractiveness of the self-help process is obvious: it sometimes permits C to recover some large portion of v without incurring the usual expenses of going to law. It is, in effect, an opportunity to use state-of-nature power to get out of the C position without having either to buy off, or buy in, the Leviathan. It is not, of course, cost free; night work with tow trucks (or even with duplicate keys) entails expense, but seldom as much as a litigated lawsuit and subsequent execution. Moreover, it is possible in this way for C to recover all. This depends on the value to C of the goods seized and the amount of the debt outstanding at the time of the repossession, but given a high enough down payment and a low enough rate of depreciation for the goods, C may be made whole solely by retaking the goods.

But there are significant additional injuries inherent in self-help or judicially-ordered repossession which do not directly inure to C's benefit but which may so increase t_D as to inhibit D's breach. First is the loss which comes solely from being deprived of a good's use. Obvious examples are the repossessed fabricating machine, the padlocked plant and (in the context of non-business "productive" assets) the automobile which is the only access to work. This harm can be visited on D even if C does not get use and/or resale rights in the item. . . . Once deprivation is coupled with a power of resale, however, D's loss increases and becomes more permanent. There are numerous things in the world that are worth much more to one person than they are to any other person. Consider the situation of a man owning a drill press which he uses in his business. He bought the press ten years ago and has used it since, not without problems, but by and large successfully. It has a slight tendency to yaw to the left and thus the operator must keep up a constant countervailing pressure. When used at top speed it tends to burn out its bearings. After some bitter experience all that is now known and integrated into its use pattern. What is also known and integrated is that it has no other material peculiarities; make but these few adjustments in use and it will drill press away to one's perfect satisfaction. The price of discovery has been spent. But put such a press up for sale, even in an honest, open auction market, and the sound buyer will discount his bid by the possibility of disaster. After all, it is a "ten-year-old press" of this manufacture and that appearance. It is a mystery. It may functionally be worth as much as it is, but it may be worth almost nothing.

When there are added to this "normal" depreciation factor all the other factors which may lower the price fetched by used goods in an aftermarket, for example, restricted buyer pools approaching minimonopsonies, insufficient sale advertising, unenthusiastic and incompetent selling, title uncertainties, and the lack of price-maximizing incentives, the potential loss in value to an ex-possessor becomes substantial. With consumer goods, yet another factor enters further to drive this value-destruction vector: non-functional depreciation, the loss in value consumer goods suffer when they move from the category "new" to the category "used." . . .

Mere value destruction, of course, cannot do C any direct good. If his claim is properly for more than the repossession-resale will realize, it may do him actual harm, assuming he cannot get the deficiency from D at all, or at least [not] without suffering unreasonable additional costs. But the threat of this value destruction is a threat of serious injury, available at a cost relatively low in comparison to the magnitude of the injury inflicted. It is an increase in the applicable t_D. . . .

Spite: A Value Secreted in "Transaction Costs"

Up to now we have assumed that all t decreases the amount of v left in the system. But this assumption is often untrue in practice; t may also represent the price of additional "goods" purchased by the two parties for themselves. The most important of these, which is as real and valuable as cars and coats, is spite.

The nature of spite can be exposed most starkly by reference to an illegal but common subspecies of collection, the protection racket. Consider an extortioner approaching a potential victim with the following proposition: "Give me $100 or I will break the plateglass window of your store." Assume that (a) the plateglass window is uninsured; (b) it can be replaced for $100; (c) the crook's cost of breaking it is $10; and (d) breaking the window once is within the crook's power, but no further depredations against the victim are. Obviously, it would make no economic difference to the victim whether he paid $100 and kept the window intact, or refused to pay, had his window broken, and replaced it at a cost of $100. In either case he is out $100.

To the extortioner, however, what the store-owner decides makes a great deal of difference. If the victim pays, his yield is $100. If the victim does not, the crook's assets may be decreased by the cost of breaking the window ($10). Thus while the cost of "cooperative" payment is zero ($100 of value "belonging to" the victim is merely transferred to the racketeer), the cost of coercion is $110, $100 of which represents absolutely destroyed value (the difference between a window and a pile of glass fragments) and $10 of which represents the crook's additional labor costs.

Naturally, the crook would strongly prefer that the store-owner pay rather than suffer the asset destruction. But to increase the chances of this outcome, the extortioner must offer some incentive. At a minimum, his proposition ought to be: "Pay me $99 or I will break your window." If the victim decides to pay, he will be $1 better off than he would be if the game were played. But the victim has the power to inflict harm on his tormentor. He can deprive the crook of any payment. If the crook decides nonetheless to engage in his game of destruction, the victim will have forced him into an actual out-of-pocket loss. Of course, to bring this about the victim must also hurt himself. He must insist upon an end-point for the confrontation more expensive for himself than the one proposed by the extortioner. What some economic analysis might overlook is that he is nevertheless very likely to do it. People do it all the time. The fulfillment of an urge to spite seems no different from the fulfillment of any other human desire. People pay to satisfy lust, hate, ambition and greed. They also pay for this. As with any other good, of course, the demand for spite is never totally inelastic, and its value is never infinite. While it has some value for almost everyone, the more it costs the less likely it is to be bought.

Power Over Transaction Costs: The Professional and the Consumer

Up to now, by using the abstract constructs "C" and "D" as if there were no material differences among the various kinds of parties to half-executed contracts, I have falsified the reality of collection practice. Abstraction, for all its usefulness, almost always does that. I shall now, so to speak, "decompose" the concepts C and D into recognizable subclasses, so as better to describe the variable impact of various collection strategies in different "typical" collection situations. . . .

The Consumer as Creditor. Picture a consumer who has just bought a color television set from a retailer for $500 in cash. The consumer takes the set home, tries it, and finds that it is defective to the tune of $50, that is, that it would cost $50 to bring the television up to warranty. In these circumstances the consumer is a creditor; the retailer has possession of $50 of parts and services "belonging to" him. The consumer approaches the retailer and asks him to repair the set. The retailer refuses. This leaves the consumer with only coercive collection if he is to recover the $50. But legal action may not be for him a realistic alternative; the t_C of such a move, the cost of hiring a lawyer, filing papers and so on, is likely to be in excess of the $50 at issue while the merchant need not until then expend any t_D.

Some consumers in this position would disregard the economic disadvantages of going to the law and would sue the seller out of spite. But even assuming that a consumer were so inclined, he would likely find his purpose frustrated, for much of the potential destructiveness locked into repossession and garnishments hardly affects professional debtors at all. A merchant faced with an adverse judgment is better able than a consumer in a similar situation to interdict property execution, by paying the judgment or posting a bond, or at least by showing up at the execution sale to restrict the frigidity of any "chill." And if a consumer were to garnish a seller because of his postjudgment obstinacy it would not be wage garnishment, with all its threatened injuries. If the garnishee is the retailer's bank (the most likely target, since it is his most solvent and notorious debtor), the ramifications on the retailer are likely to be minimal; he is unlikely to be "fired," even from his account. Thus our hypothetical businessman can allow the consumer to pursue his legal remedies with relative impunity, for none of the consumer's options can force a risk of much more than the $50 claim. That is not much satisfaction for the consumer to buy—merely preventing a windfall. It ordinarily demands "superspite," that is, infliction by the consumer of greater harm on himself than he can inflict on his enemy. It may still be done; the history of law is filled with cranks (lawyers or clients, it's often hard to tell) who spend large and unrecoverable sums to assuage feelings of outrage, moral or economic. But it is not bloody likely. We thus have a classic and unhappy creditor situation: the consumer as creditor can officially recover through judicial coercion no more than the amount by which D is in default, but to recover anything he must expend some t_C, perhaps more than the v in issue, while to cause that initial expenditure D need spend no t_D. If the ultimate t_D of a full coercive collection is also low compared to the t_C necessarily involved, and the t_C will not be borne by D even if C eventually wins, then it is usually economically rational for C just to abandon his claim.

The Professional as Creditor. When C is in the credit business, however, the impact of the factors noted above changes materially. Perhaps the largest components of t_C are the administrative and risk costs inherent in the due-process "adjudication" phase of the collection process. Naturally, it is to the creditor's advantage to avoid these if possible. Certainly he can avoid them by not playing the judicial-coercion game at all, by attempting to collect what he is owed by direct contact with the debtor. But as noted earlier, to avoid judicial coercion altogether is to fail to get the power of the State, the ultimate in civil puissance, on the creditor's side. Thus, if the creditor can get to the execution phase without the costs and risks of an adjudication, it is tempting to try to do so. This can be if done most inexpensively if the other party fails to show up and the judgment, that pass-key to judicial coercion, is procured by default. There are a number of ways to encourage default, but there are indications that consumers need very little encouragement. Largely because of the prevalence of default judgments against consumers, the cost of the initial adjudication-phase t_C is remarkably small in the vast majority of cases in which businessmen are creditors and consumers are debtors.

That does not mean that a default judgment is cost-free. But efficiency in the legal context is like efficiency elsewhere, it is in large part a function of standardization and repetition. The lawsuit will ordinarily require an attorney, but at the summons and complaint stage he needs relatively little extra time or effort to file two suits rather than one, or eight, or sixteen for that matter. In fact, in businesses which combine large volume collection with no concern about the effects of indiscriminate and excessive behavior on business reputations it may be cheaper overall to standardize all activities and "go to law" immediately in every case, without regard to the specific facts of the specific debtor's problems.

In addition to reaping benefit from specialization and standardization, professional creditors may be in a position to externalize a great many more of their costs than can consumers. First, for all Cs there is subsidization by the State of some of the costs of recourse to the judicial system. That which the State takes in payment for the use of its information-generating and force-applying mechanisms is ordinarily less than its expense in supplying those services. Second, when one gets to the point of execution one often finds instances of administrative costs being shifted to persons who had nothing to do with either the debt-generating or debt-defaulting transactions. As noted earlier, this is notably the case with respect to garnishment where the employer rather than the creditor or employee-debtor must expend the cost of organizing an installment payment plan. Third, both by "agreement" and by operation of law much of C's administrative expense, which would ordinarily be part of his t, may be transferred to D. This simultaneously decreases t_C and increases t_D, thereby giving superleverage to the move as a species of coercion. Insofar as this shift depends on contracts it will vastly favor the drafter of the contract, who is likely to be the professional.

Thus the merchant-creditor has far greater opportunities than does the consumer-creditor to decrease t_C by externalizing or shifting some portion of it, a move which under our earlier analysis should massively increase C's power to force a settlement agreeable to him. But the merchant-creditor's compensatory weapons do not stop there. He also has a very much greater power than a consumer-creditor to increase t_D. One need not here repeat all that has been said previously about the destruction-of-value aspects of property execution and garnishment. They form a clear "cost" to a D who goes through the whole coercive collection game. Even if, therefore, spite is beneath a presumably rational businessman, and credibility and reputation are ambiguous "assets" in this context, the combination of (a) the merchant-creditor's power to limit his own t_C, (b) his power to shift it, and (c) his power to increase his opponent's t_D weighs very heavily against there being any substantial practical effect, as to him, of the paradoxical disadvantage of being a creditor under American law.

Professionals on Both Ends. There is a third very common situation in which one party fully performs before the other completes his performance: transactions between businessmen. When those transactions lead to dispute, businessmen avoid the judicial-coercive system, that very flower of Western common law, like some rare Asiatic plague. They go to law only under very special circumstances and as a last resort.

For any two businessmen the components of both t_C and t_D are likely to be roughly the same. Both parties are likely already to have established mechanisms for dealing with disputes. Whatever economies come from the differential between their scales of operation are likely to be small. Except when a businessman is in extremis, those gross harms that may be visited on him by carrying out legal execution are harder to inflict; he usually has more room to maneuver and pay up prior to being sold out. The risk-of-loss-of-v factor still favors the businessman who in the particular instance occupies the D chair, but a careful C who is willing to spend for representation can usually soften the risk of total loss. In general, then, t_C and t_D are not likely to be wildly different.

But considerably more important for avoiding the coercive collection game is the availability of a strikingly effective alternative procedure, one which while not abjuring threat and coercion, avoids the costs and dangers of the official varieties. It depends not on force, but on the exchange of information. The "solution" to many of the divers disputes between businessmen takes the form of some variation on one of the following scenarios. The scene of each is simple: split screen, two telephones:

1. Buyer: Hello, Morris? Those widgets you sent us. They're breaking every minute. You want me to pay for such junk?

    Seller: Look, if your men don't know how to use widgets right, what do you want from me? They're just what you ordered, Grade A-2 stainless steel widgets.

    Buyer: Stainless steel they're not. Swiss cheese maybe, orange-crate wood, but not steel.

    Seller: Look, Kevin, maybe we've been having a little quality control problem—just temporary. Do the best you can and we'll make it up next time.

Buyer: OK, but don't forget. The noise of popping widgets my partner doesn't have to hear.

2. Seller: Hello, Kevin? So what's with our last bill?

Buyer: My bookkeeper's been sick.

Seller: Uh huh. Your hand cramps when you pick up a pen?

Buyer: Soon, Morris.

Seller: How soon? Tomorrow?

Buyer: Come on, Morris; did I make such a stink when you were shipping out those cardboard widgets?

Seller: OK, OK. Maybe I'll give you a couple of weeks more. You're not really in trouble are you?

Buyer: Absolutely not. I got plenty of orders. Go check with some of the other guys.

Seller: Don't worry. I already did. OK, take a couple of weeks. Give my get-wells to your bookkeeper.

Buyer: Hah!

3. Seller: So Kevin? Morris. Where's the money?

Buyer: Soon.

Seller: It's been soon a long time. Now it's now.

Buyer: Look Morris, I could have gotten the stuff from Acme or Nadir cheaper. I gave you the trade.

Seller: Now you're giving me the business. Now! Or there'll be trouble.

Buyer: Tell you what I'll do. I'll pay you today, right now, what I could've got the merchandise for from Acme.

Seller: Tell you what I'll do. I'll break your head is what I'll do. We got a goddam contract, Kevin, and you pay the goddam contract price.

Buyer: Morris, you don't like my deal, sue me with your contract.

Seller: I may and I may not. But one thing I know I'll do; anyone asks me if you pay your bills *the answer is no*. [The last clause is delivered in a high-pitched shriek of absolute credibility.]

Buyer: Morris? Half today, the rest at the end of the month?

Seller: OK.

Buyer: My best to Ethel.

Seller: You too. Remember me at home.

It would be folly to characterize these solutions as "coercive" or "cooperative"; they are deals and like all deals partake of both elements. But one thing is perfectly clear: this form of solution depends upon the generation, transmission, and communication of information and threats of information. And another thing is also clear: until one begins to understand this method of collection, one understands nothing. . . .

A Model in Which Suits Are Brought for Their Nuisance Value

DAVID ROSENBERG AND STEVEN SHAVELL

By a suit brought for its nuisance value, we mean a suit in which the plaintiff is able to obtain a positive settlement from the defendant even though the defendant knows the plaintiff's case is sufficiently weak that he would be unwilling or unlikely actually to pursue his case to trial. This note considers a model of the legal dispute allowing for the occurrence of such nuisance suits. . . . Specifically, the plaintiff may choose to file a claim at some (presumably small) cost. If the defendant does not then settle with the plaintiff and does not, at a cost, defend himself, the plaintiff will prevail by default judgment. If the defendant does defend himself, however, the plaintiff then may either withdraw or may, at a cost, litigate, resulting in a favorable verdict only with a probability (and a low one if his case is weak).

Given the model and the assumption that each party acts in his financial interest and realizes the other will do the same, it is easy to see how nuisance suits can arise. By filing a claim, any plaintiff, and thus the plaintiff with a weak case, places the defendant in a position where he will be held liable for the full judgment demanded unless he defends himself. Hence, the defendant should be willing to pay a positive amount in settlement to the plaintiff with the weak case—despite the defendant's knowledge that were he to defend himself, such a plaintiff would withdraw. . . .

Numerical Examples

Suppose that the cost to the plaintiff of filing a claim is $25; that the cost to the defendant of defending himself would be $200; that the cost to the plaintiff of then litigating through trial would be $100; that the probability of the plaintiff prevailing at trial would be only 1 percent (in the opinions of both parties); and that the amount the plaintiff would obtain were he to prevail at trial or by default judgment is $1,000. Thus the plaintiff's case is weak, and because his "expected" or probability discounted judgment from litigation would be only 1 percent × $1,000 or $10, which is substantially less than his litigation costs of $100, he would not litigate the case through trial.

Now let us verify that although the plaintiff would withdraw were the defendant to defend himself, the defendant would be willing to pay the plaintiff in settlement

David Rosenberg and Steven Shavell, A Model in which Suits Are Brought for Their Nuisance Value, 5 *International Review of Law and Economics* (1985), pp. 3–13. Copyright © 1985 Elsevier Science, Inc. Reprinted with permission.

any amount up to his defense costs of $200. Suppose, for instance, that the plaintiff files a claim and demands $180 in settlement. The defendant will then reason as follows. . . . If he settles, his costs will be $180. If he rejects the demand and does not defend himself, he will lose $1,000 by default judgment. If he rejects the demand and defends himself, the plaintiff will withdraw, but he will have spend $200 to accomplish this. Hence, the defendant's costs are minimized if he accepts the plaintiff's demand for $180; and the same logic shows that he would have accepted any demand up to $200.

It follows that the plaintiff will find it profitable to file his nuisance claim; indeed, this will be so whenever the cost of filing is less than the defendant's cost of defense. Because the plaintiff is able to obtain from the defendant in settlement as much as his $200 defense costs, the plaintiff will clearly decide to file if the filing costs are less than $200; the fact that they are only $25 makes filing extremely attractive.

Consider next the situation where the plaintiff's case is meritorious but he would still be unwilling to go to trial because the costs of litigation would exceed the expected judgment. This situation is qualitatively similar to that of the nuisance case in that the plaintiff may still find it profitable to file a claim. Suppose for example that the plaintiff would have a 70 percent chance of winning at trial; that the judgment amount would be $300; that his litigation costs would be $250; and that the defendant's defense costs would be $200. Then as the plaintiff's expected judgment would be only $210, he would not be willing to go to trial. But by the reasoning used above, the plaintiff nevertheless can obtain in a settlement as much as the defendant's defense cost of $200, so that he would certainly choose to file a claim if the filing cost if $25.

Now consider the situation where the plaintiff's likelihood of prevailing and the judgment amount are sufficiently high that he would be willing to engage in litigation. Suppose that not only is the likelihood of the plaintiff prevailing 70 percent, but also that the judgment amount would be $1,000, as originally assumed. Then the plaintiff's expected judgment from litigation would be $700, which exceeds his litigation costs of $250; thus the plaintiff would indeed be willing to litigate. In this situation, it is clear that the plaintiff can obtain in a settlement any amount up to the defendant's costs of defense plus the expected judgment, that is, $200 + $700, or $900. For were the defendant to reject a settlement demand and defend himself at a cost of $200, then, unlike before, the plaintiff would litigate and the defendant would bear expected liability of $700, meaning that his total expected costs would be $900. Of course, because the plaintiff's credible threat to litigate enables him to obtain a higher settlement than before, he will file a claim more often than before; he will do so whenever the cost of filing is less than the defense costs plus his expected judgment.

Last, briefly reconsider the analysis assuming that the prevailing party's cost of litigation would be shifted to the losing party, that is, that the so-called British system for the allocation of litigation costs applies. Under the British system, a plaintiff who would be unwilling to litigate would never file a claim; in particular, nuisance suits would never occur. The reason for this conclusion is that if the defendant defended himself against a plaintiff who would then withdraw, the defendant, being the winning party, would recover his defense costs of $200. Hence his defense would turn out to be costless, and this in turn means that the plaintiff would not be able to extract a settlement from him and so would not be able to profit from filing a claim.

A related point is that under the British system the willingness of the plaintiff to litigate and to file a claim will be less than under the usual American system if the likelihood of prevailing is low. The explanation for this is simply that if the likelihood of prevailing is low, then the plaintiff's expected litigation costs will be high under the British system. For instance, if the likelihood of prevailing is near 0, the plaintiff's expected litigation costs under the British system will be close to his actual (unshifted) costs plus the defendant's; thus the expected costs will exceed those under the American system, and the chance that the plaintiff will be willing to litigate will therefore be less than under the American system.

On the other hand, if the likelihood of prevailing is high, then the willingness of the plaintiff to file a claim and litigate will be greater under the British system than under the American; thus the bringing of highly meritorious suits will be encouraged. In the case where the likelihood of prevailing is near 1, for example, the plaintiff's expected litigation costs under the British system will be almost 0, so that his willingness to file and to litigate will clearly be greater than under the American system.

Note that an implication of these latter points is that use of the British system may or may not lead to a decrease in the number of claims filed, depending on the distribution of the likelihoods of the plaintiff prevailing. . . .

Concluding Remarks

1. The feature of the model of primary interest was the ability of the plaintiff cheaply to place the defendant in a position where he would lose unless he engaged in a costly defense. This feature of the model seems justified in fact. First and most obviously, a party can usually file a claim at small expense asserting that another is legally liable for a harm he has suffered. Second, it is not feasible for the courts to exercise much control over the quality of claims; thus, only the plainly frivolous claim will be disallowed. Third, if a claim goes unchallenged, the plaintiff ordinarily will prevail without further inquiry on the part of the courts. And fourth, to defeat a claim, the defendant will have to engage in actions that are frequently more expensive than the plaintiff's cost of making the claim, for the defendant will have to gather evidence supporting his contention that he was not legally responsible for harm done to the plaintiff or that no harm was actually done. Therefore, by making a claim, the plaintiff can usually do what was envisioned in the model.

It should be emphasized as well that the model does not have to be interpreted in so literal a sense for its point to be relevant. All that need be true is that the plaintiff be able to prevail with high probability unless the defendant spends a larger amount defending himself.

2. After the plaintiff has filed a claim, there will often be actions that he or the defendant can take that will impose considerable costs on the other. Notably, during discovery the plaintiff or the defendant might ask for information which the other would find expensive to prepare; or one of the parties might at some point hire an expert who asserts facts which it would be difficult for the other to refute; and so forth. It is clear that were such possibilities incorporated into a more elaborate model of litigation, we

could conclude by the logic of our simple model that whenever a party is able to impose significant costs on the other, he should be able to bargain for a relatively advantageous settlement.

3. A defendant facing the prospect of many nuisance suits (or meritorious suits which the plaintiff would not find worthwhile litigating) may in certain circumstances be able to ward them off. If the issues presented by the suits are related, then the costs of defense could be spread over many plaintiffs, making the costs per plaintiff quite low. Hence, according to the model, a plaintiff would be able to obtain in settlement only this low amount; and if it were less than his cost of filing, he would be discouraged from filing in the first place. Similarly, a defendant facing the possibility of many nuisance suits might find it worthwhile to reject a particular plaintiff's demands and to defend himself in order to acquire a reputation for "toughness," and thus to discourage other plaintiffs from filing claims.

4. We should caution that our conclusion that under the British system nuisance suits would be discouraged and highly meritorious suits encouraged should not be taken as a recommendation for adoption of the British system. For as we said, its use would affect the propensity to file claims quite generally (not just those cases clearly fitting into the category of nuisance suits or of highly meritorious suits), might also increase the volume of litigation, and thus requires a broader analysis for thorough evaluation. . . .

The Strategic Structure of Offer and Acceptance

AVERY KATZ

The purpose of this article is to promote a particular research program; namely, the use of game theory to analyze the law of contract formation. . . . At its broadest, my argument addresses all legal rules that answer two types of questions: First, which objectively verifiable actions or subjectively experienced intentions suffice to conclude a bargain and form a contractual obligation? Second, how do these actions and intentions affect the substantive content of any contract formed?

The legal doctrines governing these questions present some of the more subtle and technical problems in all of the law. Their metaphysical controversies and mechanical intricacies have puzzled countless lawyers and judges and have consigned generations of law students to torment at the hands of their professors. Yet the law of offer and acceptance has generated relatively little interest in the literature that addresses contract law from a policy perspective. Instead, commentators concerned with public policy have focused largely on the consequences of contracts after formation. . . .

The conventional wisdom's neglect of the law of bargaining is especially noteworthy given the critical importance of bargaining to the economic analysis of law.

Avery Katz, The Strategic Structure of Offer and Acceptance, 89 *Michigan Law Review* (1990), pp. 215–295. Copyright © 1990 Michigan Law Review Association. Reprinted with permission.

A chief message of law and economics, if not the chief message, is that the effect of legal rules cannot be understood properly without taking account of the incentive for private transactions. This message is most strikingly embodied in the well-known "Coase theorem," which in its strongest version claims that so long as the mechanisms of private ordering are frictionless, legal rules will have no effect on the allocation of resources. One might have expected, therefore, that the set of legal rules that regulate private ordering and determine its frictions would have occupied a more prominent place on the research agenda of law and economics.

It would be surprising in light of current theoretical understandings of bargaining, moreover, if contract formation and interpretation rules served merely a coordinating role. Most formal economic accounts of bargaining conclude that when information is imperfect or communication costly, self-interested parties generally will fail to realize the full potential surplus from exchange. Just how much is wasted will depend on the precise structure of the institutions that govern the bargaining. Different legal rules, once established, imply different institutional structures for contracting parties and may induce different forms of bargaining behavior. Hence they can have important consequences for the efficiency of exchange. . . .

The Bargaining Problem in Law and Economics

. . .

The Relation between Substantive Legal Rules and Bargaining

Many prominent contributions to the economic analysis of law suggest how substantive entitlements affect strategic behavior. Calabresi and Melamed's distinction between liability and property rules provides the classic illustration. They observed that a particular legal entitlement such as the right to undisturbed enjoyment of a given parcel of land can be protected in at least two ways. If the entitlement is protected only by a liability rule, anyone has the legal power, if not the right, to violate it provided they pay damages in compensation. In contrast, when an entitlement is protected by a property rule, no one has the power to violate it without first obtaining the permission of the holder. Such a rule might be enforced by criminal or equitable sanctions, effectively requiring that permission to make use of the entitlement be obtained in a voluntary exchange.

As Calabresi and Melamed argued, the choice among these alternatives (and a third—making the entitlement inalienable) both influences and depends on private bargaining. The different rules alter the threats and offers available to the parties. Under a liability rule, the potential infringer has the power to cut short the bargaining and force the question of the valuation of the entitlement before some public authority. Under a property rule, the entitlement holder has the power to end the bargaining without an exchange taking place. These possibilities may alter the outcome of any negotiation that occurs. . . .

Parallel arguments can be found in the economic analyses of virtually every field of the law. Much of the debate over the economics of contract remedies, for instance, turns on the commentators' differing views of ex post renegotiation. Whether specific performance is more effective than money damages in promoting the efficient level of contract breach can depend upon the strategic behavior costs it induces. One's view of the merits of the standard alternatives to measuring money damages—expectation, reliance, and restitution—will similarly be affected by one's view of bargaining. Since the efficiency of the various measures depends on the relative importance of encouraging efficient breach, reliance, or risk allocation, by affecting the relative significance of these factors ex post renegotiation can alter the ranking of the various measures. . . .

The Effect of Contract Formation Rules on Bargaining

This intellectual background makes it all the more surprising that the law of contract formation has attracted only scant attention in the law-and-economics literature. If the basic arguments of law and economics turn on the theory of bargaining, as I have argued, and if these conclusions are in fundamental dispute, we should look to the law of bargaining to help resolve the controversy. In this light, contract formation should be seen as logically prior to most if not all of the major issues in law and economics.

. . . [T]ransaction costs are affected in large part by rules of contract formation and interpretation. For example, various rules prescribe the degree of thoroughness and formality required for a contract to be enforceable. Among these are the Statute of Frauds, the definiteness doctrine (which demands that the parties specify the major terms of the agreement in requisite detail) and, in important respects, the doctrine of consideration.

Such requirements directly affect the implementation costs of agreement. To satisfy the Statute of Frauds, time and effort must be expended to create an authoritative record of the agreement; papers and perhaps the parties themselves must be transported back and forth at substantial expense and delay. To satisfy the definiteness doctrine, resources must be spent negotiating specific contract language at a time where relevant information is still unknown.

Perhaps less obviously, rules of offer and acceptance also influence strategic behavior costs. Requiring additional negotiation before a contract becomes enforceable, for instance, changes the information available to the parties at various critical moments. Parties may acquire information about the other side's likely future behavior at the time reliance investments must be sunk. This information will introduce certain possibilities of influencing the adversary's actions and will foreclose some others. For another example, rules requiring disclosure of private information or penalizing nondisclosure may, by reducing the cost of investigation and lowering the chance of error, improve the efficiency of negotiation. . . .

The network of rules that govern the mechanics of contracts concluded by correspondence provides [an example.] Suppose two individuals send each other identical offers that cross in the mail and are received simultaneously. According to blackletter law no contract is formed until one of the parties posts a responding acceptance. Mechanical rules such as these commonly are justified on the ground that they establish a benchmark around which parties can plan their affairs; any more complexity would supposedly create confusion during the bargaining. This justification would be more persuasive if parties knew at the time of bargaining whether their communications would ultimately be classified as offers, acceptances, or as some other preliminary communication such as an invitation to make an offer. In reality, however, actual communications generally are not labeled "offer" or "acceptance," and, even if they were, the label would not be legally conclusive. On the other hand, which particular mechanical rule is chosen will set the structure and sequence of bargaining, and change the time at which parties can safely rely. The choice is by no means neutral. . . .

A Full Information Model: Acceptance by Silence

[As an extended illustration], I consider the rule of acceptance by silence to show how even simple bargaining models premised on full information can help illuminate the law of offer and acceptance. While I do not mean to claim that imperfect information is unimportant in this setting, . . . the full-information analysis can reveal important strategic features of existing legal doctrine.

The Doctrinal Background

It is a basic feature of Anglo-American contract law that the person who proposes an exchange has substantial control over the structure of the bargaining—a fact captured in the maxim that "the offeror is master of his offer." . . . One important limit on the offeror's power to set the terms of the bargain arises when the offeror wishes to specify that the offeree need do nothing at all in order to accept. For instance, a seller of goods might send a letter stating that a shipment of merchandise will be sent in ten days unless the recipient sends a notice of objection. If such an offer were valid and the recipient wanted to purchase the goods at the stated price, he could merely wait for the goods to arrive. . . . As a matter of prevailing doctrine, however, failing to reply to an offer can operate as an acceptance only in certain special circumstances. . . .

With a simplified account of contract bargaining, . . . some of the functional consequences of the common law rule and its various alternatives can be examined. Before developing a formal presentation, however, let me try to provide some heuristic motivation for the modeling choices I have made. . . .

Which features of the problem are the essential ones? To begin with, the model must provide an opportunity for at least two messages to pass between the parties, to admit an alternative to silence as a mode of acceptance. Second, some positive cost should be associated with a message; otherwise all opportunities for communication would be exploited. Third, the model should admit a variety of possible sellers' costs and buyers' valuations, for otherwise full efficiency could be achieved trivially by a rule that either decreed or forbade an exchange, depending upon whether the buyer's valuation was greater or less than the seller's cost. Fourth and finally, the contract price should be determined endogenously within the model rather than prescribed at the outset, because parties can find mutual advantage in bargaining around the constraints of legal rules only by adjusting price. . . .

The Formal Model

Consider a bargain over the sale of a book. Suppose the seller's cost of providing the book, denoted as C, and the value that the buyer attaches to the book, denoted as V, are precisely measurable. I will refer to these quantities as the seller's and buyer's respective *reservation prices*. . . . Initially, the seller must decide whether to offer the book for sale at all. If she does make an offer, she must also select a specific price, which is denoted as P. It costs the seller a fixed amount, denoted as S, to send or otherwise communicate the offer to the buyer, but this cost need only be paid if an offer is actually made. If the seller makes no offer, the bargaining ends without gain or loss for either side. . . . If the seller does make an offer, on the other hand, the buyer is then faced with a decision whether to accept, reject, or ignore the offer. I assume that it does not cost the buyer anything to read and consider the offer, and that he actually does so. If the buyer simply ignores the offer after reading it, he thereby incurs no direct cost, although depending on the legal rule in force, the inaction may obligate him to purchase the book. In order affirmatively to reject or accept the offer, however, the buyer must communicate a return message to the seller; and the cost of his response is denoted as R. Following the buyer's decision, the bargaining ends, and trade takes place if and only if a contract was formed under the legal rule that happens to be in force. I assume that the established legal rule and the specific values of all relevant parameters—V, C, S, and R—are common knowledge.

The parties' respective gains or losses in the series of events that can follow an offer will depend upon the legal rule in force. [Consider] the payoffs under two alternate doctrinal regimes. The first set of payoffs, labeled (I), are those that follow under the usual common law rule, which holds that no contract is formed unless the buyer explicitly accepts. The second set of payoffs, labeled (II), correspond to a hypothetical legal regime in which silence implies acceptance.

Under the first regime, the payoffs at the end of each sequence of events are calculated in the following manner: If the buyer explicitly accepts the seller's offer, the seller receives the sales price P, and incurs production cost C and communication cost S for a net payoff of $(P - S - C)$. The accepting buyer nets $(V - R - P)$; he enjoys value V, incurs communication cost R, and must pay P to the seller. If on the other hand the buyer explicitly rejects the seller's offer, the parties' payoffs are $(-S, -R)$; no sale is concluded, but each side loses the cost of a message. Finally, if the buyer ignores the seller's offer, the buyer neither gains nor loses anything, and the seller loses S, the cost of sending her offer initially.

Under the second regime, the payoffs following any explicit response by the buyer are the same as under the common law. The only difference occurs if the buyer ignores the seller's offer, in which case the offer is accepted by silence. In this case, the seller nets $(P - S - C)$, just as if the buyer had explicitly accepted, and the buyer nets $(V - P)$ (value less price, and no costs spent responding to the offer).

Since the game is one of full information, the outcome can be found in straightforward fashion through the method of backward induction. First, one looks to see what the buyer would find rational to do if faced with an offer to buy, and then one finds the seller's optimal offer given the buyer's expected response. Focusing on the buyer, it should be obvious that he will never choose to reject under the first regime or accept under the second. In either case he can do strictly better by remaining silent and saving the cost of a response. Under the common law he will simply choose between accepting and ignoring, and will accept if and only if his net gain from doing so is positive. This will be the case if the price offered is sufficiently low; that is, if $P < V - R$.

Under the silent-acceptance regime, alternatively, the buyer chooses between rejecting and ignoring, and will reject if and only if his loss from remaining silent is greater than the cost of a response. This will be the case if the offered price is sufficiently high; that is, if $P > V + R$. Since she can anticipate the buyer's reaction, the seller when making an offer will in either case want to set her price just low enough to induce the buyer to purchase. Offering a lower price than necessary fails to maximize her profits; offering a higher price is suboptimal because she is better off saving the cost of a message than making an offer she knows the buyer will reject.

Notice that under either regime there exists some maximum amount—call it the buyer's *net reservation price*—that the seller can obtain from the buyer in a sale. The buyer's net reservation price differs under the two regimes, however, and is lower under the common law. This is because in order to get the buyer to accept under the common law, the price must be below his gross valuation V by enough to justify spending R on a response. To get buyer to accept under regime II, in contrast, the price can be above his valuation by as much as the cost of a response, since he must spend that amount in order to avoid being bound. It follows that the price charged by a rational seller will equal $V + R$ under regime II and will equal $V - R$ under regime I. The difference between the two prices will be exactly twice the buyer's cost of responding.

For the final step in the induction, we must determine whether an offer will be made at all. Because the seller can always earn a return of zero by doing nothing, and because making an offer is costly, she will want to make an offer if and only if the maximum price she can obtain will cover her combined costs of production and communication. Under regime I, this will be the case if and only if $V - R > C + S$. Under regime II, it will be the case if and only if $V + R > C + S$.

We can now summarize the equilibrium outcomes under each of the alternate legal regimes. Under the common law, the equilibrium is for no offer to be made when $V - R < C + S$, and otherwise for an exchange to be concluded at price $P = V - R$. Under the silent-acceptance rule, the equilibrium is for no offer to be made when $V + R < C + S$, and otherwise for an exchange to be concluded at price $P = V + R$.

The analysis predicts that fewer sales will take place under regime I than under regime II. Only when the gross potential surplus from trade, $(V - C)$, exceeds the sum of the two communication costs, $R + S$, will the seller offer an exchange that the buyer will accept under the common law. In contrast, for this to occur under regime II, the gross surplus from trade must exceed the difference of the two communication costs—which is necessarily less than the sum. Intuitively, a buyer of given valuation is more willing to accept a higher price under regime II; for this reason a seller facing a given production cost is more willing to make an offer initially.

Since the sum of communication costs is necessarily positive while the difference may be negative, regime II, but not the common law, permits slightly inefficient exchanges to take place. Indeed, this will occur whenever the cost of a buyer's response exceeds the seller's cost of an offer, and when production cost C is not too greatly above the buyer's value V. Grossly inefficient exchanges cannot take place even under regime II, however, because if C is too far above V, the buyer will reject any offer that the seller would want to make—and the seller, anticipating this, will make no offer.

Regime II might therefore appear inefficient. On the other hand, regime II does allow those marginally desirable exchanges that under the common law are not worth the cost of a response. It also allows those sales that would occur under either regime to be concluded with one message rather than two. A more complete accounting is necessary. . . .

One hesitates to draw any very strong conclusions from the foregoing analysis, given its illustrative character and the simplified assumptions that went into it. . . . Some might view the silent-acceptance rule as unfairly rewarding opportunistic behavior on the part of offerors, to an extent outweighing any efficiency gains that it could otherwise achieve.

Indeed, if the conclusions of the formal model are to be believed, the silent-acceptance rule is open to severe objection on just these grounds. When goods are exchanged under that regime, not only does the buyer get none of the surplus from the exchange, but he winds up losing an amount equal to his cost of sending a response. The seller's ability to force the buyer to expend resources to avoid an unwanted offer gives her a strategic advantage, with the result that the buyer actually ends up in a worse position after the exchange than if the offer had never been made. For this reason, those arrangements in which parties contract out of the common-law rule typically provide some up-front benefits to the offeree, like the introductory bonuses commonly offered by book clubs. Such loss leaders may be necessary to induce the offeree to enter into the arrangement at all, since he must be compensated ex ante for the opportunism he can expect from the offeror ex post. . . .

Conclusion

. . .

I have argued that traditional explanations of offer and acceptance doctrine, and of contract formation doctrine generally, fail to identify important incentive effects. The methods and styles of thought of noncooperative game theory, on the other hand, highlight such effects; they help to illuminate the policy consequences of the law of bargaining by drawing our attention to its strategic structure. The potential insights to be gained transcend the narrow concerns of contract lawyers and scholars. Since Coase's classic article and probably before, we have known that the opportunity for private individuals to enter into contracts can critically influence the efficiency and fairness of substantive rules and regulations in every field of the law. By devoting more attention to the specific branch of law that governs the procedures of private ordering, we may learn how better to use the law in general to promote the public interest.

An Economist's Perspective on the Theory of the Firm

OLIVER HART

. . . Introduced in Coase's famous 1937 article, transaction cost economics traces the existence of firms to the thinking, planning and contracting costs that accompany any transaction, costs usually ignored by the neoclassical paradigm. The idea is that in some situations these costs will be lower if a transaction is carried out within a firm rather than in the market. According to Coase, the main cost of transacting in the market is the cost of learning about and haggling over the terms of trade; this cost can be particularly large if the transaction is a long-term one in which learning and haggling must be performed repeatedly. Transaction costs can be reduced by giving one party authority over the terms of trade, at least within limits. But, according to Coase, this authority is precisely what defines a firm: within a firm, transactions occur as a result of instructions or orders issued by a boss, and the price mechanism is suppressed. . . .

Coase's ideas, although recognized as highly original, took a long time to catch on. There are probably two reasons for this. First, they remain to this day very hard to formalize. Second, there is a conceptual weakness, pointed out by Alchian and Demsetz, in the theory's dichotomy between the role of authority within the firm and the role of consensual trade within the market. Consider, for example, Coase's notion that an employer has authority over an employee—an employer can tell an employee what to do. Alchian and Demsetz questioned this, asking what ensures that the employee obeys the employer's instructions. To put it another way, what happens to the employee if he disobeys these instructions? Will he be sued for breach of contract? Unlikely. Probably the worst that can happen is the employee will be fired. But firing is typically the sanction that one independent contractor will impose on another whose performance he does not like. To paraphrase Alchian and Demsetz's criticism, it is not clear that an employer can tell an employee what to do, any more than a consumer can tell her grocer what to do (what vegetables to sell at what prices); in either case, a refusal will likely lead to a termination of the relationship, a firing. In the case of the grocer, this means that the consumer shops at another grocer. Thus, according to Alchian and Demsetz's argument, Coase's view that firms are characterized by authority relations does not really stand up. . . .

At the same time that doubts were being expressed about the specifics of Coase's theory, Coase's major idea—that firms arise to economize on transaction costs—was increasingly accepted. The exact nature of these transaction costs, however, remained unclear. What lay beyond the learning and haggling costs that, according to Coase, are a major component of market transactions? Professor Oliver Williamson has offered the deepest and most far-reaching analysis of these costs. Williamson recognized that transaction costs may assume particular importance in situations where

Oliver, Hart, An Economist's Perspective on the Theory of the Firm, 89 *Columbia Law Review* (1989), pp. 1757–1774. Reprinted with permission.

economic actors make relationship-specific investments—investments to some extent specific to a particular set of individuals or assets. Examples of such investments include locating an electricity generating plant adjacent to a coal mine that is going to supply it; a firm's expanding capacity to satisfy a particular customer's demands; training a worker to operate a particular set of machines or to work with a particular group of individuals; or a worker's relocating to a town where he has a new job.

In situations like these, there may be plenty of competition before the investments are made—there may be many coal mines next to which an electricity generating plant could locate or many towns to which a worker could move. But once the parties sink their investments, they are to some extent locked into each other. As a result, external markets will not provide a guide to the parties' opportunity costs once the relationship is underway. This lack of information takes on great significance, since, in view of the size and degree of the specific investment, one would expect relationships like these to be long lasting.

In an ideal world, the lack of ex post market signals would pose no problem, since the parties could always write a long-term contract in advance of the investment, spelling out each agent's obligations and the terms of the trade in every conceivable situation. In practice, however, thinking, negotiation and enforcement costs will usually make such a contract prohibitively expensive. As a result, parties must negotiate many of the terms of the relationship as they go along. Williamson argues that this leads to two sorts of costs. First, there will be costs associated with the ex post negotiation itself—the parties may engage in collectively wasteful activities to try to increase their own share of the ex post surplus; also, asymmetries of information may make some gains from trade difficult to realize. Second, and perhaps more fundamental, since a party's bargaining power and resulting share of the ex post surplus may bear little relation to his ex ante investment, parties will have the wrong investment incentives at the ex ante stage. In particular, a far-sighted agent will choose her investment inefficiently from the point of view of her contracting partners, given that she realizes that these partners could expropriate part of her investment at the ex post stage.

In Williamson's view, bringing a transaction from the market into the firm—the phenomenon of integration—mitigates this opportunistic behavior and improves investment incentives. Agent A is less likely to hold up agent B if A is an employee of B than if A is an independent contractor. However, Williamson does not spell out in precise terms the mechanism by which this reduction in opportunism occurs. Moreover, certain costs presumably accompany integration. Otherwise, all transactions would be carried out in firms, and the market would not be used at all. Williamson, however, leaves the precise nature of these costs unclear.

The Firm as a Nexus of Contracts

All the theories discussed so far suffer from the same weakness: while they throw light on the nature of contractual failure, none explains in a convincing or rigorous manner how bringing a transaction into the firm mitigates this failure.

One reaction to this weakness is to argue that it is not really a weakness at all. According to this point of view, the firm is simply a nexus of contracts, and there is therefore little point in trying to distinguish between transactions within a firm and those between firms. Rather, both categories of transactions are part of a continuum of types of contractual relations, with different firms or organizations representing different points on this continuum. In other words, each type of business organization represents nothing more than a particular "standard form" contract. One such "standard form" contract is a public corporation, characterized by limited liability, indefinite life, and freely transferable shares and votes. In principle it would be possible to create a contract with these characteristics each time it is needed, but, given that these characteristics are likely to be useful in many different contexts, it is much more convenient to be able to appeal to a "standard form." Closely held corporations or partnerships are other examples of useful "standard forms."

Viewing the firm as a nexus of contracts is helpful in drawing attention to the fact that contractual relations with employees, suppliers, customers, creditors and others are an essential aspect of the firm. . . . At the same time, the nexus of contracts approach does less to resolve the questions of what a firm is than to shift the terms of the debate. In particular, it leaves open the question of why particular "standard forms" are chosen. Perhaps more fundamentally, it begs the question of what limits the set of activities covered by a "standard form." For example, corporations are characterized by limited liability, free transferability of shares, and indefinite life. But what limits the size of a corporation—what are the economic consequences of two corporations merging or of one corporation splitting itself into two? Given that mergers and breakups occur all the time, and at considerable transaction cost, it seems unlikely that such changes are cosmetic. Presumably they have some real effects on incentives and opportunistic behavior, but these effects remain unexplained.

A Property Rights Approach to the Firm

One way to resolve the question of how [the organization of enterprise] changes incentives is spelled out in recent literature that views the firm as a set of property rights. This approach is very much in the spirit of the transaction cost literature of Coase and Williamson, but differs by focusing attention on the role of physical, that is, nonhuman, assets in a contractual relationship.

Consider an economic relationship of the type analyzed by Williamson, where relationship-specific investments are important and transaction costs make it impossible to write a comprehensive long-term contract to govern the terms of the relationship. Consider also the nonhuman assets that, in the postinvestment stage, make up this relationship. Given that the initial contract has gaps, missing provisions, or ambiguities, situations will typically occur in which some aspects of the use of these assets are not specified. For example, a contract between GM and Fisher might leave open certain aspects of maintenance policy for Fisher machines, or might not specify the speed of the production line or the number of shifts per day.

Take the position that the right to choose these missing aspects of usage resides with the owner of the asset. That is, ownership of an asset goes together with the possession of residual rights of control over that asset; the owner has the right to use the asset in any way not inconsistent with a prior contract, custom, or any law. Thus, the owner of Fisher assets would have the right to choose maintenance policy and production line speed to the extent that the initial contract was silent about these.

Finally, identify a firm with all the nonhuman assets that belong to it, assets that the firm's owners possess by virtue of being owners of the firm. Included in this category are machines, inventories, buildings or locations, cash, client lists, patents, copyrights, and the rights and obligations embodied in outstanding contracts to the extent that these are also transferred with ownership. Human assets, however, are not included. Since human assets cannot be bought or sold, management and workers presumably own their human capital both before and after any merger.

We now have the basic ingredients of a theory of the firm. In a world of transaction costs and incomplete contracts, ex post residual rights of control will be important because, through their influence on asset usage, they will affect ex post bargaining power and the division of ex post surplus in a relationship. This division in turn will affect the incentives of actors to invest in that relationship. Hence, when contracts are incomplete, the boundaries of firms matter in that these boundaries determine who owns and controls which assets. In particular, a merger of two firms does not yield unambiguous benefits: to the extent that the (owner-)manager of the acquired firm loses control rights, his incentive to invest in the relationship will decrease. In addition, the shift in control may lower the investment incentives of workers in the acquired firm. In some cases these reductions in investment will be sufficiently great that nonintegration is preferable to integration.

Note that, according to this theory, when assessing the effects of integration, one must know not only the characteristics of the merging firms, but also who will own the merged company. If firms A and B integrate and A becomes the owner of the merged company, then A will presumably control the residual rights in the new firm. A can then use those rights to hold up the managers and workers of firm B. Should the situation be reversed, a different set of control relations would result in B exercising control over A, and A's workers and managers would be liable to holdups by B.

It will be helpful to illustrate these ideas in the context of the Fisher Body–General Motors relationship. Suppose these companies have an initial contract that requires Fisher to supply GM with a certain number of car bodies each week. Imagine that demand for GM cars now rises and GM wants Fisher to increase the quantity it supplies. Suppose also that the initial contract is silent about this possibility, perhaps because of a difficulty in predicting Fisher's costs of increasing supply. If Fisher is a separate company, GM presumably must secure Fisher's permission to increase supply. That is, the status quo point in any contract renegotiation is where Fisher does not provide the extra bodies. In particular, GM does not have the right to go into Fisher's factory and set the production line to supply the extra bodies; Fisher, as owner, has this residual right of control. The situation is very different if Fisher is a subdivision or subsidiary of GM, so that GM owns Fisher's factory. In this case, if Fisher management refuses to supply the extra bodies, GM always has the option to fire management and hire someone else to supervise the factory and supply extra bodies (they could even run Fisher themselves on a temporary basis). The status quo point in the contract renegotiation is therefore quite different.

To put it very simply, if Fisher is a separate firm, Fisher management can threaten to make both Fisher assets and their own labor unavailable for the uncontracted-for supply increase. In contrast, if Fisher belongs to GM, Fisher management can only threaten to make their own labor unavailable. The latter threat will generally be much weaker than the former.

Although the status quo point in the contract renegotiation may depend on whether GM and Fisher are one firm rather than two, it does not follow that the outcomes after renegotiation will differ. In fact, if the benefits to GM of the extra car bodies exceed the costs to Fisher of supplying them, we might expect the parties to agree that the bodies should be supplied, regardless of the status quo point. However, the divisions of surplus in the two cases will be very different. If GM and Fisher are separate, GM may have to pay Fisher a large sum to persuade it to supply the extra bodies. In contrast, if GM owns Fisher's plant, it may be able to enforce the extra supply at much lower cost since, as we have seen in this case, Fisher management has much reduced bargaining and threat power.

Anticipating the way surplus is divided, GM will typically be much more prepared to invest in machinery that is specifically geared to Fisher bodies if it owns Fisher than if Fisher is independent, since the threat of expropriation is reduced. The incentives for Fisher, however, may be quite the opposite. Fisher management will generally be much more willing to come up with cost-saving or quality-enhancing innovations if Fisher is an independent firm than if it is part of GM, because Fisher management is more likely to see a return on its activities. If Fisher is independent, it can extract some of GM's surplus by threatening to deny GM access to the assets embodying these innovations. In contrast, if GM owns the assets, Fisher management faces total expropriation of the value of the innovation to the extent that the innovation is asset-specific rather than management-specific, and GM can threaten to hire a new management team to incorporate the innovation.

So far, we have discussed the effects of control changes on the incentives of top management. But workers' incentives will also be affected. Consider, for example, the incentive of someone who works with Fisher assets to improve the quality of Fisher's output by better learning some aspect of the production process. Suppose further that GM has a specific interest in this improvement in car body quality, and that none of Fisher's other customers cares about it. There are many ways in which the worker might be rewarded for this, but one important reward is likely to come from the fact that the worker's value to the Fisher-GM venture will rise in the future and, due to his additional skills, the worker will be able to extract some of these benefits through a higher wage or promotion. Note, however, that the worker's ability to do this is greater if GM controls the assets than if Fisher does. In the former case, the worker will bargain directly with GM, the party that benefits from the worker's increased skill. In the latter case, the worker will bargain with Fisher, who only receives a fraction of these benefits, since it must in turn bargain with GM to parlay these benefits into dollars. In consequence, the worker will typically capture a lower share of the surplus, and his incentive to make the improvement in the first place will fall.

In other words, given that the worker may be held up no matter who owns the Fisher assets—assuming that he, himself, does not—his incentives are greater if the number of possible hold-ups is smaller rather than larger. With Fisher management in control of the assets, there are two potential hold-ups: Fisher can deny the worker access to the assets, and GM can decline to pay more for the improved product. As a result, we might expect the worker to get, say, a third of his increased marginal product (supposing equal division with Fisher and GM). With GM management in control of the Fisher assets, there is only one potential hold-up, since the power to deny the worker his increased marginal product is concentrated in one agent's hands. As a result, the worker in this case might be able to capture half of his increased marginal product (supposing equal division with GM).

The above reasoning applies to the case in which the improvement is specific to GM. Exactly the opposite conclusion would be reached, however, if the improvement were specific to Fisher, such as the worker learning how to reduce Fisher management's costs of making car bodies, regardless of Fisher's final customer (a cost reduction, furthermore, which could not be enjoyed by any substitute for Fisher management). In that event, the number of hold-ups is reduced by giving control of Fisher assets to Fisher management rather than GM. The reason is that with Fisher management in control, the worker bargains with the party who benefits directly from his increased productivity, whereas with GM management in control, he must bargain with an indirect recipient; GM must in turn bargain with Fisher management to benefit from the reduction in costs.

Up to this point we have assumed that GM management will control GM assets. This, however, need not be the case; in some situations it might make more sense for Fisher management to control these assets—for Fisher to buy up GM. One thing we can be sure of is that if GM and Fisher assets are sufficiently complementary, and initial contracts sufficiently incomplete, then the two sets of assets should be under common control. With extreme complementarity, no agent—whether manager or worker—can benefit from any increase in his marginal productivity unless he has access to both sets of assets (by the definition of extreme complementarity, each asset, by itself, is useless). Giving control of these assets to two different management teams is therefore bound to be detrimental to actors' incentives, since it increases the number of parties with hold-up power. This result confirms the notion that when lock-in effects are extreme, integration will dominate nonintegration.

These ideas can be used to construct a theory of the firm's boundaries. First, as we have seen, highly complementary assets should be owned in common, which may provide a minimum size for the firm. Second, as the firm grows beyond a certain point, the manager at the center will become less and less important with regard to operations at the periphery in the sense that increases in marginal product at the periphery are unlikely to be specific either to this manager or to the assets at the center. At this stage, a new firm should be created since giving the central manager control of the periphery will increase hold-up problems without any compensating gains. It should also be clear from this line of argument that, in the absence of significant lock-in effects, nonintegration is always better than integration—it is optimal to do things through the market, for integration only increases the number of potential hold-ups without any compensating gains.

Finally, it is worth noting that the property rights approach can explain how the purchase of physical assets leads to control over human assets. To see this, consider again the GM-Fisher hypothetical. We showed that someone working with Fisher assets is more likely to improve Fisher's output in a way that is specifically of value to GM if GM owns these assets than if Fisher does. This result can be expressed more informally as follows: a worker will put more weight on an actor's objectives if that actor is the worker's boss, that is, if that actor controls the assets the worker works with, than otherwise. The conclusion is quite Coasian in spirit, but the logic underlying it is very different. Coase reaches this conclusion by assuming that a boss can tell a worker what to do; in contrast, the property rights approach reaches it by showing that it is in a worker's self-interest to behave in this way, since it puts him in a stronger bargaining position with his boss later on.

To put it slightly differently, the reason an employee is likely to be more responsive to what his employer wants than a grocer is to what his customer wants is that the employer has much more leverage over his employee than the customer has over his grocer. In particular, the employer can deprive the employee of the assets he works with and hire another employee to work with these assets, while the customer can only deprive the grocer of his custom and as long as the customer is small, it is presumably not very difficult for the grocer to find another customer. . . .

NOTES AND QUESTIONS

1. There is by now a burgeoning literature on strategic issues in law. An excellent general survey of the subject is Douglas Baird, Robert Gertner, and Randal Picker, *Game Theory and the Law* (Cambridge, Mass.: Harvard University Press, 1994). For readers interested in the basic principles of game theory, the branch of applied economics that deals with strategic behavior, a more formal introduction can be found in Eric Rasmusen, *Games and Information: An Introduction to Game Theory*, 2d ed. (Cambridge, Mass.: B. Blackwell, 1994). A recent lively and informal survey of strategic issues can be found in Avinash Dixit and Barry Nalebuff, *Thinking Strategically: The Competitive Edge in Business, Politics, and Everyday Life* (New York: Norton, 1991). The classic informal discussion is Thomas C. Schelling, *The Strategy of Conflict*, 2d ed. (Cambridge, Mass.: Harvard University Press, 1980). Additional applications, beyond those covered here, include family law [Robert Mnookin and Lewis Kornhauser, Bargaining in the Shadow of the Law: The Case of Divorce, 88 *Yale Law Journal* 950 (1979)]; intellectual property [Suzanne Scotchmer, Standing on the Shoulders of Giants: Cumulative Research and the Patent Law, 5 *Journal of Economic Perspectives* 29 (1991)]; and corporate law [Lucien Bebchuk, The Case for Facilitating Competing Tender Offers, 95 *Harvard Law Review* 1028 (1982)].

2. Cooter critiques the Coase theorem from the perspective of the model of market failure, in that he argues that the costs of strategic behavior justify limits on freedom of contract. A good Coasian would respond, however, that Cooter has only told half the story. There are strategic costs in government institutions as well, and to make the case for state regulation it is necessary to show that strategic behavior is less of a problem in the governmental setting. Consider in this regard Cooter's discussion of the doctrine of duress, typified by the case of *Austin v. Loral.* What sorts of strategic behavior might arise from granting the courts discretion to invalidate contracts on grounds of duress? Are the costs of such behavior greater or less than the costs that would result from abandoning duress as a legal defense?

3. Consider Cooter's suggestion that if Hobbesian rather than Coasian behavior is the norm, legal institutions should be designed to minimize the total social costs of breakdowns in

bargaining. It is not obvious how best to promote this goal, however, given that the cost of a breakdown can influence its probability. If the parties know that the consequences of failing to agree are very severe, for instance, they may act in a more cooperative fashion than otherwise. Whether shielding individual bargainers from the consequences of an impasse reduces social costs thus depends on whether the probability of an agreement increases more or less than proportionately with the severity of the resulting loss. Which aspect of bargaining inefficiency do you think is likely to be dominant: probability or severity? Is it better to protect the parties from the costs of bargaining failure, or to charge them with those costs?

4. The outcome of the Cooter/Marks/Mnookin model illustrates what game theorists call *Nash equilibrium,* after the mathematical economist John Nash. A Nash equilibrium arises when each party to a strategic situation chooses a strategy that is a best response to the other parties' strategies, given the information available. In the case of single-offer sealed bids, hedging between a tough and soft offer is a best response given the variety of persons one might be negotiating against. The informational assumption is critical, however, for if one knew that the opponent had turned in a tough bid, the optimal response would be to turn in a soft one, and vice versa. Accordingly, while Nash equilibrium is generally regarded as a likely outcome in situations where the parties move simultaneously and in ignorance of each others' choices, it may not be so likely if one party is able to choose first, or to commit in advance to a particular strategy. For this reason, a substantial amount of strategic behavior in negotiation may be devoted to jockeying for temporal advantage. A classic example of this is provided by the game colloquially called "Chicken," in which victory goes to the party who can persuade his opponent to be the first one to yield in the face of some impending mutual disaster. For a general introduction to the Nash criterion, as well as a discussion of some of the refinements that have been suggested for it, see David Kreps, Nash Equilibrium, in *The New Palgrave: A Dictionary of Economics,* ed. J. Eatwell, M. Milgate, and P. Newman (London: MacMillan, 1987), vol. 3, 584–88. For a discussion of the game of "Chicken," see Baird, Gertner, and Picker, supra note 1, pp. 43–44.

5. Strategic issues can also arise when one person's decision about precaution influences the costs and benefits of another's; a good example of this can be found in the case of joint tortfeasors. For instance, in *Summers v. Tice,* 33 Cal. 2d 80, 199 P.2d 1 (1948), Summers was injured while shooting quail when Tice and a third hunter simultaneously fired in Summers' direction. Summers recovered damages even though he could not show whose shot had hit him. Does this rule of liability give defendants in a joint venture the best incentives for precaution? Does it matter if the tortfeasor who is sued is entitled to contribution from the one who is not? See William Landes and Richard Posner, Joint and Multiple Tortfeasors: An Economic Analysis, 9 *Journal of Legal Studies* 517 (1980); Lewis Kornhauser and Richard Revesz, Settlements under Joint and Several Liability, 68 *New York University Law Review* 427 (1993).

6. Why precisely are Leff's hypothetical business adversaries able to settle their differences without resort to legal process? Are such methods available to individual consumers in their disputes with merchants or with each other? Why or why not? See Lisa Bernstein, Opting Out of the Legal System: Extralegal Contractual Relations in the Diamond Industry, 21 *Journal of Legal Studies* 115–57 (1992) [discussing roles of reputation and of anticipated future dealings within mercantile community].

7. Note that in the strategic situations described by Leff and by Rosenberg and Shavell, a great deal turns on timing. For instance, in the debtor-creditor context, if the creditor can commit to bringing suit before debtor defaults and if this commitment can be made public knowledge, the debtor will have the incentive to perform fully. Similarly, in the Rosenberg/ Shavell model, if defendants can commit publicly to take all cases to trial, frivolous suits will lose their profitability. Can you think of ways in which disputing parties could make such

commitments, and thus protect themselves against opportunism from the other side? Are there risks to making such commitments? Are they subject to strategic problems of their own? See Marc Galanter, Why the "Haves" Come Out Ahead: Speculations on the Limits of Legal Change, 9 *Law and Society Review* 95 (1974) [outlining various ways in which repeat players have advantages over one-shot participants in litigation.]

8. Issues of timing are also important in tort law, which provides special doctrines to deal with sequential interactions. For instance, under the doctrine of "last clear chance," a person who fails to take reasonable precautions to avoid an accident is liable for the damages that result, even if others could have avoided the accident by taking earlier, cheaper precautions. Is this doctrine efficient, or would it be better to apply the usual rule of comparative negligence? See Steven Shavell, Torts in Which Victim and Injurer Act Sequentially, 26 *Journal of Law and Economics* 589 (1983).

9. The incentive problem that Leff describes in the debtor-creditor context is potentially present in any contractual setting, since in most cases it will not be worthwhile to sue over a small shortfall in performance. Given this, should a buyer be entitled to reject goods that fail to conform in any respect, however trivial, to the contract description (the "perfect tender" rule of the common law of sales), or should a substantial defect be required to justify rejection? See George Priest, Breach and Remedy for the Tender of Non-conforming Goods under the Uniform Commercial Code: An Economic Approach, 91 *Harvard Law Review* 960 (1978) [suggesting that the perfect tender rule is an efficient response to the threat of seller opportunism].

10. As we introduce additional refinements to the economic approach, readers should reflect on whether the policy conclusions derived from simpler models need to be adjusted or abandoned. Consider in this regard the effect of strategic behavior on the optimal rule of liability. For instance, a major issue in the economic literature on contract law has been the idea of efficient breach—that is, that the law of damages should be designed to encourage promisors to breach their promises only when the benefits of doing so exceed the costs. The conventional wisdom has been that a rule of expectation damages does this most effectively, on the theory that if a promisor knows that she will have to make her promisee whole, she has the correct incentives to weigh all the relevant costs and benefits of a breach. Critics of this view have pointed out that if a contract is to be breached the parties can and ought to negotiate a cancellation, since this will better ensure that the promisee's interests are fully considered. Given Leff's account of post-contractual bargaining, which side in this debate do you think has the better argument? How will the rules for allocating the costs and benefits following breach affect any bargaining that takes place? Would your answer be any different in the tort or property setting—for example, in a case like *Boomer v. Atlantic Cement Co.*?

Refining the Model II:
Risk and Insurance

In the discussion so far, we have implicitly assumed that economic actors are risk neutral, that is, that they view an uncertain prospect as equivalent to its *expected value:*
the amount of any potential gain or loss multiplied by the probability of its occurrence. For instance, this assumption underlay Posner's discussion of the Hand formula for negligence; the conclusion that a burden B should be undertaken if and only
if it would prevent a loss L with probability p presumes that the expected value $p \times$
L is an appropriate measure of the potential cost of an accident. In general, however,
most people are somewhat averse to risk and willing to pay a premium to avoid it.
Accordingly, a reasonable person might prefer to incur a small cost of $10 in order to
avoid a 1 percent chance of a $900 accident, even though the accident has expected
value of only $9. For this reason, we need to refine our model further to take account
of the social fact that risk is a cost and its reduction is a benefit. Indeed, the efficient
allocation of risk is just as important to social welfare as the production and allocation of material goods.

The first reading in this chapter, by Steven Shavell, presents the economic theory
of risk allocation and insurance. Shavell shows that the phenomenon of risk aversion
can be understood as a consequence of the diminishing marginal utility of wealth; because rational individuals attend to their material needs in order of importance, they
get proportionately less value out of additional increments of money. As a result, the
downside portion of a risky prospect, which threatens the loss of relative necessities,

carries more weight than the upside portion, which merely promises the gain of relative luxuries. Because individuals differ in the extent of their aversion to risk, however, a bargain in which a less risk-averse person acts as insurer to a more risk-averse
one can make both better off ex ante. Similarly, individuals can improve their expected welfare by pooling the diverse risks they face. Shavell also shows that because
insurance undermines incentives to take precautions against insured-against dangers,
it is often optimal for insurers to provide less than full coverage. This tradeoff between efficiency in risk-bearing and efficiency in precaution incentives is called
moral hazard.

The advantages of risk spreading and insurance helps explain many legal and
commercial institutions. In tort law, insurance considerations are commonly offered
to justify strict liability for business defendants—what is sometimes called *enterprise
liability.* The standard argument for enterprise liability maintains that large tort defendants are likely to be less risk-averse than individual plaintiffs, better able to
spread losses over repeated transactions, and in a better position to monitor risks—
even when such defendants have acted without fault. The reading by Jon Hanson and
Kyle Logue surveys these standard arguments and offers a further rationale: that business defendants are better able to obtain accident insurance than are individual plaintiffs. The reason for this, in the authors' view, is that the transaction costs of moral
hazard and risk screening are greater in the market for first-party accident insurance,
which covers personal losses of the insured, than in the market for third-party liability insurance, which covers damages paid out in tort.

Richard Posner and Andrew Rosenfield apply the economic theory of risk-
bearing to the impossibility and impracticability doctrines of contract law. Insurance
is a fundamental issue in contract law, as the rule of expectation damages effectively
makes promisors into insurers against the risk of breach. Posner and Rosenfield argue that, just as privately marketed insurance policies typically exclude or limit coverage for certain specified types of risks, an optimal contract liability regime should
also contain limitations and exclusions. The doctrine of impossibility, in this view,
serves to exclude certain risks—those that are large, unforeseen, and out of the
promisor's control—from the standard insurance policy provided by expectation
damages. The authors argue that such an exclusion is efficient: because the risk is
large, it is best shared among the parties; because it is unforeseen and outside the
promisor's control, excusing him from responsibility for it poses little problem of
moral hazard.

Louis Kaplow examines the subject of compensation for governmental takings.
Kaplow presents the takings issue as a specific instance of the general category of legal transitions, which includes changes in tax, regulatory, and expenditure policy. In
his view, the functional problem in all these areas is that unanticipated changes in the
legal regime impose risks on private citizens, ranging from the physical taking of real
property to the losses suffered by owners of capital assets when tax deductions are
cut back. He argues that, from an economic viewpoint, risk resulting from govern-

mental actions is no different from other types of risk. The goal of transition policy, accordingly, ought to be to make sure that citizens have adequate access to insurance. Such insurance, however, must take account of moral hazard, in order to make sure that persons contemplating investments take proper account of the possibility of future legal reforms. Because insurance is ordinarily available on the private market, and because private insurance generally takes account of moral hazard while government compensation often does not, Kaplow concludes that the efficiency justification for governmentally provided transitional relief is weak.

Finally, A. Mitchell Polinsky and Steven Shavell discuss the consequences of risk aversion for the economics of punishment. In their article, Polinsky and Shavell analyze Gary Becker's claim that it is efficient to impose high fines with relatively low probability when enforcing a legal standard, rather than the other way around, in order to conserve on the costs of apprehending and prosecuting offenders. They point out that this arrangement is rarely observed in reality, and suggest that the explanation stems from the fact that individuals are risk averse. Risk aversion implies that large, infrequent fines impose social costs on those who are punished—costs that are not recouped by the state when it collects the amounts assessed. The authors conclude that if costs to offenders are considered as part of social welfare (which seems plausible, at least in the case of civil infractions), certain punishment may be more efficient than severe punishment, in addition to being more just.

5.1 THEORY

The Allocation of Risk and the Theory of Insurance
STEVEN SHAVELL

Risk Aversion and the Allocation of Risk

Assumption of Risk Aversion

In contrast to risk-neutral parties, risk-averse parties care not only about the expected value of losses, but also about the possible magnitude of losses. Thus, for instance, risk-averse parties will find a situation involving a 5 percent chance of losing 20,000 worse than a situation involving a 10 percent chance of losing 10,000, and this situation, in turn, they will find worse than a situation involving a sure loss of 1,000—

Steven Shavell, The Allocation of Risk and the Theory of Insurance, from *Economic Analysis of Accident Law* (Cambridge: Harvard University Press, 1987), pp. 186–199. Copyright © 1987 The President and Fellows of Harvard University. Reprinted with permission.

even though each of the situations involves the same expected loss of 1,000. (Risk-neutral parties would not find any one of the situations worse than any other.) Risk-averse parties, in other words, dislike uncertainty about the size of losses per se.

The assumption that a party is risk-averse turns out to be equivalent to a simple assumption concerning the utility the party attaches to his wealth. In particular, suppose that while the party's utility increases with the level of his wealth, it does so at a decreasing rate. . . . [Such a party] will especially dislike bearing the risk of large losses, for such losses will evidently matter to him disproportionately in terms of utility. To be precise, . . . [the] party is assumed to evaluate a risky prospect by measuring its effect on his expected utility. Expected utility is obtained by multiplying the utility of each possible consequence—here the utility of each possible level of wealth—by its probability. Calculations will show that [the party's] expected utility will be lower if he faces the 5 percent chance of a 20,000 loss than if he faces the 10 percent chance of a 10,000 loss, because a loss of 20,000 will result in more than twice the diminution in utility than will follow from a loss of 10,000. . . .

Importance of Risk Aversion with Regard to Individuals and Firms

The importance of risk aversion will ordinarily depend on the size of risk in relation to an individual's assets and to his needs. Thus it may make sense to think of a person with assets of $10,000 as quite averse to a risk of a $5,000 loss, especially if he will soon want to use (say, for medical or educational purposes) the greater part of his $10,000. But where a person with assets of $300,000 faces a $5,000 risk, risk aversion will likely be an unimportant factor, and it will usually do no harm to consider the person as risk neutral (although risk aversion would probably become relevant if the magnitude of the risk he faced was $200,000).

The attitude toward risk of firms will reflect the attitudes towards risk of their managers, employees, and shareholders. To the extent that the managers and employees of a firm are risk averse and that their rewards (or positions) are tied to the firm's performance, they will want the firm to behave in a risk-averse way. One would therefore expect there to be some tendency for firms to avoid risks jeopardizing their profitability or their assets. However, to the extent that shareholders hold well-diversified portfolios, they will not be much concerned about the risks borne by a firm (since the risks of different firms in a portfolio will tend to cancel one another). Consequently, shareholders will often wish firms to be operated in an approximately risk-neutral manner, and firms will be operated in that way insofar as shareholders exercise control over managers and employees.

Risk Aversion, the Allocation of Risk, and Social Welfare

The presence of risk-averse parties means that the distribution or allocation of risk will itself affect social welfare. Specifically, and assuming for convenience that social welfare is the sum of parties' expected utilities, the shifting of risks from the risk averse to the risk neutral or, generally, from the more to the less risk averse will raise social welfare. This is because the bearing of risk by the more risk averse would result in a greater reduction in their expected utility than will the bearing of risk by the less risk averse or by the risk neutral. Indeed, for this reason, it is always possible for the more risk averse to pay the less risk averse or the risk neutral to assume risk, so as to leave both better off in terms of expected utility. . . .

Social welfare is raised not only by the complete shifting of risks from the more to the less risk averse or to the risk neutral, but also by the sharing of risks among risk-averse parties. Sharing risks reduces the magnitude of the potential loss that any one of them might suffer. . . . [In general,] some unequal sharing of losses can always be shown to be beneficial where the parties are not equally risk averse. Also, . . . social welfare can in theory always be enhanced by bringing additional parties into a risk-sharing agreement (here we ignore the "transaction" costs of so doing), because this further reduces the size of the losses each might face. . . .

Remarks on the Allocation of Risk and Social Welfare

1. The proper allocation of risk raises social welfare not only directly, by reducing the risk borne by the risk averse, but also indirectly, by making the risk averse willing to engage in socially desirable, risky activities. Thus, for example, an individual may decide to undertake a promising business venture only because he has partners with whom he can share the risk.

2. Protection of the risk averse against risk is socially beneficial for reasons quite distinct from those appealing to the desirability of equity in the distribution of wealth. This is apparent, for instance, from the point that two risk-averse parties with equal levels of wealth (and therefore about whom there can be no questions concerning lack of distributional equity) may each be made better off, ex ante, by arranging to share a risk. That there should be a tendency to conflate the issue of distributional equity with that of the allocation of risk is no doubt engendered by the fact that after a party has suffered a loss he will be in a disadvantageous position relative to others in the absence of any risk-bearing agreement.

3. It should also be emphasized that the allocation of risk is in principle just as important a determinant of social welfare as the production of goods and services or the reduction of accident losses. The impression some may have that the conventional normative economic calculus is concerned only with the latter, "real" elements is incorrect.

The Theory of Insurance

Assumptions of the Theory

Under the arrangement known as insurance, parties referred to as insureds pay premiums to an insurer in exchange for protection against possible future losses. The insurer is obligated to pay insureds an amount specified by an insurance policy if the insureds make claims for losses they suffer.

In the analysis of insurance here it will be assumed that there are many risk-averse insureds facing identical, independent risks of loss and that there are essentially no administrative expenses associated with the insurer's operations. This assumption implies that the insurer can be virtually sure of covering its costs by collecting from each insured the expected value of the amount it will have to pay him. If, for instance, each insured faces a 5 percent risk of losing 10,000 and will be paid that amount in the event of a loss under the insurance policy, the insurer can cover its costs by collecting premiums of 500.

In the following sections the theory of insurance will be examined in two situations: where insureds cannot affect risks and where they can.

Expected Utility Maximizing Insurance Policies
Where Insureds Cannot Influence Risks

There are many types of loss the probability or severity of which insureds can do relatively little to alter. Consider, for instance, property damage suffered by insureds caused by objects dropped from airplanes or injuries sustained while insured[s] are under general anesthesia. Keeping such examples in mind, assume here that insureds have no influence over risks, so that their ownership of insurance policies cannot itself lead to a change in risks.

Although in principle insureds may own any type of policy—such as a policy covering only a percentage of losses, one imposing a low ceiling on payments, or one supplying greater than full coverage—the policy offering exactly full coverage against loss is optimal for an insured: the full-coverage policy will yield an insured higher expected utility than any other type of policy. The explanation for this result is twofold. First, obviously, a policy offering less than complete coverage will not protect an insured against risk, which, being risk averse, he will wish to avoid. Second, and perhaps not obviously, a policy offering more than complete coverage will require, in effect, that an insured make an added expenditure (in the form of a higher premium) to engage in a gamble (where "winning" will occur if he suffers a loss), and being risk averse, the insured will not wish to engage in such a gamble. . . .

Expected Utility Maximizing Insurance Policies
Where Insureds Can Influence Risks

Insured parties may be able to cause losses, and for this reason, in addition to the element of gambling mentioned above, it will be disadvantageous to insureds for insurance coverage to exceed losses. Were coverage to exceed losses, there would be an incentive for insureds to create losses (and an enhanced incentive to falsify claims), as where an owner of a building worth 20,000 arranges to burn it down to collect against a fire damage policy in the amount of 25,000. (A similar incentive would operate were someone who is not the owner of a building—someone who does not have an "insurable interest" in the building—able to insure it.) This behavior will raise insureds' premiums undesirably.

On the other hand, insured parties are, of course, also often able to lower the probability or the magnitude of the losses they might suffer. They may be able to reduce the risk of fire by purchasing safety devices or by exercising various precautions; they may be able to limit the losses arising from an accident by taking remedial actions; and so forth. The question therefore arises whether insureds will in fact act in these ways, and in particular whether their ownership of insurance might not dull their incentives to do so. (Such dulling of incentives is called the *moral hazard* in the insurance literature.)

To answer this question, suppose first that insurers can obtain perfect information at no cost about insureds' risk-reducing actions (whether an insured installed fire extinguishers). In this case there will be no problem of adverse incentives created by insureds' ownership of coverage because insurers will be able to link the terms of policies to insureds' actions: insurers may make premiums depend on insureds' risk-reducing actions (as where premiums are lowered for those who install extinguishers); and insurers may limit or deny coverage in the event of a loss if insurers determine that insureds' actions did not conform to policy requirements (as where coverage is not paid to those who failed to install extinguishers or who did not prevent greater losses by moving valuable property away from a fire). . . .

Assume now that insurers are not able to obtain information about insureds' risk-reducing actions (that insurers cannot tell whether insureds discard oily rags and take other precautions to prevent fire). In this case insurers plainly cannot make premiums or other policy terms depend directly on insureds' actions, and insureds therefore cannot be rewarded with premium reductions for reducing risk nor be punished by denial of their claims for failing to do so. Hence insureds' ownership of insurance will affect their incentives to reduce risk. If insureds possess complete coverage, the problem will be most serious, for they will then have no reason to avoid losses (here we ignore nonpecuniary losses). Thus premiums for coverage will be high and insureds' utility low. This problem may often be ameliorated if insureds have only partial coverage, for because insureds will bear a fraction of their losses and have some motive to avoid losses, insurers' costs and premiums charged will be lower. . . . In general, the level of coverage provided in policies that maximize expected utility under present assumptions are those that strike the best balance between the advantage and the disadvantage of lowering coverage from full coverage, namely, those that balance the creation of incentives to reduce risk on the one hand and the increased exposure to risk on the other. . . .

Further Remarks on Expected Utility Maximizing Policies

1. It should be emphasized that expected utility maximizing insurance policies are the policies that will be sold in a competitive insurance market, since parties will prefer to purchase these policies over any alternatives. In addition, they are the policies that will be sold by a self-financed social insurer that desires to maximize the welfare of insureds.

2. It is clear from the analysis in [the preceding subsection] that the expected utility of insureds will be higher where insurers have information about their actions than where insurers do not, for in the former but not the latter case full coverage can be provided while at the same time insureds can be supplied incentives to reduce risk. Thus insureds will want insurers to obtain information (by means of inspections and the like) about their risk-reducing activity and should be willing to pay (directly or in the form of a higher premium) for that to be done if the cost is not too high. The cost, however, will vary. One suspects that as a general matter it will often be relatively

easy for insurers to learn about safety features and other "fixed" physical characteristics that affect risk, but it will frequently be difficult for them to monitor effectively
insureds' precautionary behavior (such as drivers' attention to the road) since this is
so readily modified.

A more complete analysis of the foregoing would recognize that information
about insureds' behavior may be inaccurate. (Despite appearances, was the fire really
due to the insured's failure to put out his cigarette?) Such analysis would conclude
that under the expected utility maximizing policy, use of imperfect information will
be limited (evidence pointing to the cigarette will result in only a moderate reduction
in payment to the insured), because if a policy were to depend on possibly erroneous
information in a substantial way (notably, so as to lead to complete denial of payment)
insureds would be subjected to too great a risk due to the chance that their true behavior would be mistakenly assessed. Similarly, an insured's loss history may be a
faulty indicator of the care he has taken (he can have suffered his losses through bad
luck). If so, the use of loss history in the expected utility maximizing policy will be
restricted.

Another point related to the analysis in [the preceding subsection] is that insureds
as a group should be willing to pay insurers to develop information about risk (as
where fire insurers determine the flammability of different types of furnishings and
building materials and provide advice to insureds). This will allow insureds to benefit through premium reductions and, if they are not fully insured, through a reduction
in their expected out-of-pocket losses. Of course, insureds could develop the information on an individual basis (each insured could ascertain for himself the flammability of various furnishings), but it will generally be more efficient for insurers to
carry out this function for the whole group of insureds.

3. If it is supposed, realistically, that there are positive administrative costs associated with the supply of insurance, then the premium must include a component to
cover these costs and the nature of expected utility maximizing insurance policies will
change. Specifically, the policies will involve less coverage than will otherwise be
desirable and, often, will include a deductible feature (according to which the insured
receives only the excess of his loss over the deductible amount). Policies with a deductible have the virtue that they reduce insurers' administrative costs because losses
falling below the deductible do not lead to claims. At the same time, the policies protect insureds against the major portion of large losses, which is what insureds care
about most, being risk averse.

4. If it is supposed, contrary to what was true in the analysis, that the risks facing
different insureds are correlated with each other or large (as with earthquakes, nuclear
war, design defects affecting most units of a widely distributed product, or major
changes in liability law), then insurers cannot be confident of meeting their costs by
charging premiums equal to the expected value of the payments they have to make.
Insurers will therefore wish to charge higher premiums, presuming some degree of
risk aversion on their part, and insureds' expected utility will be maximized if they
obtain less coverage than otherwise.

5. Where the correlation or size of risks, administrative costs, or problems with
incentives are sufficiently important, the expected utility maximizing amount of coverage may be none at all. That is, it may turn out that parties do not have any insur-

ance against certain types of loss. For example, they do not generally have insurance against business losses, because of the problems with incentives that would arise were they to own such insurance.

6. If insureds or insurers misperceive risks, the amount of insurance coverage purchased may be inappropriate. Insureds who overestimate risks will tend to buy too much coverage (because premiums will appear low to them), and those who underestimate risks will buy too little. Similarly, insurers who underestimate risks will charge too little for coverage so insureds will tend to buy too much, and those who overestimate risks will charge too much so insureds will buy too little. The latter possibility may help to explain (along with the correlation of risks and the risk aversion of insurers) why insurance is sometimes not sold where the risks are hard for insurers to estimate.

Actual Insurance in View of the Theory

The character of insurance policies and of insurance practice seems to comport with the theory presented in this part. When parties have no real influence over risks (as, for instance, with the possible loss of goods during transport by others), insurance policies are relatively simple and parties frequently purchase complete coverage. When parties can influence risks, insurers often supply advice about risk reduction and include in policies a great variety of features that serve to induce insured parties to lower risk. Moreover, as the discussion in [the preceding subsection] should have suggested, a detailed examination of these policy features (where premiums and conditions of payment are based on information about insureds; where coverage is partial or is excluded; and so forth) would further confirm the theory. . . .

5.2   APPLICATIONS

The First-Party Insurance Externality: An Economic Justification for Enterprise Liability
JON HANSON AND KYLE LOGUE

. . . Law-and-economics scholars generally agree that an efficient products liability regime would accomplish two principal economic goals. First, it would encourage parties to prevent all preventable accidents (the "deterrence" goals). Second, it would efficiently allocate the risk of unprevented accident costs (the "insurance" goal).

. . . Consumers are insured through first-party mechanisms against most of the

Jon Hanson and Kyle Logue, The First-Party Insurance Externality: An Economic Justification for Enterprise Liability, 76 *Cornell Law Review* (1990), pp. 129–196. Copyright © 1990 Cornell University. Reprinted with permission.

risks of product accidents. However, first-party insurers rarely and imperfectly adjust premiums according to an individual consumer's decisions concerning exactly what products she will purchase, how many of those products she will purchase, and how carefully she will consume them. . . . This failure by first-party insurers to adjust premiums according to consumption choices gives rise to a first-party insurance externality. . . .

Optimal Product Risk Insurance: First-Party Insurance versus Enterprise Liability

The Insurance Goal: Allocating Risk

A tort regime's ability to allocate the risks of unprevented product accidents may be as important a determinant of that regime's overall efficiency as is its ability to deter product accidents. The proper allocation of risk increases social welfare directly by reducing the risk borne by the risk averse. It also increases social welfare indirectly by permitting the risk averse to engage in socially desirable, risky activities and by permitting them to use the assets they would otherwise reserve to offset potential losses.

If transaction costs were zero and if insurers and consumers were perfectly informed as to each individual consumer's expected accident costs, risk-averse consumers would willingly pay an actuarially fair insurance premium to protect themselves fully against the uncertain costs of product accidents. Under these assumptions, insurers could write and monitor fully specified policies that would induce insureds to invest optimally in accident prevention. Under the more realistic assumptions of positive transaction and information costs, however, the provision of full insurance against product risks will, for several reasons, create a loss of social welfare.

Moral Hazard. First, to the extent that insurers do not perfectly adjust premiums to reflect the actual risks generated by each insured's consumption choices, insureds will not fully internalize expected accident costs and, consequently, will not invest efficiently in prevention. An individual insured will tend to consider only what she must pay out-of-pocket rather than the total costs. In the long run, the total costs will be included in raised premiums, but the extra cost will be spread over all policyholders. . . . Therefore, all insureds end up paying more for insurance and having more accidents than they would if they were required to weigh the full costs to themselves of an accident-prone activity against the benefits they receive. . . .

Cross-Subsidization. To the extent that insureds who present different levels of risk (expected damages) are charged the same premium and lumped into the same insurance pool, low-damage insureds will cross-subsidize high-damage insureds. Consequently, the former will pay more (and the latter will pay less) than the efficient amount towards insurance. An allocatively efficient insurance regime would charge insureds competitively according to their individual risk.

Adverse Selection and Unravelling. Another potential welfare loss associated with full insurance occurs when potential insureds who know that they pose above average risk "self-select" into insurance pools. This "adverse selection" is a function of (1) the insurer's being unable to classify insureds perfectly according to each insured's expected damages, and (2) the insureds' knowing how their own expected damages compare to the average expected damages of the insurance pool. It depends, in other words, on asymmetrical information regarding expected damages. Under these circumstances, an insurer cannot perfectly control the variance of risk pools. The greater the heterogeneity or variance of risks allowed in an insurance pool, the greater the tendency will be of high-damage individuals to opt into the pool and of low-damage individuals to opt out. Adverse selection creates a social welfare loss by raising the pool's average risk and thereby forcing low-risk individuals to choose between paying disproportionately high premiums or foregoing insurance. . . .

Meeting the Insurance Goal: First-Party Insurance versus Enterprise Liability

Copayment Features. To limit possible risk-pool inefficiencies, first-party insurers often introduce copayment features, such as deductibles and coinsurance, into insurance arrangements. Those copayment features, by requiring insureds, ex ante, to include some portion of the accident costs in their decision calculus, help to align the incentives of insureds and insurers and thereby mitigate risk-pool inefficiencies. Deductibles require insureds to pay up to some set portion of their accident expenses (for example, the first $175) before the insurer will pay all or some fraction of the remainder. Coinsurance arrangements, on the other hand, provide that the insurer pays only a fraction (for example, seventy-five percent) of an insured's total losses. Note, however, that first-party insurance contracts typically include "out-of-pocket limits," which place a cap on the amount of copayments an insured can be required to pay per year. Recent critics of enterprise liability have argued that moral hazard would abound under such a system because manufacturers supposedly cannot employ copayment arrangements. Enterprise liability, however, contains features that approximate copayment arrangements; consumers, under enterprise liability, bear some portion of their product-accident losses. For example, consumers making consumption choices will discount potential product liability awards by the probability of losing a suit and will subtract from their expected awards the costs of litigation.

Suppose a consumer, in deciding whether to take optimal or suboptimal care when using a particular product, faces an expected accident cost of $10,000 should she take suboptimal care. Suppose further that she considers her chances of winning an enterprise liability suit, should she actually be injured, to be eighty percent, and the cost of litigating such a suit to be $3000. Under those circumstances, our consumer would be insured for up to $5600 of her expected product-accident cost. Therefore, she would, ex ante, internalize $4400 in copayments. Furthermore, she will be even less willing to take inefficient risks because she is uncertain of the actual amount of compensation she will receive and is risk averse. Finally, to the extent that consumers are not fully compensated for nonpecuniary losses, the risk of such losses will also mitigate moral hazard. . . .

Risk Classification: The Insurance Implications of the First-Party Insurance Externality. As already explained, the broader the risk pools, the greater will be the cross-subsidization and the tendency toward adverse selection and unravelling. Hence, the more narrowly insurers can define risk pools, the more efficient the insurance mechanism will be. In choosing between first-party insurance and enterprise liability qua insurance the goal should be to pick the scheme that—other things equal—best classifies insureds according to their expected accident costs. Insurers, to earn higher profits, have an incentive to classify insureds according to expected accident costs and to adjust premiums accordingly. By doing so, an insurer can offer low-damage insureds lower premiums, while charging high-damage insureds higher premiums.

An insured's expected accident costs are a function of her consumption choices and individual-risk characteristics. We examine each of these in turn.

Classifying According to Consumption Choices. . . . [F]irst-party insurers make little or no effort to segregate insureds according to consumption choices. . . . It is true that some first-party insurers for some types of insurance segregate consumers who use certain exceptional products. Some insurers attempt to classify according to whether applicants smoke cigarettes, abuse alcohol, use controlled drugs, or engage in extremely dangerous activities (e.g., scuba diving below forty feet or hang gliding). These are consumption choices whose expected costs are high enough to significantly affect an insured's total expected losses; they also are products whose use correlates highly with other risky activities. . . .

The extent to which this sort of classification narrows risk pools, however, is de minimis for several reasons. First, only a small percentage of insurers attempts to classify in this way. Most health insurance, for example, is provided through private-sector group insurers such as Blue Cross/Blue Shield, through prepaid plans such as health maintenance organizations, or through public programs such as Medicaid or Medicare. For these forms of insurance, insurers virtually never require smokers to pay higher premiums than non-smokers. Second, those insurers who do attempt to classify insureds according to whether they smoke do not classify according to how often, how long, or what brand of cigarette an insured has smoked. . . . Even if these general classifications serve a segregative purpose, the ease with which insureds can falsify responses further attenuates their usefulness. More important, other than for these few exceptional products, insurers do not even attempt to classify insureds according to their consumption choices. Thus, insurers do not raise premiums for insureds who have particularly hard and slippery bath tubs nor lower premiums for those who use clothes irons that shut off automatically.

There are several reasons why first-party insurers fail to classify insureds efficiently according to consumption choices, notwithstanding the putative benefits of risk classification. First, an insurer will only classify risks when the benefits to the insurer of further classification exceed the costs. The cost to first-party insurers of analyzing the safety characteristics of every consumer product and the consumption choices of every insured may be high enough alone to prevent such classification. By contrast, manufacturers have relatively inexpensive access to and control over information regarding the risks inherent in their products. For this reason, it may be more efficient to put the insurance burden on manufacturers by adopting an enterprise liability regime than to put it on first-party insurers through a negligence regime. . . .

Furthermore, even when classifying insureds according to product use is cost-justified for insurers, actual investments in determining product risks will be inefficiently low. Individual insurers would be unable to reap the full return on their investments in information because the information they generate cannot be protected. The information will, by necessity, be revealed in the insurer's applications and policies. Suppose, for example, Insurer One learns through costly investments in research that consumers who use Type A lawnmowers face an increase in expected accident costs of $50 per policy period over consumers who use Type B lawnmowers and a $100 increase over those consumers who avoid using lawnmowers altogether. When the insurer offers a $50 discount to insureds who use Type B lawnmowers and a $100 discount to insureds who do not use lawnmowers at all, Insurer One's competitors will simply replicate Insurer One's policy discounts. The insurer cannot force its competitors to pay for the valuable information, so the insurer gives away the fruits of its costly research. Free-riding competitors, because they can costlessly acquire the classification information without the investor's consent and without compensating the investor, can offer insureds comparatively low rates. The problem of free-riding competitors would likely be much less severe for manufacturers under an enterprise liability regime.

There is an important, practical reason why insurers do not adjust premiums according to consumption choices. Suppose insurers knew that the expected accident costs to an insured of consuming, say, a 6.5 ounce can of chunk light tuna packed in spring water was ten cents. How could insurers incorporate that information into their policies? The insurer might, through its application, ask the insured how many of those cans of tuna she expects to eat during the policy period. The insurer could then adjust the premium accordingly. If the consumer is a tuna lover (or just a big eater) and expects to eat 100 cans, her premium will be raised by $10. But if her premium is raised just because she loves tuna, then she will have an incentive to understate her expected level of tuna consumption. Because the insurer has no way of cost-justifiably monitoring the insured's consumption of tuna, the information about the expected accident costs of tuna will have little or no value to the insurer. . . .

Notice, however, that an enterprise liability regime could adjust insurance premiums through the market price of the product at a much lower cost—indeed, at practically no cost. Manufacturers would simply add 10¢ to the cost of every 6.5 ounce can of chunk light tuna packed in spring water, and consumers would automatically pay an adjusted insurance premium with every can they buy.

Classifying According to Individual-Risk Characteristics. We have argued that first-party insurance fails to classify insureds according to the products they consume, how many of those products they consume, and how carefully they consume them. We have also offered several reasons to explain why risk pools under first-party insurance, unlike those that would exist under enterprise liability, are inefficiently broad with regard to consumption choices. . . .

One could respond to this conclusion by arguing that first-party insurance, unlike enterprise liability qua insurance, segregates consumers according to other relevant individual-risk characteristics such as age or income. A consumer's income is relevant in the products liability context because, for the same product-caused disability, the lost-income component—typically one of the two largest components of products liability damages—may be greater for consumers with higher incomes. Under an enterprise liability regime, there is a potential inefficiency: low-income consumers (or more generally, consumers who would have relatively low damages if injured) would, in buying a product, subsidize high-income (or high-damage) consumers. Such cross-subsidies would not only be inefficient, they would also be distributively unjust. The poor and low-income person would be required to pay the same price up front but would, on average, receive less compensation were they later injured by the product. . . .

The first problem [with this conclusion] is that most first-party insurers do not make a significant effort to discriminate among insureds according to their individual-risk characteristics (again, these are characteristics besides consumption choices that affect the expected injury costs of insureds). A majority of health insurance policies, for instance, are provided through large group policies as an employment benefit. Those group policies provide the same coverage for the same premium to very large groups of individuals. Thus, there is likely a significant variance of risk characteristics among the insureds of such pools. The same is likely true of life insurance, roughly forty percent of which also is provided through employer-employee (and other general) group plans. Absent the necessary empirical support, scholars are wrong to merely assert that first-party insurers can (or do) optimally segregate insureds according to their individual-risk characteristics. . . .

Enterprise Liability: More Efficient than Scholars Assert. . . . The conventional wisdom is that enterprise liability is inefficient because it leads to cross-subsidies and adverse selection and that it is unjust because, in essence, it requires low-income consumers to pay a regressive tax when they buy a product. This view depends entirely on two (typically unstated) assumptions: (1) that a manufacturer cannot design, package and market its product so that the product will be consumed by individuals with relatively homogeneous individual-risk characteristics; and (2) that a manufacturer must charge the same price to all consumers regardless of the consumers' individual-risk characteristics. These assumptions, however, are based on an erroneous oversimplification of consumer-product markets.

For starters, it seems unimaginable that any product would ever be consumed by a random selection of the general population. Casual empiricism suggests that most products are consumed by relatively homogeneous subsets of the population—as defined by income, gender, age, and the like. Consider the injuries—whatever they might be—that bowling balls and golf balls cause their consumers. . . . For instance, it is probably true, though there are undoubtedly exceptions, that golfers tend to have higher incomes than bowlers. Furthermore, even among golfers, different brands of golf balls attract still narrower groups of consumers. Relatively low-income golfers tend to use Wal-Mart Floaters, whereas high-income golfers tend to use Titlest [sic] X-100 Blacks. Similarly, female golfers would be more likely to use Lady Titlests, whereas male golfers would likely use high-compression Golden Rams. This point can be stated more generally: Because demand for any product is to a greater or lesser extent a function of consumer characteristics such as income, age, gender, and taste, most products will, in terms of those characteristics, be consumed by some nonrandom portion of the population at large.

It should be remembered that not only do products naturally attract relatively homogeneous consumer groups, but manufacturers actually design their products with a view to attracting particular sets of consumers. Manufacturers do not produce products that are perfectly homogeneous and perfectly substitutable. Instead, each manufacturer differentiates its product or brand of product from that of other manufacturers. . . . Manufacturers can, through market research, collect much the same data concerning consumers' risk characteristics that first-party insurers collect through insurance applications. Therefore, manufacturers can with at least some success segregate consumers by income, age, gender, taste, etc., such that a reasonably homogeneous group of consumers uses each distinct product. . . . [Furthermore,] through the use of rebates or coupons manufacturers not only can target very narrow subsets of consumers but also can offer larger or smaller discounts to different consumers according to those consumers' individual-risk characteristics.

Another way in which manufacturers can segregate consumer risk pools is through the use of warnings. By alerting those consumers especially susceptible to the product's risks, a manufacturer may be able to discourage high-risk consumers from using their product. Consider, for example, the warnings to pregnant women regarding the particular dangers posed by their smoking.

Finally, it is worth observing . . . that manufacturers not only can segregate consumers according to risk but also can lower the risk facing consumers by improving their products' safety. To enhance safety, manufacturers could, for instance, alter their manufacturing processes or the designs of their products. Because manufacturers would be liable for all the costs under an enterprise liability regime, they would have optimal incentives to ascertain the costs and benefits of alternative designs.

Summary

We have argued that, contrary to the received wisdom, first-party insurers fail, for several reasons, to segregate consumers efficiently according to consumption choices. Additionally we have argued further that first-party insurance is probably less effective, and enterprise liability is probably more effective, at segregating according to individual-risk characteristics than has previously been supposed. Whether enterprise liability qua insurance or first-party insurance actually creates narrower risk pools (with regard to individual-risk characteristics) is an empirical question that is beyond the scope of this Article. Indeed, many of the relevant factors do not, at least for the time being, lend themselves to empirical evaluation. The overall efficiency of enterprise liability qua insurance is currently indeterminate; we have argued simply that it is plausible (especially given the failure of first-party insurers to adjust premiums according to consumption choices) that enterprise liability will be superior in this regard. . . .

Product Accident Deterrence

The Deterrence Goal

The deterrence goal of products liability law is to minimize the costs of product accidents, including the costs of preventing accidents. Consistent with standard models in the economics of tort law, we assume that the costs of product accidents are a function of three variables: the care taken by producers in manufacturing a product ("manufacturer care levels"); the care taken by consumers in using a product ("consumer care levels"); and the quantity of a product purchased and sold ("activity levels"). These variables will be optimized (i.e., the costs of product accidents will be minimized) only when both manufacturers and consumers internalize the total accident costs that a product causes. Only then will both manufacturers and consumers have incentive to invest in lowering accident costs—by increasing care levels and/or decreasing activity levels—up to the point at which the marginal costs to society of such investments equal the marginal benefits. . . .

A "No Liability" Regime Assuming Perfect First-Party Insurance

. . . How would it affect the deterrence analysis to assume, more realistically, that consumers are covered for these risks through first-party mechanisms?

Assume initially that, although contracting costs exist, first-party insurers have free access to product risk information and that they are therefore able costlessly to segregate risk pools. Under these assumptions, insurers will perfectly adjust premiums according to every consumption choice of each insured. Consumers will then be induced through first-party insurance premiums to internalize the expected accident costs associated with each of their consumption choices. Optimal deterrence will obtain.

Note that by adjusting premiums according to consumption choices (i.e., what products an individual consumes, how many of those products she consumes, and how much care she takes when consuming them) the insurer sees to it that all elements of deterrence (i.e., manufacturer care, activity levels, and consumer care, respectively) are optimized. For any product, each consumer will internalize the product's total price by paying to the manufacturer the product's market price and by paying to the insurer (in the form of an insurance premium) the expected accident cost resulting from the consumer's consumption choices. As before, consumers internalizing a product's total price will prefer to pay the lowest possible total price. Again, consumers will make all cost-justified investments in consumer care. Thus consumer care levels will be optimized. Manufacturers, competing for consumers, will help lower the total price of their products by making all cost-justified investments in their products' safety. Thus manufacturer care levels will be optimized. Just as they did in the absence of insurance, consumers—now through both the market price and insurance premiums—will internalize the total costs of consuming each additional unit of a product. Consequently, they will consume until the marginal cost of the product equals the marginal benefit. Thus activity levels will be optimized. In sum, perfect first-party insurance against product risks would facilitate, not impede, optimal deterrence.

A "No Liability" Regime Given the First-Party Insurance Externality

. . . The ability of consumers to externalize the costs of product accidents significantly undermines deterrence goals. First, consumer care levels will be suboptimal. Consumers will be unwilling to invest in consumer care because they will be compensated by first-party insurers for their injuries; taking care would impose upon consumers a cost with no offsetting benefit. Consumers, therefore, will have no incentive to take care. For the same reasons, consumers will be unwilling to pay any additional price to compensate manufacturers for investments in care. Competitive manufacturers will refrain from making cost-justified investments in care. Thus manufacturer care levels will be suboptimal. Finally, because consumers externalize accident costs, they will consume beyond the point at which the marginal costs of the product equal the marginal benefits. Activity levels will therefore be too high. The failure of first-party insurers to classify according to insureds' consumption choices, then, ensures that none of the deterrence objectives will be met: activity levels will be too high, care levels too low. . . .

The Optimal Rule: Enterprise Liability without a Contributory Negligence Defense

Some scholars have argued that enterprise liability in conjunction with the defense of contributory negligence will optimize all deterrence objectives, including consumer care levels. Assuming courts can accurately determine when a consumer has been contributorily negligent, and that consumers fully internalize losses imposed on them by courts, enterprise liability plus contributory negligence would indeed be the optimal products liability rule. Under such a rule, manufacturers would prevent initially preventable accidents and insure consumers against unpreventable accidents, and consumers would prevent residually preventable accidents. Scholars supporting such a regime, however, have failed to consider that consumers will still be insured by first-party insurers against these risks. Therefore, even if found contributorily negligent by a court, consumers will be compensated by their insurers. As a result, the consumer care problem will persist. That is, the use of contributory negligence does not alter the fact that first-party insurance allows consumers to externalize residually preventable accidents. Contributory negligence, then, serves little or no useful function because of the first-party insurance externality. . . .

Enterprise liability without the defense of contributory negligence, on the other hand, imposes upon manufacturers the costs of residually preventable accidents. Manufacturers will therefore internalize these costs and will prevent the subset of residually preventable accidents that are preventable by manufacturers. Put differently, those accidents that are not initially preventable but are nonetheless (cost-justifiably) preventable by increasing manufacturer care will be prevented if, and only if, there is no defense of contributory negligence. Given that consumer care levels cannot be optimized, disallowing a contributory negligence defense optimizes manufacturer care levels. The first-party insurance externality essentially raises the optimal care level of manufacturers above what it would be in the absence of first-party insurance externality.

For these reasons, enterprise liability without contributory negligence is the liability rule that most efficiently deters product accidents. . . .

Impossibility and Related Doctrines in Contract Law: An Economic Analysis

RICHARD POSNER AND ANDREW ROSENFIELD

Ordinarily the failure of one party to a contract to fulfill the performance required of him constitutes a breach of contract for which he is liable in damages to the other party. But sometimes the failure to perform is excused and the contract is said to be discharged rather than breached. This study uses economic theory to investigate three closely related doctrines in the law of contracts that operate to discharge a contract: "impossibility," "impracticability," and "frustration." . . .

Impossibility and Related Doctrines: Basic Principles

As Perceived in Conventional Legal Scholarship

Conventional legal scholars who have dealt with the discharge cases have indicated a pervasive dissatisfaction with the prevailing doctrinal articulations. Their uneasiness may reflect an inability to develop a coherent positive theory consistent with the typical outcomes in the recurring cases, and may be responsible for the tendency of courts and commentators alike to treat the field as too broad and diverse to be adequately understood within a single theoretical framework. Thus, in an effort to make descriptive treatment of the case law more manageable, the subject has been subdivided. The principal subdivisions are "impossibility of performance," "frustration of purpose," and "extreme impracticability," and each is thought to merit separate analytical treatment. "Impossibility" is the rubric used when the carrying out of a promise is no longer "physically possible," and "frustration of purpose" when performance of the promise is physically possible but the underlying purpose of the bargain is no longer attainable. Impracticability—a catch-all for any discharge case that does not fit snugly into either the impossibility or the frustration pigeonhole—is the term used when performance of the promise is physically possible and the underlying purpose of the bargain achievable but as a result of an unexpected event enforcement of the promise would entail a much higher cost than originally contemplated.

This categorization is unhelpful. It is true that some contracts could be performed only at infinite cost, and these, though rare (Professor Corbin's use of a promise to supply a trip to the moon to illustrate impossibility showed a certain want of imagination), are cases of true physical impossibility. But whether the cost is infinite, or merely prohibitive relative to the gains from performance (as in the impracticability and frustration cases), is a distinction without relevance to the purposes of contracts

Richard Posner and Andrew Rosenfeld, Impossibility and Related Doctrines in Contract Law: An Economic Analysis, 6 *Journal of Legal Studies* (1977), pp. 83–118. Copyright © 1977 University of Chicago Law School. Reprinted with permission.

or contract law. There is thus no functional distinction between impossibility and frustration cases on the one hand and impracticability cases on the other. In every discharge case the basic problem is the same: to decide who should bear the loss resulting from an event that has rendered performance by one party uneconomical. . . .

Economic Principles

. . . The typical case in which impossibility or some related doctrine is invoked is one where, by reason of an unforeseen or at least unprovided-for event, performance by one of the parties of his obligations under the contract has become so much more costly than he foresaw at the time the contract was made as to be uneconomical (that is, the costs of performance would be greater than the benefits). The performance promised may have been delivery of a particular cargo by a specified delivery date—but the ship is trapped in the Suez Canal because of a war between Israel and Egypt. Or it may have been a piano recital by Gina Bachauer and she dies between the signing of the contract and the date of the recital. The law could in each case treat the failure to perform as a breach of contract, thereby in effect assigning to the promisor the risk that war, or death, would prevent performance (or render it uneconomical). Alternatively, invoking impossibility or some related notion, the law could treat the failure to perform as excusable and discharge the contract, thereby in effect assigning the risk to the promisee.

From the standpoint of economics—and disregarding, but only momentarily, administrative costs discharge should be allowed where the promisee is the superior risk bearer; if the promisor is the superior risk bearer, nonperformance should be treated as a breach of contract. "Superior risk bearer" is to be understood here as the party that is the more efficient bearer of the particular risk in question, in the particular circumstances of the transaction. Of course, if the parties have expressly assigned the risk to one of them, there is no occasion to inquire which is the superior risk bearer. The inquiry is merely an aid to interpretation.

A party can be a superior risk bearer for one of two reasons. First, he may be in a better position to prevent the risk from materializing. This resembles the economic criterion for assigning liability in tort cases. It is an important criterion in many contract settings, too, but not in this one. Discharge would be inefficient in any case where the promisor could prevent the risk from materializing at a lower cost than the expected cost of the risky event. In such a case efficiency would require that the promisor bear the loss resulting from the occurrence of the event, and hence that occurrence should be treated as precipitating a breach of contract.

But the converse is not necessarily true. It does not necessarily follow from the fact that the promisor could not at any reasonable cost have prevented the risk from materializing that he should be discharged from his contractual obligations. Prevention is only one way of dealing with risk; the other is insurance. The promisor may be the superior insurer. If so, his inability to prevent the risk from materializing should not operate to discharge him from the contract, any more than an insurance company's inability to prevent a fire on the premises of the insured should excuse it from its liability to make good the damage caused by the fire. . . .

The factors relevant to determining which party to the contact is the cheaper insurer are (1) risk-appraisal costs and (2) transaction costs. The former comprise the costs of determining (a) the probability that the risk will materialize and (b) the magnitude of the loss if it does materialize. The amount of risk is the product of the probability of loss and of the magnitude of the loss if it occurs. Both elements—probability and magnitude—must be known in order for the insurer to know how much to ask from the other party to the contract as compensation for bearing the risk in question.

The relevant transaction costs are the costs involved in eliminating or minimizing the risk through pooling it with other uncertain events, that is, diversifying away the risk. This can be done either through self-insurance or through the purchase of an insurance policy (market insurance). To illustrate, a corporation's shareholders might eliminate the risk associated with some contract the corporation had made by holding a portfolio of securities in which their shares in the corporation were combined with shares in many other corporations whose earnings would not be (adversely) affected if this particular corporation were to default on the contract. This would be an example of self-insurance. Alternatively, the corporation might purchase business-loss or some other form of insurance that would protect it (and, more important, its shareholders) from the consequences of a default on the contract; this would be an example of market insurance. Where good opportunities for diversification exist, self-insurance will often be cheaper than market insurance.

The foregoing discussion indicates the factors that courts and legislatures might consider in devising efficient rules for the discharge of contracts. An easy case for discharge would be one where (1) the promisor asking to be discharged could not reasonably have prevented the event rendering his performance uneconomical, and (2) the promisee could have insured against the occurrence of the event at lower cost than the promisor because the promisee (a) was in a better position to estimate both (i) the probability of the event's occurrence and (ii) the magnitude of the loss if it did occur, and (b) could have self-insured, whereas the promisor would have had to buy more costly market insurance. As we shall see, not all cases are this easy.

The Analysis Applied. Two hypothetical cases will illustrate the nature of the economic analysis of a discharge case.

1. A, a manufacturer of printing machinery, contracts with B, a commercial printer, to sell and install a printing machine on B's premises. As B is aware, the machine will be custom-designed for B's needs and once the machine has been completed its value to any other printer will be very small. After the machine is completed, but before installation, a fire destroys B's premises and puts B out of business, precluding B from accepting delivery of the machine. The machine has no salvage value and A accordingly sues for the full price. B defends on the ground that the fire, which the fire marshal has found occurred without negligence on B's part, indeed (the same point, in an economic sense), which could not have been prevented by B at any reasonable cost, should operate to discharge B from its obligations under the contract.

The risk that has materialized, rendering completion of the contract uneconomical, is that a fire on B's premises would prevent B from taking delivery of the machine at a time when the machine was so far completed (to B's specifications) that it would have no value in an alternative use. The fact that the fire occurred in premises under B's control suggests that B had the superior ability to prevent the fire from occurring. This consideration is entitled to some weight even though the fire marshal found that B could not, in fact, have prevented the fire (economically); the fire marshal might be wrong. Certainly as between the parties B had the superior ability to prevent the fire. But in light of the fire marshal's finding, ability to prevent cannot weigh too heavily in the decision of the case.

Turning to the relative ability of the parties to insure against the machine's loss of value as a result of the fire, we note first that while B was in a better position to determine the probability that a fire would occur, A was in a better position to determine the magnitude of the relevant loss (the loss of the resources that went into making the machine) if the fire did occur. That loss depended not only on the salvage value of the machine if the fire occurred after its completion but also on its salvage value at various anterior stages. A knows better than B the stages of production of the machine and the salvage value at each stage.

Assuming the actuarial value of the risk has been computed, there remains the question which of the parties could have obtained insurance protection at lower cost. Depending on the volume of A's production and on A's prior experience with contingencies such as occurred in the contract with B, A may be able to eliminate the risk of such contingencies simply by charging a higher price—in effect, an insurance premium—to all of its customers; A may in short be able to self-insure. B is less likely to be able to do so: the magnitude of its potential liability to A in the event of a default may greatly exceed any amount it could hope to pass on to its customers in the form of higher prices. As for market insurance, it seems unlikely that B could obtain for a reasonable price a fire insurance policy that protected it not only against the damage to its premises (and possibly to its business) caused by a fire but also against its contractual liability to A which, as mentioned, depends on the stage in the production of the machine at which the fire occurs, a matter within the private knowledge of A.

We are inclined to view A as the superior risk bearer in these circumstances and thus to discharge B. This inclination would be strengthened if it turned out that A was a publicly held, and B a closely held, corporation, for then the owners of A could eliminate the risk of the loss of the machine's value simply by combining their shares in A with shares of other companies in a suitably diversified portfolio. It is generally more difficult for the owners of a closely held corporation to diversify away the risks associated with their holdings in the corporation, for often those holdings represent a large fraction of their net assets.

2. For our second hypothetical case, let C be a large and diversified business concern engaged in both coal mining and the manufacture and sale of large coal-burning furnaces. C executes contracts for the sale of furnaces to D, E, F, etc. in which it also agrees to supply coal to them for a given period of time at a specified price. The price, however, is to vary with and in proportion to changes in the consumer price index.

A few years later the price of coal unexpectedly quadruples and C repudiates the coal-supply agreements arguing that if forced to meet its commitments to supply coal at the price specified in its contracts it will be bankrupted. Each purchaser sues C seeking as damages the difference between the price of obtaining coal over the life of C's commitment and the contract price. C argues that the rise in the price of coal was unforeseeable and ought to operate to discharge it from its obligations.

On these facts the case might be decided against C simply on the ground that the contract explicitly assigned to it all price risks (except those resulting from changes in the value of money). If, however, C were able to convince a court that the risk had not been specifically assigned in the instrument (either on the theory that C was really contracting for the sale of a furnace and the coal provisions of the contract were incidental, or that the source or magnitude of the price change that occurred was not within the parties' contemplation), it would then be necessary to determine whether C or the purchasers were the superior risk bearer(s).

With regard to the parties' relative abilities to forecast the consequences for contract performance of a steep change in the price of coal, two factors seem critical: the amount of coal that C has contracted to deliver forward at a fixed price, and the degree of C's exposure to coal price changes. C's exposure depends on the amount of coal sold forward that is not covered either by C's existing coal stocks or by its forward purchase contracts, multiplied by the spread in price between the average forward sale price and the average forward purchase price. Thus the magnitude of the potential loss from the price increase is simply C's net short position, and C is in a better position than any of its (typical) customers to estimate this magnitude since only C has precise knowledge of its own net asset position and contractual commitments.

The likelihood (as distinct from magnitude) of loss in this case appears to depend crucially on the probability of a large movement in the price of coal, a movement which C may have no greater ability to predict than its purchasers. But the appearance is deceptive. The critical variable is again the extent of C's exposure. If C had a perfectly neutral hedge position in coal, no movement in the price of coal could affect it. The closer C is to a neutral hedge position, the less impact a given movement in price will have on its balance sheet. Hence the ability to forecast the relevant probability here depends ultimately on knowing C's net coal position, and C knows it best.

Moreover, C can readily insure against the contingency involved in this case. Its owners can self-insure against the financial risks to C of having to make good on its coal-supply commitments at the price specified in the contract either by holding a diversified portfolio of common stocks or by purchasing shares just in the firms that are on the buying side of C's contracts. To be sure, the shareholders of C's customers may be able to insure themselves in the same fashion and at no greater cost. But an additional factor is that, as suggested above, C can self-insure by purchasing covering contracts to perfect a neutral hedge. Since there are transactional economics of scale in making forward contracts, C could execute the hedge at lower cost than each of its purchasers.

To complicate the analysis, observe that while C has better knowledge of its total exposure, it doesn't really know its potential liability to its customers since that depends on the steps they might be required by the contract doctrine of mitigation of damages to take to minimize the net cost of the unexpected increase in the price of coal. For example, at the current price of coal, oil or natural gas might become an economical substitute. If so, the measure of damages would not be the difference between the contract and current prices of coal; it would be the difference between the contract price of coal and the current price of a substitute fuel, adjusting for any differences in the quality of the substitute. Nonetheless, it seems reasonably clear to us that, everything considered, C is the superior risk bearer and its claim of discharge should fail.

The Costs of Particularized Inquiry. In the discussion of our two hypothetical cases, we applied the standard of efficient discharge developed earlier directly to the facts of the case. This is not necessarily the optimum approach. A broad standard makes it difficult to predict the outcome of particular cases. If the purpose of contract law (so far as relevant here) is to supply standard contract terms in order to economize on negotiation, it will be poorly served by a legal standard so vague and general that contracting parties will encounter great difficulty in trying to ascertain the judicially implied terms of their contract; if the allocation of risks in the contract is unclear, neither party will know which risks he should take steps to prevent or insure against because he will be held liable if they materialize.

Our second hypothetical case is a particularly good illustration of the dangers of a broad standard. The contract seemed on its face to allocate the risk of coal price changes (save those due to inflation) to C; if the allocation is instead to depend on how a court decides years later who the superior risk bearer was, the apparent definiteness of the contract terms evaporates. One way of avoiding this result in the coal hypothetical is to deem the case outside the scope of the discharge defense by noting the absence of any showing that performance under the contract was uneconomical. We assume the coal company's position is not that it could not comply with the contract at an economically reasonable price (it could buy on the open market all of the coal that it needed to fulfill its contractual obligations), but that compliance would bankrupt it. This is tantamount to arguing that a breach of contract should be excused when the breaching party for some reason lacks the resources to make good the other party's damages.

Another approach one can take in the coal case to rule out discharge is to reason that when a contract explicitly assigns the risk of price changes to one party, discharge should not be allowed simply because the price change is greater than anticipated, regardless of which party is the superior bearer of the unanticipated portion of the change. The theory here would be that since the parties must negotiate with regard to price anyway, they can, at the same time and at little additional negotiating cost, place a limit on the promisor's price exposure. If they do not do so, the court will not do it for them.

The proper use of the sort of general standard developed earlier in this section is to guide not the decision of particular cases but the formulation of rules to decide groups of similar cases. . . .

Remedial Consequences of Discharge

Thus far we have treated the discharge issue dichotomously, as if the only choice were between on the one hand enforcing the contract and awarding full contract damages, and on the other hand discharging the contract and awarding no damages. This dichotomous treatment is not inevitable. Intermediate solutions are possible and have seemed appealing to many commentators on the impossibility doctrine. . . .

Many legal commentators are distressed by this [standard doctrinal] outcome because it places the entire loss on [one party]. Their failure to understand the efficiency basis of the discharge doctrine leads them to view the remedy issue as separable from that of liability and to argue that it is "only fair" to apportion reliance losses. . . .

The proposals to change the remedial outcome to one where reliance losses are shared result from a misplaced emphasis on ex post loss distribution rather than ex ante risk bearing. Viewed from the risk-bearing perspective the refusal to apportion reliance losses is well founded; it creates an incentive for the more efficient risk bearer to adopt cost-justified risk-avoidance or risk-minimization techniques (diversification, market insurance, or whatever). The existing rule whereby discharge results in zero damages (save where restitution is appropriate) is analogous to the approach of the tort law. There too the focus is on determining which party, the plaintiff or defendant, is the lower-cost avoider of the event in question, and it is clearly not proper to approach the question by attempting to ascertain which party is in a better position ex post to bear the loss.

The ex post approach to risk assignment is in fact circular. Often the party who ex post is the superior risk bearer is the party who purchased an insurance policy covering the risk in question. But if this were the criterion of liability, it would give parties an incentive not to insure, for by not insuring they would increase the possibility that the court would assign the risk to the other party to the contract. So neither party might insure; or, fearing this possibility, both parties might insure. Neither result would be optimal.

The 50–50 loss-sharing approach favored by the commentators is at least superficially appealing in the case where it is difficult for the court to determine which of the parties is the superior risk bearer. If the parties are equally good (or bad) risk bearers, and each is risk averse, then each will prefer ex ante a solution that reduces the variance of the expected outcome of the contract. Splitting the loss does this, so long as each party estimates the same probability that the court will assign him the risk. For in that case instead of each party facing some probability of bearing all of the costs of nonperformance each has a certainty of bearing 50 percent of those costs. Of course, if neither party is risk averse, there is no gain from such a rule. And if one party is a superior risk bearer, the entire loss should be placed on him in order to encourage future parties similarly situated to insure or take other measures to minimize the economic consequences of nonperformance.

Whether carving out a class of indeterminate risk allocations and applying a rule of 50–50 loss sharing to them can be justified in terms of the administrative costs of seeking in each case (or class of cases) to determine the superior risk bearer may be doubted. It is easy to exaggerate those costs. The cost of deciding a question in a lawsuit ought properly to be apportioned over all future transactions that will be controlled by the rule declared in the suit. Even if it is very expensive to decide in the first case how to assign the risk of some event that renders contract performance uneconomical, once the question is decided and a rule declared, the rule will be available to guide the behavior of future contracting parties and will thus reduce the costs of future contract negotiations. To define a class of cases where the proper assignment of risk is treated as indeterminate could stifle the evolution of rules of risk allocation designed to enhance the efficiency of the contract process.

While 50–50 loss sharing seems an unappealing alternative to the law's solution of placing the loss entirely on one party or the other, in a world where the costs of using the legal system to fill in contract terms were very low a principle of flexible loss sharing might be superior to either approach. Loss sharing is a common characteristic of contractual arrangements: one thinks of deductibles and coinsurance in formal insurance contracts, and of joint ventures of all sorts, including sharecropping. Sometimes loss sharing is designed to deal with the moral-hazard problem in insurance; sometimes it is itself the method of insurance, as in sharecropping. But it seems hardly feasible to design a legal rule that will imitate voluntary transacting in all its variety yet be administrable at reasonable cost. Here as elsewhere the law prefers the dichotomous to the continuous solution, presumably because of administrative-cost considerations, thus illustrating the second-best character of legal compared to market resource allocation. . . .

The remedial questions pertinent to impossibility generally arise in cases of partial performance. . . . [One] form of partial performance is part payment. If there has been part payment prior to delivery and delivery is prevented by some catastrophe, the question may arise whether the payee should be required to return all or part of the payment. The answer ought to depend, in an economic analysis, on whether the purpose of the prepayment was to compensate the payee for bearing the risk that something might happen to prevent him from making delivery, or whether the prepayment was unrelated. Suppose in our hypothetical repair case that the homeowner had advanced the repair contractor part of the payment before the latter completed the repairs. If the purpose of the advance was to compensate the repairer for bearing the risk that fire or some other event would prevent him from completing the repairs, then clearly he is entitled to keep the advance—it is his insurance premium. But the advance may be completely unrelated to such risks. It could be intended to finance the purchase of supplies by the contractor (the homeowner may be the cheapest source of the necessary capital) or to protect the contractor against the risk that the homeowner might be insolvent or otherwise difficult to collect the contract price from. It could be intended simply to avoid, for tax or other financial reasons, a large, lumpy payment at the completion of construction. Given the number of plausible reasons for prepayment that are unrelated to the provision of insurance against an event that may prevent completion of performance, there can be no presumption that prepayment is intended to compensate the performing party for the risk of such an event. Accordingly, in the absence of other evidence that the payor is the superior risk bearer, his prepayments should be returned to him. This is the generally prevailing legal rule.

Where it is the performing party to the contract who has partially completed performance, rather than the paying party, there will often be a valid claim for restitution (for example, where a contract to deliver 10,000 widgets is terminated after 1,000 have been delivered), even if the performer is deemed the superior risk bearer. But where there is no basis for restitution—no value conferred on the other party to the contract by partial performance—neither is there any good economic argument for redistributing the loss between the parties. The loss should then fall entirely on the superior risk bearer.

A concept related to restitution which courts have also used in these cases is that of the "divisible contract." If an event giving rise to discharge prevents the completion of a performance that has begun, payment of the completed portion of the performance will be held to be due if that portion is deemed divisible, and the recovery may include a pro rata portion of the performing party's expectation interest as well as its reliance loss. For example, in one well-known case a steamship company had contracted to supply a given number of trips at a fixed rate per trip to be paid monthly. When the steamship was destroyed by fire, the contract was held to be discharged but the steamship company was allowed to recover the agreed-upon rate for each of the trips that had already been supplied. The basic rationale in such cases seems restitutionary, and appears to make good sense. But it would be cleaner to drop the fiction of divisibility and simply to view the contract as discharged, and then decide what compensation is due the performing party for having conferred benefits on the other party to the contract before the event giving rise to the discharge occurred. . . .

An Economic Analysis of Legal Transitions

LOUIS KAPLOW

Most normative legal analysis is devoted to determining which procedures or policies society should prefer. Any divergence between proposed solutions and the current legal regime raises the question of how the gains and losses caused by the transition to the more desirable system should be addressed. In particular, the government chooses among such options as compensation, grandfathering, phase-ins, and simply providing no relief. This article uses economic analysis to evaluate these competing transition policies. The analysis challenges much of the conventional wisdom concerning reliance arguments, constitutional protections of private property, hostility toward retroactivity, and, more generally, the desirability of transitional relief. . . .

Consider a factory that emits pollution that harms the health of a community. A legislature might enact a statute making the operation of such factories illegal, including those built and operating prior to the date of enactment. Alternatively, a common law court might extend the law of nuisance to cover the newly discovered problem. A third possibility is that a regulatory agency with jurisdiction to protect the health of the population might prohibit the operation of such factories or might impose a pollution tax rendering operation of the factory unprofitable. Finally, the government might determine that none of these policies could be enforced because the operation of such factories at night is difficult to detect, leaving no alternative but to condemn the factory.

From the point of view of both the factory owner and society as a whole, all of these actions are virtually indistinguishable, assuming that each has its intended effect. Each policy effects a transition from a regime that permitted operation of the factory to one that does not; each costs the owner the value of the factory and benefits society by protecting health. Because each of these four approaches has the same effects, the choice among them should not alter one's conclusion about the appropriateness of various forms of transitional relief, such as compensation, grandfathering of factories already in operation, or some other mitigation of the reform's effect on the factory owner. Simply recognizing the strong similarities among these transition contexts motivates critical analysis of prevailing views because virtually identical issues arising in the different settings are often approached and resolved quite differently.

The point of departure for this study is that all these transition contexts have very much in common and thus can usefully be analyzed together. The central feature of any transition situation is the existence of uncertainty concerning future government policy prior to the government action. . . .

An Economic Analysis of Transitions

The Trade-Off between Risk and Incentives

The Failure of Reliance and Expectations Arguments. One of the most commonly noted arguments against allowing private actors to bear losses resulting from changes in government policy is that they reasonably relied on preexisting law in making investment decisions. Many legal theorists, however, have long recognized the circularity of such arguments, which implicitly assume that it is reasonable to expect laws never to change—a particularly perverse assumption given that laws change quite frequently, and often in predictable ways. Even if this assumption were plausible, it could easily be overcome by decreeing that all legal rules are subject to change, in which case there could never be a contrary claim of entitlement based upon positive law. . . .

More importantly, even if actors rationally expect that legal change of a given type is unlikely, there is still the question of whether they have a compelling normative claim to fulfillment of that expectation. For example, consider a product previously assumed to be safe—an assumption rationally based on existing evidence, but by no means indicating with certainty that the product might not prove dangerous in the future. It is then discovered that the product has caused the illness or deaths of thousands of people. In such circumstances, it is generally believed to be just that the product be banned, without compensating those whose investments are rendered less valuable, precisely because there does not exist a legitimate expectation of continuing to profit from such activity.

The logic of this illustration applies to any change in preexisting law. The issue that varies from case to case is the legitimacy of the expectation of future profit. But the issue of legitimacy presents a normative question that demands an analysis of the desirability of compensation or other relief. Direct invocations of a "reliance interest" beg this question. . . .

The Effects of Transitional Relief on Risk and Incentives. Many of the deficiencies in previous investigations of transition policy result directly from deficiencies in commentators' attempts to define the problem. The analysis here focuses on the two consequences of changes in government policy: imposition of risk and modification of incentives to engage in affected activities. Uncertainty concerning future government action imposes risk, something generally considered undesirable. Insurance is one of the more common techniques for mitigating risk; government compensation has a similar effect. For example, from the owner's point of view, losing a home to a taking is very much like losing it to a fire. Just as homeowners typically purchase fire insurance, the requirement of just compensation for takings can be seen as government-sponsored takings insurance in that compensation similarly spreads the risk of an event adverse to the homeowner. Diversification through the financial markets is the other primary market mechanism for spreading risks.

One can be seriously misled by considering the risk aspect of changes in government policy without also carefully taking into account effects on economic incentives. As previously discussed, the prospect that legal provisions favoring an investment might be repealed makes the investment less attractive. Yet, contrary to the commonly held view, it is generally desirable to discourage such investment. The efficient level of investment is that induced when investors bear all real costs and benefits of their decisions. Therefore, the encouragement resulting from the assurance that compensation or other protection will be provided in the event of change results in overinvestment.

Two simple, parallel examples illustrate the desirability of exposing investors to the full costs and benefits of their decisions. Suppose there is a substantial chance that land will be taken and leveled for a highway and that a product will be found hazardous and will therefore be banned. Should these events occur, investments in improvements on the land and in manufacturing equipment to produce the product would be rendered worthless. Accordingly, ex post, it might well have been socially preferable for the landowner and the manufacturer not to have made the investments in the first place. The opposite preference would exist if the events were not to occur. As a result, it is just as socially desirable for the landowner and the manufacturer to take both possibilities into account ex ante as it is for investors to take into account the possibility that the production resulting from their investments might be rendered obsolete by superior competing products or changes in demand. A familiar example outside the context of government reform is that investors in land or equipment located in a high-risk earthquake or flood zone should take into account the prospect that their project will be destroyed in determining whether the investment is worthwhile.

These examples illustrate that the incentives question is in essence an application of the analysis typically used in connection with externalities. . . . Simply put, government compensation creates an externality that otherwise would not be present. Compensation shifts part of the long-run cost of private investment to the government and thus distorts an otherwise efficient decisionmaking process. It is socially desirable for investors to take into account the prospects for government reform; compensation eliminates this incentive by insulating investors from an important element of downside risk. . . .

The Similarity of Government- and Market-Created Risks. Most risks in society are not attributable to uncertainty concerning future government action. More commonly considered risks—which I will refer to as "market risks"—arise from uncertainty with respect to such factors as the level of future demand (the general level of demand as well as demand for particular products), technological change, behavior of competitors, and prices of other goods (such as inputs) that in turn depend upon a variety of factors. The government generally does not mitigate market risks. . . . Previous discussions of tax policy transitions, as well as analyses of government takings, have occasionally noted the similarity of market risks, where there generally is no compensation or other direct government mitigation, and government risks, where there often is.

Most commentators who recognize this similarity nonetheless go on to defend mitigation of government risks, but not of market risks. Yet none of the distinctions they offer for treating government and market risks differently withstands scrutiny. Except for institutional factors . . . , there is little to distinguish losses arising from government and market risk. For purposes of analyzing risk and incentive issues, the source of the uncertainty is largely irrelevant. A private actor should be indifferent as to whether a given probability of loss will result from the action of competitors, an act of government, or an act of God, except to the extent that the source of the risk will affect the likelihood of compensation or other relief. But whether the source of risk should determine the availability of compensation is precisely the issue this analysis is designed to address. Investors do not care whether the probability that an investment will prove worthless arises from it being leveled for a highway or being leveled by a storm; in either case, investors will invest less as the likelihood that the investment will go sour increases.

If this analogy between market and government risks is accepted, transition policy should vanish as a separate concern. Transitional relief can be seen as an instance of the more general issue concerning when government action can improve upon traditional market responses to the risk and incentive problems posed by uncertainty. If one accepts the common belief that government relief of market risk is generally undesirable because the market usually operates efficiently, or at least that government relief would be no better, then one should conclude that mitigation of government risks is economically unjustified.

The government does occasionally mitigate private risk, as through the provision of disaster relief, income maintenance programs, and the subsidization of risky undertakings. Although there are far more such instances than most would immediately recognize, such programs are the exception in the vast universe of market risks. For example, there is no general government compensation for new products that fail, production facilities that prove more costly than anticipated, or people who earn less than they had expected due to a variety of unfortunate circumstances. The adoption of mechanisms to deal with all conceivable market risks would be tantamount to government displacement of the market economy. . . .

Market versus Government Solutions in the Presence of Market Failure

. . .

Moral Hazard and Incentives. . . . Consider, for example, an insurance contract covering a manufacturer of potentially hazardous products against the risk that its products will be banned because they are too dangerous. Full insurance coverage would produce a serious moral hazard problem by decreasing the manufacturer's incentives to design safe products. A monitoring solution would be infeasible because it would require the insurance company to assess the level and composition of investment in safety research as well as choices in product design. This does not mean, however, that no risk-spreading arrangements are available in such circumstances. In addition to obtaining partial-coverage insurance contracts, firms often spread their risks widely through the financial markets; investors in these companies typically base their decisions to buy securities on the reputations of the firms, expectations concerning the companies' likely future performances, and their estimation of their ability to induce management to behave in their interests. The market will often be unable to maintain appropriate incentives while simultaneously spreading all the risk. Yet even in contexts in which the moral hazard problem represents an important market imperfection, there is no a priori reason to believe that the market will not make the best possible trade-off between risk spreading and incentive maintenance in light of the costs of monitoring behavior. . . .

Direct government compensation differs from private insurance in a number of important respects: it is provided by the government, it is typically considered to be full (whereas many insurance contracts or other diversification channels only partially spread the risk), no conditions for efficient behavior are bargained for, and no premium need be paid. The government determines all but the first feature; it could offer insurance rather than compensation by requiring the payment of premiums, and the insurance contracts could in principle duplicate any features offered in private insurance policies. The most general question thus becomes whether the government can improve on imperfect private markets, and, if so, through what mechanisms. . . .

Private arrangements, as we have seen, often provide full risk spreading without distorting incentives, and, when this is impossible, balance risk spreading and maintenance of incentives in the most efficient manner. It follows that further insurance coverage or compensation provided by the government—beyond what would have been offered by the market—is undesirable. If the benefits of additional risk spreading were sufficiently great to justify the distortions of incentives that would result, private parties would pay the higher price for more complete insurance coverage. Full compensation is therefore undesirable whenever the market would not have provided full protection.

For the same reason, partial compensation is undesirable whenever it exceeds the level of mitigation that would have been provided by the market. In fact, even providing compensation for a smaller portion of the loss than market insurance would cover in the absence of compensation is undesirable. The failure to charge a premium, and impose conditions where relevant, induces excessive total coverage—taking into account the market's response—and inefficient investment behavior. Thus, any compensation will be inefficient.

Imperfect Information and Adverse Selection. The probability that undesirable events will occur is often not known with great certainty. This does not, however, always cause private insurance or other forms of diversification to fail in spreading such risks. Insurance companies often can devise profitable policies by making reasonable approximations of risks—for example, by averaging risk factors over groups that cannot be more precisely differentiated without significant cost. Thus, for example, fire insurance premiums may be higher in some regions than in others, and on some buildings than on others, when there are readily detectable differences among them. . . .

A more serious problem arises, however, if the difficulty is not that average probability estimates are uncertain, but that the probability of loss varies among individuals in a way known to such individuals but not ascertainable by insurance companies. This informational asymmetry—referred to by economists as the adverse selection problem—can undermine the operation of insurance markets. Suppose an insurance company charges a premium based on the average probability of loss. Those with high probabilities of loss would gladly purchase such insurance, whereas those with lower than average probabilities might not purchase coverage because they would be forced to pay more than the expected value of their losses. If a significant number of low-risk individuals refused coverage, the average probability of loss to those remaining in the insurance pool would rise, thus raising premiums. Higher premiums might then induce the lower-risk individuals of those remaining in the insurance pool to drop out, and so on. In some instances, no insurance would be offered; in others, a significant number of individuals might purchase less insurance than they otherwise would, or even none at all.

Adverse selection might not, however, be a significant problem in many contexts involving uncertainty concerning future government action. The two conditions necessary for adverse selection problems to arise are (1) that the probability of loss differ significantly among individuals, who are themselves aware of these differences, and (2) that insurance companies or other institutions be unable to detect those differences at a sufficiently low cost. Much government action, however, is of general impact. For example, the repeal of a tax benefit is a reform having a common probability for all taxpayers who receive that benefit. Differences may arise if only a partial repeal is expected, leaving a subset of investors exempt. Even then, however, the second necessary condition for the adverse selection problem would often fail. Information concerning who would be exempt from repeal would not be the private, personal knowledge of the affected individuals, but would be more generally known by the relevant investment community, and thus by the financial institutions and investors that might spread such a risk. In contrast, adverse selection problems are typically thought to arise in instances such as auto insurance—where individuals have different driving abilities and habits that cannot readily be observed—and life and health insurance—where an insurance company cannot determine the health of individuals at the time of enrollment as well as the individuals can. . . .

In contrast to moral hazard, which the government is generally unable to remedy, it is more plausible that the government would be in a better position than the market to counteract adverse selection or other information problems. The government might, for example, prevent an "unraveling" of the market by compelling the purchase of insurance, thereby preventing low-risk individuals from dropping their coverage. Such action is not without efficiency costs of its own. Nevertheless, if adverse selection were a significant problem in some instances, contrary to the above discussion, there might be some room for the government to improve on the situation. Even then, the incentive benefits of requiring the payment of a premium ex ante would, as in the moral hazard context, lead to a preference for insurance over compensation.

Administrative and Transaction Costs. Insurance companies must charge more than actuarially fair premiums to stay in business because they must incur the costs of collecting information upon which to base premiums, writing policies, and determining the amount owed under the contract in the event of a loss. The greater such administrative costs are, the greater the price of insurance coverage will be, and the less insurance coverage risk bearers will purchase. In some instances, these costs are likely to be quite small, and in others, individuals will be sufficiently risk-averse that they will purchase significant insurance coverage in spite of these costs. The widespread purchase of homeowner's insurance (even when not required by banks) and life insurance suggests that administrative costs often do not make insurance prohibitive. On the other hand, many hazards remain uninsured, although it is possible that other factors explain this result.

For losses having particularly low probabilities of occurrence, up-front administrative costs may represent a large fraction of total insurance costs and thus discourage insurance protection. Similarly, when the losses imposed are relatively small (which also indicates that there is less to be gained from spreading risk), the administrative costs of determining the amount of a particular loss may be a substantial portion of the total amount at stake, making insurance unattractive. Of course, the failure of consumers to purchase insurance under these conditions will be efficient if the administrative costs reflect social costs and are otherwise unavoidable.

The existence of administrative problems has varied implications for the desirability of government transitional relief. In general, these problems do not suggest any preference for government insurance over private insurance, because the former faces all the administrative requirements of the latter, and perhaps additional costs for some of the reasons to be noted in connection with compensation. As between compensation and insurance, however, there may be administrative reasons for preferring the former. The primary administrative difference is that compensation requires administrative expense only in the actual event of a loss, and thereby avoids the costs associated with creating an insurance policy and collecting the premiums. This savings would be particularly important in the case of low-probability losses, the insurance of which involves high contracting costs relative to the stakes. Of course, to the extent that the market spreads risk through diversification of ownership, the administrative costs of relying solely on the market might be lower than those of compensation, because losses would automatically be shared.

This advantage of compensation over insurance must be balanced against both the distortion of incentives caused by compensation and the possibly higher cost of administering government compensation payments rather than insurance recoveries. The cost of administering compensation might be higher because our government institutions, at least as currently constituted, rely upon review procedures that typically are more costly than those used by private parties in similar contexts. Administrative proceedings to determine the decrease in value resulting from the government's actions are likely to be particularly costly. Potential recipients have strong incentives to argue for high valuations. The market avoids such administrative costs when risk is spread through financial mechanisms such as diversified stock ownership. In addition, such costs will often be less with private insurance because value is agreed upon in advance when the insured decides what level of coverage to purchase. Thus, the savings that compensation offers through not contracting in advance might be offset by the additional costs of determining the level of compensation to be awarded. Furthermore, to the extent that the administrative costs of compensating for losses are substantial relative to the loss itself, as is true for many losses, it may be preferable to spread risks through financial institutions, or even not at all, rather than using either conventional insurance or compensation. . . .

General Implications of the Economic Analysis of Transitions
. . .

The Symmetry of Gains and Losses. Like most previous studies of transition issues, the analysis thus far has been conducted as though future government policy (as well as market uncertainty) caused many losses but no gains. This implicit assumption, however, is particularly inappropriate for government-imposed risk because policy change is presumably undertaken in order to generate net gains in social welfare. A few examples outline the general scope of the argument that transitions produce gains and losses in a symmetrical fashion. Just as the repeal of a tax exemption creates losers, the enactment of an exemption, or any other subsidy for that matter, creates winners. Promulgation of restrictive regulations may create losers, and repeal may create winners (or vice versa). Regulations also have indirect and often opposite effects on other groups. Regulated businesses may be losers, while consumers, workers, breathers of cleaner air, or competitors of such businesses may be winners. Holders of assets that have lost their subsidies and tax preferences are losers; holders of competing assets will typically be winners. Owners of land taken for a highway are losers; owners of land near the new highway may be even bigger winners because of changes in land use patterns. And, as some of these examples illustrate, gains, like losses, can be concentrated as well as dispersed.

. . . All of the analysis offered in connection with losses applies, in a symmetric fashion, to gains. Risk concerns apply to gains as well as losses. For example, a group of investors may be discouraged from financing projects with uncertain although very high possibilities of profit if they must bear too much of the risk, and, for a given level of investment, they will be less well off if they bear the full risk. The incentives analysis also fully applies to both gains and losses. If a change may occur that will make an investment more profitable, investors should invest more in that project than in a similar one that lacks the additional upside potential. The more the upside risk is borne by others, the smaller is the incentive to invest, mirroring the investment distortion that results in the case of downside risk.

Despite this strong similarity between gains and losses, virtually all investigations of transition issues have ignored the issue of how to treat windfall gains. In particular, those advocating mitigation of windfall losses virtually never recommend taxation of similar windfall gains. In the takings context, it is not very surprising that commentators generally focus only on losses, because the takings clause of the Constitution is limited to a requirement that losses be compensated. Yet even authors who explicitly offer a general policy analysis of takings do not evaluate the parallel case for taxation of windfall gains. . . .

The similarity between the arguments for and against compensation of losses and taxation of gains suggests that advocacy of one result for losses should be accompanied by advocacy of a corresponding result for gains. If the common view in many contexts is that losses should be compensated or otherwise mitigated, while gains can be ignored, the question is which position should be abandoned. The analysis presented in this article suggests a general preference for ignoring both gains and losses. . . .

Extending and Applying the Economic Analysis of Transition Policy

. . .

Applications . . .

. . .

Retroactive Application of Newly Evolved Common Law Tort Doctrine. Suppose that a common law court decides that tort doctrine should be interpreted, extended, or changed to broaden the scope of liability for injuries caused to third parties by the manufacture or use of a product. Assume that this change is made because it has become apparent that harm can be caused in ways not adequately dealt with by prior doctrine. For example, tolling rules of a statute of limitations might be modified to account for long latency periods of diseases caused by toxins, or causality rules might be adjusted to account for the nature of state-of-the-art epidemiological evidence. The transition policy question, then, would be whether the new rule should be applicable (1) to future production from factories built prior to the announcement of the new rule and (2) to harms arising from production prior to the announcement. The analysis of incentives presented [in the preceding section] suggests that the new rule should apply to both situations and that no other transitional relief should be provided.

Both of the applications impose losses on owners of the factories. Although in response to the new liability rule owners might modify future production to result in less harm than was caused by past production—or halted altogether if that is not possible—either alternative will be more costly to the owners than if there were no liability. The liability for past production can neither be avoided nor mitigated, and it will be costly as well. That both applications have effects that are qualitatively similar is a direct implication of the analysis presented [above].

Standard deterrence analysis in tort theory suggests that liability rules should internalize costs in order to achieve efficiency in terms of the combined objectives of maximizing safety and minimizing production costs. For the most part, this discussion will assume that the new rule was adopted precisely to affect behavior in that manner. In the transition context, however, the behavior at issue is not only future behavior, but also behavior that occurred prior to announcement of the new rule.

Economic behavior is most obviously implicated by transition policy in terms of the first question: whether the rule should apply to future production from preexisting factories. The desirability of the rule suggests that such application is appropriate, for harm would otherwise be produced that might well exceed the costs of avoiding it. The second question concerns harms that arise prior to announcement of the new rule. Although behavior prior to the rule change cannot be altered after the fact, transition policy can influence such behavior ex ante. In this example, firms making initial construction and product design decisions will have made earnings projections that took into account expected liabilities. If it was known that tort liability would ensue when a product or production process caused substantial harm, decisions might have been substantially different than if it was likely that the firm would be immune. Transition policy that exempts or otherwise gives relief to past investments confers such an immunity. Thus, when the risk of tort liability depends in important ways on the evolution of tort law, as might often be the case with toxic substances and products liability, the expectation that future evolution in the law will be made applicable to harms arising, and factories built, prior to the announcement of new rules will have a desirable effect on behavior. . . . In contrast, a pattern of exempting prior activity and investments would more likely lead subsequent investors facing uncertainty concerning future evolutions in tort law to expect that they too would be exempted. Such expectations would induce inefficient investment decisions. . . .

Just Compensation for Takings. [The preceding section] used examples involving takings to illustrate the incentives argument against transitional relief and the likely market response to the risks posed by the possibility that land would be taken. The basic economic framework suggests that, as a matter of economic efficiency, compensation is unwise. Nonetheless, a number of additional factors pertaining to the economic analysis raise doubts about this initial conclusion.

One factor is that takings are often thought to be low-probability events. This factor would not alter the general economic argument, but the transaction costs involved would be more likely to impede efficient operation of the market: it may not be worth the cost of arranging for private insurance if the event is sufficiently unlikely. The transaction costs, however, may not be so high as to impede the market from spreading risk efficiently. For firms that have diversified ownership, special insurance for takings may be unnecessary. And for individuals, comprehensive homeowner's policies, which would be purchased in any event, might deal with the problem. . . .

Another reason why insurance markets could fail is that individuals might underestimate the probability that their homes will be taken. . . . [T]his possibility strengthens the case for some relief, although it does not necessarily imply that compensation would be desirable. [Rather,] compulsory insurance would be preferable to transitional relief. Translated to the takings context, the approach might involve imposing on homeowners annual taxes equal to what their market insurance premiums would otherwise be, in addition to providing compensation after-the-fact. Because these charges would be higher if additional investments were made on property when takings were more probable, investment incentives would be preserved. This system would entail administrative costs additional to those required with pure compensation because it would require ex ante assessments in all instances in which the probability of taking was sufficiently high. On the other hand, most localities already administer property taxes, which appraise property values and attempt periodically to account for improvements. One suspects that at some point the probability of a taking would be sufficiently high that such a scheme would be worthwhile in terms of the incentive benefits, despite its administrative costs. This scheme has the added virtue that it might be permissible to apply it even in cases in which any compensation was inefficient, but in which compensation was nonetheless constitutionally required. . . .

Concerns about abuse of power are potentially far more important in the context of takings than in most other transition contexts precisely because takings often single out individuals or groups. Again, the increased likelihood of private insurance in the absence of a compensation requirement, or the compulsory insurance alternative that provides full compensation but taxes insurance premiums ex ante, would do much to alleviate this problem. Of course, if affected groups tended to be those least likely to have made adequate market provision, for any of the reasons discussed [above], and if compensation under a compulsory insurance alternative was not required, for example, as by a constitutional provision, the problem would be that compensation might be denied in precisely those instances involving an abuse of power. . . . [Thus, a]rguments concerning political feasibility, and particularly those addressed to potential abuse of power, seem to emerge as the strongest reasons in favor of compensation. . . .

The Optimal Tradeoff between the Probability and Magnitude of Fines

A. MITCHELL POLINSKY AND STEVEN SHAVELL

Fines are used in a variety of situations to control activities which impose external costs. Examples of such activities include polluting the air, speeding or double parking, evading taxes, littering a highway, and attempting to monopolize an industry. If it were costless to "catch" or "observe" individuals (or firms) when they engage in an externality creating activity, presumably everyone would be caught and fined an amount equal to the external cost of the activity. This is simply the traditional Pigouvian tax solution. Individuals would then engage in the activity only if their private benefits exceed the external cost.

However, in most situations it is difficult or costly to catch individuals who impose external costs. If, as a result, individuals are caught with probability less than one, it is often observed that the fine could be raised to a level such that, as in the Pigouvian solution, individuals would engage in the activity only if their private gains exceed the external cost. Since this can apparently be done for any given probability of catching individuals and since it is normally costlier to catch a larger fraction of those engaging in the activity, it is frequently argued that the probability should be as low as possible. The only constraint on lowering the probability that is recognized is the inability of individuals to pay the fine; thus, the optimal fine implied by this argument equals an individual's wealth. Note that this reasoning applies regardless of the external cost of the activity.

This view of the optimal tradeoff between the probability and magnitude of fines does not seem realistic. Individuals are rarely if ever fined an amount approximating their wealth, especially for activities which impose relatively small external costs.

A. Mitchell Polinsky and Steven Shavell, The Optimal Tradeoff between the Probability and the Magnitude of Fines, 69 *American Economic Review* (1979), pp. 880–891. Copyright © 1979 The American Economic Association. Reprinted with permission.

The present paper points out an error in this view and suggests an explanation of the choice of the probability and fine which seems consistent with reality. The mistake is that the view does not properly take into account the possibility that individuals may be risk averse. The possibility of risk aversion does not imply that individuals cannot be induced to make the same decision about engaging in the activity as they would under the Pigouvian solution. (For any given probability, the fine can be lowered from the level at which its expected value equals the external cost to a level, reflecting risk aversion, such that only those individuals for whom the private gains exceed the external costs engage in the activity.) Rather, the error is that the view does not take into account the risk that is borne by those who do engage in the activity. This risk is present whenever those individuals have to pay a fine with a (positive) probability less than one. . . .

[The logic of the argument is as follows. I]f individuals are risk neutral, the optimal probability is as low as possible and the fine is as high as possible. However, the assumption of risk neutrality seems implausible in situations in which individuals would face the risk of losing all of their wealth. [Two major conclusions follow] if individuals are risk averse. First, if the cost of catching individuals is sufficiently small, the optimal probability equals one. This is true because the disutility from bearing the risk of being caught and fined outweighs the potential savings from a reduction in the probability. Also, when the optimal probability is one, the optimal fine equals the private gain of those who engage in the activity. Second, if it is optimal to control the activity at all, then, regardless of how costly it is to catch individuals, it may never be optimal to catch them with a very low probability and to fine them much more than the external cost. This is true because doing so would lower utility due to risk bearing and could more than offset the benefits from controlling participation in the activity. [Furthermore,] it could not be beneficial to allow individuals to reduce the risk of being fined by the purchase of insurance against fines. This is true because the government could achieve the same result by lowering the fine. . . .

Although this paper does not take considerations of justice into account, it is obvious that they are relevant to determining the desired probability and fine. For example, fining individuals far in excess of the external cost they impose on society may be thought of as unfair, or catching only a small fraction of those who impose the external costs may be seen as arbitrary. These considerations appear to complement the conclusions of this paper if individuals are risk averse. . . .

Double Parking—A Numerical Example

The optimal tradeoff between the probability and magnitude of fines if individuals are risk averse may be illustrated by a numerical example. The example concerns the control of double parking in a hypothetical city in which the private gains to residents from double parking usually are exceeded by the congestion costs imposed, but occasionally are not. The example is described by the following data:

$100,000$ = population of the city
$10,000$ = number of locations where a resident could double park

$$25 = \text{number of locations which each policeman can check per hour}$$

(and, if necessary, at which he must write a ticket)

s = wage of a policeman per hour

$\log w$ = utility of wealth w for each resident

$10,000 = initial wealth of each resident

$5.00 = congestion costs from double parking

$4.50, $50.00 = possible private gains to each resident from double parking

.90 = probability that the private gain is $4.50.

In order to catch individuals who double park with probability p, $10,000p$ locations must be checked, $10,000p/25 = 400p$ policemen must be employed, and a total of $400ps$ must be spent on enforcement. If the fine is f, then total fine revenue is pf times the number of residents who double park. Per capita taxes equal total expenditures on enforcement minus total fine revenue, all divided by 100,000. Per capita congestion costs are $5.00 times the number of people who double park, divided by 100,000. Given his wealth net of taxes and congestion costs, his private gain, and the probability and magnitude of the fine, each resident decides whether to double park. The expected utility of a typical resident may then be calculated.

The probability and fine which maximize expected utility of the residents were computed for various values of the policemen's wage [see Table 1]. The results may be summarized as follows:

a. If the wage is less than $2.83, the optimal probability is 1 and the optimal fine is $50. This result . . . shows that when policemen can be cheaply hired, it is best to employ a sufficient number to catch all individuals who double park. Total expenditures on enforcement may be as high as $1,132 (when the wage is $2.83). In contrast, the government could use the maximum possible fine of $10,000 with the threshold probability of .00006. This is the lowest probability which would deter those who would gain $4.50 from double parking; however, those who would gain $50 would double park. Although this system of enforcement would involve total expenditures of at most $0.07, it is not optimal because of the high risk imposed on those who double park.

b. As the wage rises above $2.83, the optimal probability rapidly declines and the fine increases. Table 1 reports the results for several reasonable values of a policeman's wage. As the wage increases from $3 to $25, the optimal probability falls from .09 to .03, the optimal fine rises from $50.10 to $149.00, and total enforcement expenditures increase from $108 to $300. In contrast, if the fine were incorrectly set at its maximum $10,000, and the probability were set at the threshold .00006, then total expenditures on enforcement would never exceed $0.58.

c. As the wage increases to extremely high levels, the optimal probability declines to a value slightly above the threshold probability and the optimal fine rises to approximately $9,000. Only when the wage exceeds $200,000 per hour is the optimal policy to do nothing about double parking. . . . [W]hen it is optimal to control double parking, the optimal fine never reaches $10,000 and the optimal probability never becomes as low as the threshold probability. If the parameters of the example were different, the wage at which it first becomes optimal to do nothing about double parking might be much lower; as a result, the maximum optimal fine might be much lower and the minimum positive optimal probability much higher. For instance, this would

Table 1. The Control of Double Parking

Wage (per hour)	Optimal Probability	Optimal Fine	Total Enforcement Expenditures at Optimal Probability	Threshold Probability	Total Enforcement Expenditures at Threshold Probabilty
$3.00	.09	$50.10	$108.00	.00006	$0.07
$5.00	.07	$64.10	$140.00	.00006	$0.12
$7.50	.06	$74.80	$180.00	.00006	$0.18
$10.00	.05	$89.80	$200.00	.00006	$0.23
$15.00	.04	$111.90	$240.00	.00006	$0.35
$25.00	.03	$149.00	$300.00	.00006	$0.58

be the case if the lower private gain from double parking were just below the congestion costs imposed.

Concluding Remarks

The model used here abstracted from a variety of considerations relevant to the determination of the optimal probability and fine. Several of these are now mentioned.

It was assumed that the private gains from engaging in the externality creating activity could not be observed. However, in some contexts the private gains might be identifiable at little cost. If this is the case, the fine (and possibly the probability) could depend on the private gain. Those who would gain less than the external cost could be discouraged from engaging in the activity by setting their fine sufficiently high, and those who would gain more than the external cost could be induced to engage in the activity by setting their fine sufficiently low, possibly at zero. Therefore, the optimal probability can be lowered since disutility due to risk bearing can be reduced.

It was also assumed in the model that individuals who did not engage in the activity were never mistakenly fined. If this possibility had been taken into account, the conclusions of this paper would be reinforced, since all individuals would then bear the risk of paying a fine.

Furthermore, it was assumed that if individuals engaged in the activity at all, they did so at a particular level. A more general model would allow individuals to engage in the activity at varying levels. Similarly, it was assumed that the distribution of private gains was discrete, whereas in general the distribution might be continuous. Neither of these extensions would affect the basic results of this paper. All that is required for the results is that, given the optimal probability and fine, some risk-averse individuals generate externalities at a positive level and are subject to the risk of having to pay a fine.

Finally, it was assumed that individuals had the same level of wealth. However, differences in wealth may be important in many situations. Suppose that absolute risk aversion decreases with wealth and that the probability and fine cannot be made to depend on wealth. Then, ceteris paribus, any given probability and fine would be less likely to discourage a wealthy individual from engaging in the activity than a poor one. As a result, in the optimal solution, some wealthy individuals might be under-deterred—induced to engage in the activity even though their private gains are less than the external cost—and some lower income individuals who are able to pay the fine might be overdeterred. However, some poor individuals who are unable to pay the fine might be underdeterred.

NOTES AND QUESTIONS

1. As Shavell explains, attitudes toward risk can be explained by a theory of *expected utility maximization*. According to this theory, a rational individual will accept a gamble if and only if the expected value of the gamble, measured in terms of utility, is positive. Thus, the reason that you are unwilling to risk $1,000 on the flip of a coin is that the $1,000 you stand to gain would provide less utility than the $1,000 you may lose, so that the expected utility is negative. Conversely, those who attach increasing utility to additional increments of money, such as someone who would get great satisfaction out of a large purchase that is currently just out of his reach, can be expected to be risk-preferring, and hence willing to gamble at unfair odds. The theory also explains why owners of an insolvent or barely insolvent firm might behave in a risk-preferring manner; because they are entitled to the benefits of upside gains while their creditors must bear the downside losses, they get substantial utility from increasing their wealth and lose relatively little utility from decreasing their wealth. For a discussion of the implications of this idea for the law of bankruptcy, see Michael Bradley and Michael Rosenzweig, The Untenable Case for Chapter 11, 101 *Yale Law Journal* 1043 (1992).

2. While much of the intuitive appeal of expected utility comes from its association with classical utilitarianism, it is possible to develop an account of choice under uncertainty without relying on any notion of measurable utility in the classical sense. Indeed, economists often refer to expected utility as *Von Neumann/Morgenstern utility,* in order to indicate that, unlike classical utility, it is not to be understood as comparable across individuals or between pecuniary and nonpecuniary goods. For formal accounts of expected utility theory, see John Von Neumann and Oskar Morgenstern, *The Theory of Games and Economic Behavior* (Princeton: Princeton University Press, 1953); L. J. Savage, *The Foundations of Statistics* (New York: Wiley, 1954). A more classical interpretation, however, is necessary for certain applications. Consider, for example, the question of optimal insurance against nonpecuniary loss. A risk-averse person who would be happy to pay $150 for an insurance policy against a 1 percent chance of losing $10,000, might not be willing to buy such a policy covering a nonpecuniary loss that she otherwise regarded as comparable to a $10,000 loss (say, for instance, the loss of a beloved household pet). This is because the two losses do not have the same effect on the marginal utility of wealth; monetary loss raises the individual's marginal utility of wealth, so that the utility pro-

duced by replacing the lost $10,000 in the event of the accident more than makes up for the sacrifice of the $150 premium. Nonpecuniary loss, in contrast, need have no effect on the marginal utility of wealth, so that the utility of an extra $10,000 in the event of an accident does not warrant a $150 sacrifice. Does this analysis make sense to you? Would you buy life insurance for your pet if it were available? See Samuel Rea, Nonpecuniary Loss and Breach of Contract, 11 *Journal of Legal Studies* 35 (1982) [arguing that optimal insurance should not cover nonpecuniary loss]; Steven Croley and Jon Hanson, The Nonpecuniary Costs of Accidents: Pain-and-Suffering Damages, 108 *Harvard Law Review* 1787 (1995) [arguing that most consumers would want insurance coverage for nonpecuniary loss].

3. Damages for nonpecuniary loss have been controversial among both economic and more traditional legal commentators. One reason for this is that the issue exposes a tension among goals of deterrence, compensation, and insurance. Consider, for example, the question of damages for pain, suffering, and emotional distress in tort. From the economic viewpoint, pain is a social bad, and it is worth spending material resources to lessen the risk that it will be suffered. Liability for pain and suffering, therefore, is necessary to encourage optimal precaution, since if injurers do not have to pay for the pain they cause, there will be an externality. On the argument of the previous note, however, optimal risk-bearing suggests that there should be no liability. What then is the proper policy? Can you think of any ways to resolve the tradeoff? For one suggestion, see Robert Cooter, Towards a Market in Unmatured Tort Claims, 75 *Virginia Law Review* 383 (1989).

4. Another transaction cost of insurance, mentioned in several of the preceding selections and discussed in more detail in the next chapter, is *adverse selection*. Adverse selection can occur whenever some unobservable characteristic of the insured party affects the insurer's expected costs. For example, sellers of health insurance will find it more expensive on average to provide coverage to persons who are sickly or who have preexisting medical conditions. If the purchase of insurance is voluntary, and if insurers cannot observe who is sickly and who is not, they will find that their customers are disproportionately drawn from the ranks of the sickly, since at any given price a sickly person will have a higher demand for insurance than a healthy one. In general, the cost of insurance is higher in markets with significant adverse selection, because the insurer must raise its price to reflect the costs of covering a relatively risky pool of insureds; this helps to explain why short-term health insurance is more expensive than long-term insurance, and why group insurance is cheaper than nongroup insurance. Insurers can adopt strategies of risk management to reduce the costs of moral hazard and adverse selection, including audits and inspections, experience rating, discounts for specific precautions such as airbags or a car alarm, exclusion of particular risks such as those resulting from a preexisting medical condition, copayments, deductibles, and policy caps. Similarly, contracting parties whose relationships include a significant insurance element often provide for analogous terms in their contracts. Such provisions, however, raise the administrative costs of insurance and limit the extent of its coverage. See generally Kenneth Arrow, Insurance, Risk, and Resource Allocation, in *Essays in the Theory of Risk-Bearing* (Chicago: Markham, 1971).

5. Do you agree with Hanson and Logue that sellers of first-party insurance have

little ability or incentive to monitor consumers' risk-taking behavior, or that the existence of first-party insurance makes contributory negligence irrelevant? Can you think of any situations in which first-party insurers would be better able than third-party insurers to deal with moral hazard, adverse selection, or the other transaction costs of insurance? See Richard Epstein, Products Liability as an Insurance Market, 14 *Journal of Legal Studies* 645 (1985); George L. Priest, Can Absolute Manufacturer Liability Be Defended?, 8 *Yale Journal on Regulation* 1 (1990) [both arguing that efficient tradeoff between deterrence and risk allocation requires that consumers bear some of the risks of defective products].

6. How would you decide Posner and Rosenfield's second hypothetical case, which deals with an unanticipated increase in the price of coal? Would it ever be efficient to excuse contractual performance in response to unforeseen price increases? Consider in this regard the case of *Alcoa v. Essex,* 499 F. Supp. 53 (W.D. Pa. 1980), in which an aluminum refining contract containing a complex and elaborately negotiated price index nonetheless failed to anticipate the rise of the Organization of Petroleum Exporting Countries (OPEC) as an effective cartel and the resultant increase in worldwide energy prices. Compare Paul Joskow, Commercial Impossibility, The Uranium Market, and the Westinghouse Case, 6 *Journal of Legal Studies* 119 (1977) [discussing efficient allocation of risks resulting in the uranium market from OPEC-led price increases].

7. The economic theory of risk-bearing sheds light not just on the doctrinal excuses of impossibility and mistake, but on contract damages generally, since the various measures of contract damages allocate risks differently. Expectation damages, for instance, when properly assessed, give the promisee full insurance against the promisor's inability to perform. To the extent that they are less than the expectation, reliance and restitution damages provide less than full insurance. Depending on the circumstances, though, it may not be optimal for the promisor to provide full insurance against nonperformance. Even if there is no question of moral hazard on the promisee's part, efficient risk allocation implies that the less risk-averse party should insure the more risk-averse one. If the two parties are equally averse to risk, it may be optimal to share or split the costs resulting from nonperformance. This will result in some moral hazard on the promisor's part, to be sure, but if he is sufficiently risk averse, this cost may be worth bearing. Indeed, if the promisee is risk-neutral, it may even be second-best for the promisee to insure the promisor against breach, in exchange for a reduction in the contract price. One way to implement this arrangement may be though liquidated damages or a deposit. See A. Mitchell Polinsky, Risk Sharing through Breach of Contract Remedies, 12 *Journal of Legal Studies* 427 (1983), Victor Goldberg, An Economic Analysis of the Lost-Volume Retail Seller, 57 *Southern California Law Review* 283 (1985).

8. As Kaplow admits, his argument against compensating the losers from governmental policies depends on the claim that government- and market-created risks are functionally equivalent. Are these risks in fact equivalent? Does his solution to the problem of legal transitions provide proper incentives to governmental as well as private actors? See William A. Fischel and Perry Shapiro, Takings, Insurance, and Michelman: Comments on Economic Interpretations of "Just Compensation" Law,

17 *Journal of Legal Studies* 269 (1988) [arguing that compensation is necessary to prevent excessive government use of eminent domain].

9. Many legal systems regard certain categories of insurance as against public policy. For instance, under the legal regime of the former Soviet Union, liability insurance covering negligence in tort was prohibited. Similarly, various American jurisdictions restrict a person from obtaining insurance against his own crimes or intentional torts, and in some cases, recklessness. Is there any economic reason to prohibit or to restrict the private purchase of insurance? As a specific example, suppose that a corporation wishes to offer its officers and directors, as part of their compensation package, an insurance policy that covers their losses if they are sued for insider trading or for other breaches of their fiduciary duty. Is there any reason to prohibit or limit such an arrangement? Would the situation be different if the insurance covered liability for torts committed in the corporation's service? See Steven Shavell, The Judgment Proof Problem, 6 *International Review of Law and Economics* 45 (1986).

10. Readers will find it a useful exercise to try to develop their own applications of the theory of risk-bearing in other fields of law. For instance, in the field of professional responsibility, considerations of risk and insurance appear central to devising and regulating systems of attorney compensation. A client who hires an attorney on an hourly fee bears most of the risks of litigation, upside and downside. By using a fixed fee, the attorney can undertake the risk that litigation will be more complicated and burdensome than expected; and through a contingent fee she may further assume some of the risks of an adverse outcome. The allocation of risks between attorney and client, however, will affect their incentives to invest in litigation. See, for example, Patricia Danzon, Contingent Fees for Personal Injury Litigation, 14 *Bell Journal of Economics* 213 (1983); Murray Schwartz and Daniel Mitchell, An Economic Analysis of the Contingent Fee in Personal-Injury Litigation, 22 *Stanford Law Review* 1125 (1970). Other applications include plea bargaining [Robert E. Scott and William Stuntz, Plea Bargaining as Contract, 101 *Yale Law Journal* 1909 (1992)], warranties [George Priest, A Theory of the Consumer Product Warranty, 90 *Yale Law Journal* 1297 (1981)], and bankruptcy [Susan Rose-Ackerman, Risk Taking and Ruin: Bankruptcy and Investment Choice, 20 *Journal of Legal Studies* 277 (1991)].

6

Refining the Model III: Information

In the preceding chapters, we have for the most part abstracted from problems of incomplete or imperfect information, and instead analyzed economic incentives under the assumption that the relevant parties knew everything they needed to know to make rational decisions. Our discussion of moral hazard and adverse selection in the market for insurance, however, demonstrated how imperfect information can substantially raise the transaction costs of economic exchange. In this chapter, therefore, we explicitly take up the economics of information, and consider how our previous conclusions might be refined in light of its teachings.

Like other departures from the classical model of perfect competition, imperfect information can be evaluated from either the Pigouvian or the Coasian perspective. From the standpoint of the model of market failure, imperfect information is a market failure in and of itself. It leads people to miscalculate costs and benefits and thus to make incorrect choices; and it gives them incentives to waste resources trying to take advantage of each others' ignorance. In this view, public institutions can improve the efficiency of private exchange by preventing fraud and by encouraging disclosure. From the standpoint of the model of cooperation, however, imperfect information is simply another kind of transaction cost, present in private and public institutions alike. Because information is both valuable and costly to acquire, it is not surprising that people are less well off with less of it; but this is true of all scarce resources and does not by itself establish an argument either for government regulation or for public provision. Instead, the optimal policy toward information is the one that best provides incentives for its efficient acquisition and dissemination.

In the first excerpt in this chapter, George Stigler sets out a Coasian account of imperfect information, arguing both that information is an economic good and that many important private institutions can be understood as responding to the need for its production and distribution. In particular, Stigler suggests that the scarcity of information helps explain phenomena such as price dispersion in markets for homogenous goods, advertising, brand names, and the existence of specialized dealers and market-makers. In his view, because information is valuable, it pays to engage in efforts to acquire it, but because such search is costly, it is both rational and efficient to stop searching before learning everything that might be relevant to a decision. Accordingly, the fact that informed traders do better in exchange than do uninformed traders, rather than being a sign of market failure, shows that markets are working properly; the informed traders' profits are necessary to preserve proper incentives to engage in search. Anecdotal evidence that consumers invest greater efforts in search when making large and important purchases is, in his view, consistent with the conclusion that their behavior is efficient.

The Pigouvian view of imperfect information as market failure is represented here by George Akerlof's article on the market for used cars. In this article, Akerlof shows how quality variations among the cars offered for sale, together with buyers' rational reluctance to purchase cars of low quality, or "lemons," can impede market exchange. Generalizing from this example, he argues that market outcomes will be inefficient in any situation where buyers have poorer information than sellers about the items that are being sold. This is because a buyer who cannot tell with whom she is dealing will reasonably assume she is dealing with an average-quality seller, and will pay no more for the goods she is purchasing than she would pay for average-quality goods. If sellers of high-quality goods cannot somehow communicate their high quality to buyers, they will be unable to recover the full value of their wares; if they are unwilling or unable to sell at a loss, they will be forced either to lower their quality or to leave the market. As the highest-quality goods are driven from circulation, furthermore, average market quality falls, putting pressure on those sellers who remain to lower quality further. In the end, it may be that only the lowest-quality goods can be traded. Akerlof shows that this problem is a type of adverse selection, and that it is potentially present in any market in which traders have asymmetric information about the value of exchange. In general, at least some mutually beneficial bargains will be missed in such markets, the bargains that are concluded may be at suboptimal terms, and in extreme cases, the market can collapse entirely.

The remaining readings apply these basic concepts to various topics in tort, contract, and procedure. William Bishop suggests that the contract doctrine of foreseeability, as embodied in the classic case of *Hadley v. Baxendale,* is best understood as responding to a "lemons" problem in the market for contract damages. Specifically, if a contracting party cannot tell whether his promisee will suffer high or low damages in the event of breach, and if he is held liable for all damages that result in either case, then he will have to factor into his costs an implicit insurance premium that covers high-risk promisees as well as low-risk ones. This premium will make the underlying contract relatively attractive to parties who expect their damages in event of breach to be high, and unattractive to those who expect damages to be low. Low-risk parties therefore must choose either to subsidize their high-risk counterparts or to abstain from the transaction; and the resulting adverse selection can hinder exchange. The foreseeability doctrine, in contrast, forces high-cost parties to identify themselves if they want full insurance for their losses, thus obliging them to pay for the actuarially fair costs of such coverage.

The article by Sanford Grossman argues that asymmetric information about quality need not lead to market failure so long as there is some mechanism for the informed parties to reveal their information cheaply and credibly before exchange. This is because parties with relatively favorable information will want to reveal it in order to get credit for being a high-quality contractual partner. Once the highest-quality parties reveal themselves, those of the next level of quality will have a similar incentive to reveal, and so on; and in this way any informational asymmetry will unravel of its own accord. In particular, Grossman suggests that warranty contracts provide just such a mechanism in markets for the sale of goods. Because sellers of high-quality goods can afford to supply warranties at relatively low cost while sellers of low-quality goods cannot, the voluntary offer of a warranty is a cheap and effective signal that the underlying good is reliable. Enforcing warranties through the law of contracts, therefore, may help solve the lemons problem.

Finally, Louis Kaplow discusses the role of information in the adjudicatory process, focusing on its private value for litigants and its public value for the legal system as a whole. In Kaplow's account, additional information is valuable only when it helps courts and litigants make more accurate substantive decisions. For instance, if it is uncertain whether the loss L resulting from an accident is low or high, then a court applying the Hand formula will have to make a negligence determination based on some estimate that will at best be correct on average. Similarly, individual citizens will make imperfect estimates of their expected liability when deciding how much care to take. Such errors are socially costly because uninformed courts may sanction nonnegligent behavior and fail to sanction negligence, thus providing inefficient incentives for precaution, and because uninformed individuals may fail to take care even when it is cost-justified to do so. Kaplow argues that there is no social value to investing in accuracy at trial if individuals cannot anticipate the court's findings at the time that they decide how to act, because in such circumstances increased precision ex post will not alter primary behavior. If private individuals can invest in information acquisition ex ante, however, accuracy at trial may be socially valuable.

6.1 THEORY

The Economics of Information

GEORGE STIGLER

One should hardly have to tell academicians that information is a valuable resource: knowledge is power. And yet it occupies a slum dwelling in the town of economics. Mostly it is ignored: the best technology is assumed to be known; the relationship of commodities to consumer preferences is a datum. And one of the information-producing industries, advertising, is treated with a hostility that economists normally reserve for tariffs or monopolists.

There are a great many problems in economics for which this neglect of ignorance is no doubt permissible or even desirable. But there are some for which this is not true, and I hope to show that some important aspects of economic organization take on a new meaning when they are considered from the viewpoint of the search for information. In the present paper I shall attempt to analyze systematically one important problem of information—the ascertainment of market price.

The Nature of Search

Prices change with varying frequency in all markets, and, unless a market is completely centralized, no one will know all the prices which various sellers (or buyers) quote at any given time. A buyer (or seller) who wishes to ascertain the most favorable price must canvass various sellers (or buyers)—a phenomenon I shall term "search."

The amount of dispersion of asking prices of sellers is a problem to be discussed later, but it is important to emphasize immediately the fact that dispersion is ubiquitous even for homogeneous goods. . . . Price dispersion is a manifestation—and, indeed, it is the measure—of ignorance in the market. Dispersion is a biased measure of ignorance because there is never absolute homogeneity in the commodity if we include the terms of sale within the concept of the commodity. Thus, some automobile dealers might perform more service, or carry a larger range of varieties in stock, and a portion of the observed dispersion is presumably attributable to such differences. But it would be metaphysical, and fruitless, to assert that all dispersion is due to heterogeneity.

At any time, then, there will be a frequency distribution of the prices quoted by sellers. Any buyer seeking the commodity would pay whatever price is asked by the seller whom he happened to canvass, if he were content to buy from the first seller.

Table 1. Distribution of Hypothetical Minimum Prices by Numbers of Bids Canvassed

Number of Prices Canvassed	Probability of Minimum Price of		Expected Minimum Price
	$2.00	$3.00	
1	.5	.5	2.5
2	.75	.25	2.25
3	.875	.125	2.125
4	.9375	.0625	2.0625
∞	1	0	2

But, if the dispersion of price quotations of sellers is at all large (relative to the cost of search), it will pay, on average, to canvass several sellers. Consider the following primitive example: let sellers be equally divided between asking prices of $2 and $3. Then the distribution of minimum prices, as search is lengthened, is shown in Table 1. The buyer who canvasses two sellers instead of one has an expected saving of 25 cents per unit, etc. . . .

Whatever the precise distribution of prices, it is certain that increased search will yield diminishing returns as measured by the expected reduction in the minimum asking price. This is obviously true of the rectangular [i.e., uniform] distribution, with an expected minimum price of $1/(n + 1)$ with n searches, and also of the normal distributions. In fact, if a distribution of asking prices did not display this property, it would be an unstable distribution for reasons that will soon be apparent.

For any buyer the expected savings from an additional unit of search will be approximately the quantity (q) he wishes to purchase times the expected reduction in price as a result of the search. . . . The expected saving from given search will be greater, the greater the dispersion of prices. The saving will also obviously be greater, the greater the expenditure on the commodity. Let us defer for a time the problem of the time period to which the expenditure refers, and hence the amount of expenditure, by considering the purchase of an indivisible, infrequently purchased good—say, a used automobile.

The cost of search, for a consumer, may be taken as approximately proportional to the number of (identified) sellers approached, for the chief cost is time. This cost need not be equal for all consumers, of course: aside from differences in tastes, time will be more valuable to a person with a larger income. If the cost of search is equated to its expected marginal return, the optimum amount of search will be found.

Of course, the sellers can also engage in search and, in the case of unique items, will occasionally do so in the literal fashion that buyers do. In this—empirically unimportant—case, the optimum amount of search will be such that the marginal cost of search equals the expected increase in receipts, strictly parallel to the analysis for buyers.

With unique goods the efficiency of personal search for either buyers or sellers is extremely low, because the identity of potential sellers is not known—the cost of search must be divided by the fraction of potential buyers (or sellers) in the population which is being searched. If I plan to sell a used car and engage in personal search, less than one family in a random selection of one hundred families is a potential buyer of even a popular model within the next month. As a result, the cost of search is increased more than one hundredfold per price quotation. . . .

Advertising is, of course, the obvious modern method of identifying buyers and sellers: the classified advertisements in particular form a meeting place for potential buyers and sellers. The identification of buyers and sellers reduces drastically the cost of search. But advertising has its own limitations: advertising itself is an expense, and one essentially independent of the value of the item advertised. The advertising of goods which have few potential buyers relative to the circulation of the advertising medium is especially expensive. We shall temporarily put advertising aside and consider an alternative.

The alternative solution is the development of specialized traders whose chief service, indeed, is implicitly to provide a meeting place for potential buyers and sellers. A used-car dealer, turning over a thousand cars a year, and presumably encountering three or five thousand each of buying and selling bids, provides a substantial centralization of trading activity. Let us consider these dealer markets, which we shall assume to be competitive in the sense of there being many independent dealers.

Each dealer faces a distribution of (for example) buyers' bids and can vary his selling prices with a corresponding effect upon purchases. Even in the markets for divisible (and hence non-unique) goods there will be some scope for higgling (discrimination) in each individual transaction: the buyer has a maximum price given by the lowest price he encounters among the dealers he has searched (or plans to search), but no minimum price. But let us put this range of indeterminacy aside, perhaps by assuming that the dealer finds discrimination too expensive, and inquire how the demand curve facing a dealer is determined.

Each dealer sets a selling price, p, and makes sales to all buyers for whom this is the minimum price. . . . The number of buyers from a dealer increases as his price is reduced, and at an increasing rate. Moreover, with the uniform distribution of asking prices, the number of buyers increases with increased search if the price is below the reciprocal of the amount of search. We should generally expect the high-price sellers to be small-volume sellers.

The stability of any distribution of asking prices of dealers will depend upon the costs of dealers. . . . With economies of scale, the competition of dealers will eliminate the profitability of quoting very high selling and very low buying prices and will render impossible some of the extreme price bids. On this score, the greater the decrease in average cost with volume, the smaller will be the dispersion of prices. Many distributions of prices will be inconsistent with any possible cost conditions of dealers, and it is not evident that strict equalities of rates of return for dealers are generally possible.

If economies of scale in dealing lead to a smaller dispersion of asking prices than do constant costs of dealing, similarly greater amounts of search will lead to smaller dispersion of observed selling prices by reducing the number of purchasers who will pay high prices. Let us consider more closely the determinants of search.

Determinants of Search

. . . If the correlation of asking prices of dealers in successive time periods is perfect (and positive!), the initial search is the only one that need be undertaken. In this case the expected savings of search will be the present value of the discounted savings on all future purchases, the future savings extending over the life of the buyer or seller (whichever is shorter). On the other hand, if asking prices are uncorrelated in successive time periods, the savings from search will pertain only to that period, and search in each period is independent of previous experience. If the correlation of successive prices is positive, customer search will be larger in the initial period than in subsequent periods. . . .

As a rule, positive correlations should exist with homogeneous products. The amount of search will vary among individuals because of differences in their expenditures on a commodity or differences in cost of search. A seller who wishes to obtain the continued patronage of those buyers who value the gains of search more highly or have lower costs of search must see to it that he is quoting relatively low prices. In fact, goodwill may be defined as continued patronage by customers without continued search (that is, no more than occasional verification).

A positive correlation of successive asking prices justifies the widely held view that inexperienced buyers (tourists) pay higher prices in a market than do experienced buyers. The former have no accumulated knowledge of asking prices, and even with an optimum amount of search they will pay higher prices on average. Since the variance of the expected minimum price decreases with additional search, the prices paid by inexperienced buyers will also have a larger variance.

If a buyer enters a wholly new market, he will have no idea of the dispersion of prices and hence no idea of the rational amount of search he should make. In such cases the dispersion will presumably be estimated by some sort of sequential process, and this approach would open up a set of problems I must leave for others to explore. But, in general, one approaches a market with some general knowledge of the amount of dispersion, for dispersion itself is a function of the average amount of search, and this in turn is a function of the nature of the commodity:

1. The larger the fraction of the buyer's expenditures on the commodity, the greater the savings from search and hence the greater the amount of search.
2. The larger the fraction of repetitive (experienced) buyers in the market, the greater the effective amount of search (with positive correlation of successive prices).
3. The larger the fraction of repetitive sellers, the higher the correlation between successive prices, and hence, by condition (2), the larger the amount of accumulated search.
4. The cost of search will be larger, the larger the geographical size of the market.

An increase in the number of buyers has an uncertain effect upon the dispersion of asking prices. The sheer increase in numbers will lead to an increase in the number of dealers and, ceteris paribus, to a larger range of asking prices. But, quite aside from advertising, the phenomenon of pooling information will increase. Information is pooled when two buyers compare prices: if each buyer canvasses s sellers, by combining they effectively canvass 2*s* sellers, duplications aside. Consumers compare prices of some commodities (for example, liquor) much more often than of others (for example, chewing gum)—in fact, pooling can be looked upon as a cheaper (and less reliable) form of search.

Sources of Dispersion

One source of dispersion is simply the cost to dealers of ascertaining rivals' asking prices, but even if this cost were zero the dispersion of prices would not vanish. The more important limitation is provided by buyers' search, and, if the conditions and participants in the market were fixed in perpetuity, prices would immediately approach uniformity. Only those differences could persist which did not remunerate additional search. . . .

The maintenance of appreciable dispersion of prices arises chiefly out of the fact that knowledge becomes obsolete. The conditions of supply and demand, and therefore the distribution of asking prices, change over time. There is no method by which buyers or sellers can ascertain the new average price in the market appropriate to the new conditions except by search. Sellers cannot maintain perfect correlation of successive prices, even if they wish to do so, because of the costs of search. Buyers accordingly cannot make the amount of investment in search that perfect correlation of prices would justify. The greater the instability of supply and/or demand conditions, therefore, the greater the dispersion of prices will be.

In addition, there is a component of ignorance due to the changing identity of buyers and sellers. There is a flow of new buyers and sellers in every market, and they are at least initially uninformed on prices and by their presence make the information of experienced buyers and sellers somewhat obsolete.

The amount of dispersion will also vary with one other characteristic which is of special interest: the size (in terms of both dollars and number of traders) of the market. As the market grows in these dimensions, there will appear a set of firms which specialize in collecting and selling information. They may take the form of trade journals or specialized brokers. Since the cost of collection of information is (approximately) independent of its use (although the cost of dissemination is not), there is a strong tendency toward monopoly in the provision of information: in general, there will be a "standard" source for trade information.

Advertising

Advertising is, among other things, a method of providing potential buyers with knowledge of the identity of sellers. It is clearly an immensely powerful instrument for the elimination of ignorance—comparable in force to the use of the book instead of the oral discourse to communicate knowledge. A small $5 advertisement in a metropolitan newspaper reaches (in the sense of being read) perhaps 25,000 readers, or fifty readers per penny, and, even if only a tiny fraction are potential buyers (or sellers), the economy they achieve in search, as compared with uninstructed solicitation, may be overwhelming. . . .

The effect of advertising prices, then, is equivalent to that of the introduction of a very large amount of search by a large portion of the potential buyers. It follows from our discussion [above] that the dispersion of asking prices will be much reduced. Since advertising of prices will be devoted to products for which the marginal value of search is high, it will tend to reduce dispersion most in commodities with large aggregate expenditures.

Conclusions

The identification of sellers and the discovery of their prices are only one sample of the vast role of the search for information in economic life. Similar problems exist in the detection of profitable fields for investment and in the worker's choice of industry, location, and job. The search for knowledge on the quality of goods, which has been studiously avoided in this paper, is perhaps no more important but, certainly, analytically more difficult. Quality has not yet been successfully specified by economics, and this elusiveness extends to all problems in which it enters.

Some forms of economic organization may be explicable chiefly as devices for eliminating uncertainties in quality. The department store, as Milton Friedman has suggested to me, may be viewed as an institution which searches for the superior qualities of goods and guarantees that they are good quality. "Reputation" is a word which denotes the persistence of quality, and reputation commands a price (or exacts a penalty) because it economizes on search. When economists deplore the reliance of the consumer on reputation—although they choose the articles they read (and their colleagues) in good part on this basis—they implicitly assume that the consumer has a large laboratory, ready to deliver current information quickly and gratuitously.

Ignorance is like subzero weather: by a sufficient expenditure its effects upon people can be kept within tolerable or even comfortable bounds, but it would be wholly uneconomic entirely to eliminate all its effects. And, just as an analysis of man's shelter and apparel would be somewhat incomplete if cold weather is ignored, so also our understanding of economic life will be incomplete if we do not systematically take account of the cold winds of ignorance.

The Market for Lemons: Qualitative Uncertainty and the Market Mechanism

GEORGE AKERLOF

This paper relates quality and uncertainty. The existence of goods of many grades poses interesting and important problems for the theory of markets. On the one hand, the interaction of quality differences and uncertainty may explain important institutions of the labor market. On the other hand, this paper presents a struggling attempt to give structure to the statement: "Business in underdeveloped countries is difficult";

George Akerlof, The Market for Lemons: Qualitative Uncertainty and the Market mechanism, 84 *Quarterly Journal of Economics* (1970), pp. 488–500. Copyright © 1970 The President and Fellows of Harvard College. Reprinted with permission.

in particular, a structure is given for determining the economic costs of dishonesty. Additional applications of the theory include comments on the structure of money markets, on the notion of "insurability," on the liquidity of durables, and on brand-name goods.

There are many markets in which buyers use some market statistic to judge the quality of prospective purchases. In this case there is incentive for sellers to market poor quality merchandise, since the returns for good quality accrue mainly to the entire group whose statistic is affected rather than to the individual seller. As a result there tends to be a reduction in the average quality of goods and also in the size of the market. It should also be perceived that in these markets social and private returns differ, and therefore, in some cases, governmental intervention may increase the welfare of all parties. Or private institutions may arise to take advantage of the potential increases in welfare which can accrue to all parties. By nature, however, these institutions are nonatomistic, and therefore concentrations of power—with ill consequences of their own—can develop.

The automobile market is used as a finger exercise to illustrate and develop these thoughts. It should be emphasized that this market is chosen for its concreteness and ease in understanding rather than for its importance or realism.

The Model with Automobiles as an Example

The Automobiles Market

The example of used cars captures the essence of the problem. From time to time one hears either mention of or surprise at the large price difference between new cars and those which have just left the showroom. The usual lunch table justification for this phenomenon is the pure joy of owning a "new" car. We offer a different explanation. Suppose (for the sake of clarity rather than reality) that there are just four kinds of cars. There are new cars and used cars. There are good cars and bad cars (which in America are known as "lemons"). A new car may be a good car or a lemon, and of course the same is true of used cars.

The individuals in this market buy a new automobile without knowing whether the car they buy will be good or a lemon. But they do know that with probability q it is a good car and with probability $(1 - q)$ it is a lemon; by assumption, q is the proportion of good cars produced and $(1 - q)$ is the proportion of lemons.

After owning a specific car, however, for a length of time, the car owner can form a good idea of the quality of this machine; i.e., the owner assigns a new probability to the event that his car is a lemon. This estimate is more accurate than the original estimate. An asymmetry in available information has developed: for the sellers now have more knowledge about the quality of a car than the buyers. But good cars and bad cars must still sell at the same price—since it is impossible for a buyer to tell the difference between a good car and a bad car. It is apparent that a used car cannot have the same valuation as a new car—if it did have the same valuation, it would clearly be advantageous to trade a lemon at the price of new car, and buy another new car, at a higher probability q of being good and a lower probability of being bad. Thus the owner of a good machine must be locked in. Not only is it true that he cannot receive the true value of his car, but he cannot even obtain the expected value of a new car. . . .

Asymmetrical Information

It has been seen that the good cars may be driven out of the market by the lemons. But in a more continuous case with different grades of goods, even worse pathologies can exist. For it is quite possible to have the bad driving out the not-so-bad driving out the medium driving out the not-so-good driving out the good in such a sequence of events that no market exists at all.

One can assume that the demand for used automobiles depends most strongly upon two variables—the price of the automobile p and the average quality of used cars traded, μ. . . . Both the supply of used cars and also the average quality will depend upon the price. . . . And in equilibrium the supply must equal the demand for the given average quality. . . . As the price falls, normally the quality will also fall. And it is quite possible that no goods will be traded at any price level.

Examples and Applications

Insurance

It is a well-known fact that people over 65 have great difficulty in buying medical insurance. The natural question arises: why doesn't the price rise to match the risk?

Our answer is that as the price level rises the people who insure themselves will be those who are increasingly certain that they will need the insurance; for error in medical check-ups, doctors' sympathy with older patients, and so on make it much easier for the applicant to assess the risks involved than the insurance company. The result is that the average medical condition of insurance applicants deteriorates as the price level rises—with the result that no insurance sales may take place at any price. This is strictly analogous to our automobiles case, where the average quality of used cars supplied fell with a corresponding fall in the price level. This agrees with the explanation in insurance textbooks. . . .

The statistics do not contradict this conclusion. While demands for health insurance rise with age, a 1956 national sample survey of 2,809 families with 8,898 persons shows that hospital insurance coverage drops from 63 per cent of those aged 45 to 54, to 31 per cent for those over 65. And surprisingly, this survey also finds average medical expenses for males aged 55 to 64 of $88, while males over 65 pay an average of $77. While noninsured expenditure rises from $66 to $80 in these age groups, insured expenditure declines from $105 to $70. The conclusion is tempting that insurance companies are particularly wary of giving medical insurance to older people.

The principle of "adverse selection" is potentially present in all lines of insurance. The following statement appears in an insurance textbook written at the Wharton School:

> There is potential adverse selection in the fact that healthy term insurance policy holders may decide to terminate their coverage when they become older and premiums

mount. This action could leave an insurer with an undue proportion of below average risks and claims might be higher than anticipated. Adverse selection "appears (or at least is possible) whenever the individual or group insured has freedom to buy or not to buy, to choose the amount or plan of insurance, and to persist or to discontinue as a policy holder."

Group insurance, which is the most common form of medical insurance in the United States, picks out the healthy, for generally adequate health is a precondition for employment. At the same time this means that medical insurance is least available to those who need it most, for the insurance companies do their own "adverse selection."

This adds one major argument in favor of medicare. On a cost benefit basis medicare may pay off: for it is quite possible that every individual in the market would be willing to pay the expected cost of his medicare and buy insurance, yet no insurance company can afford to sell him a policy—for at any price it will attract too many "lemons." The welfare economics of medicare, in this view, is exactly analogous to the usual classroom argument for public expenditure on roads.

The Employment of Minorities

The Lemons Principle also casts light on the employment of minorities. Employers may refuse to hire members of minority groups for certain types of jobs. This decision may not reflect irrationality or prejudice—but profit maximization. For race may serve as a good statistic for the applicant's social background, quality of schooling, and general job capabilities.

Good quality schooling could serve as a substitute for this statistic; by grading students the schooling system can give a better indicator of quality than other more superficial characteristics. . . . An untrained worker may have valuable natural talents, but these talents must be certified by "the educational establishment" before a company can afford to use them. The certifying establishment, however, must be credible; the unreliability of slum schools decreases the economic possibilities of their students.

This lack may be particularly disadvantageous to members of already disadvantaged minority groups. For an employer may make a rational decision not to hire any members of these groups in responsible positions—because it is difficult to distinguish those with good job qualifications from those with bad qualifications. This type of decision is clearly what George Stigler had in mind when he wrote, "in a regime of ignorance Enrico Fermi would have been a gardener, Von Neumann a checkout clerk at a drugstore."

As a result, however, the rewards for work in slum schools tend to accrue to the group as a whole—in raising its average quality—rather than to the individual. Only insofar as information in addition to race is used is there any incentive for training.

An additional worry is that the Office of Economic Opportunity is going to use cost-benefit analysis to evaluate its programs. For many benefits may be external. The benefit from training minority groups may arise as much from raising the average quality of the group as from raising the quality of the individual trainee; and, likewise, the returns may be distributed over the whole group rather than to the individual.

The Costs of Dishonesty

The Lemons model can be used to make some comments on the costs of dishonesty. Consider a market in which goods are sold honestly or dishonestly; quality may be represented, or it may be misrepresented. The purchaser's problem, of course, is to identify quality. The presence of people in the market who are willing to offer inferior goods tends to drive the market out of existence—as in the case of our automobile "lemons." It is this possibility that represents the major costs of dishonesty—for dishonest dealings tend to drive honest dealings out of the market. There may be potential buyers of good quality products and there may be potential sellers of such products in the appropriate price range; however, the presence of people who wish to pawn bad wares as good wares tends to drive out the legitimate business. The cost of dishonesty, therefore, lies not only in the amount by which the purchaser is cheated; the cost also must include the loss incurred from driving legitimate business out of existence.

Dishonesty in business is a serious problem in underdeveloped countries. Our model gives a possible structure to this statement and delineates the nature of the "external" economies involved. In particular, in the model economy described, dishonesty, or the misrepresentation of the quality of automobiles, costs $\frac{1}{2}$ unit of utility per automobile; furthermore, it reduces the size of the used car market from N to 0. We can, consequently, directly evaluate the costs of dishonesty—at least in theory.

There is considerable evidence that quality variation is greater in underdeveloped than in developed areas. For instance, the need for quality control of exports and State Trading Corporations can be taken as one indicator. In India, for example, under the Export Quality Control and Inspection Act of 1963, "about 85 per cent of Indian exports are covered under one or the other type of quality control." Indian housewives must carefully glean the rice of the local bazaar to sort out stones of the same color and shape which have been intentionally added to the rice. Any comparison of the heterogeneity of quality in the street market and the canned qualities of the American supermarket suggests that quality variation is a greater problem in the East than in the West.

In one traditional pattern of development the merchants of the pre-industrial generation turn into the first entrepreneurs of the next. The best-documented case is Japan, but this also may have been the pattern for Britain and America. In our picture the important skill of the merchant is identifying the quality of merchandise; those who can identify used cars in our example and can guarantee the quality may profit by as much as the difference between [low-quality] traders' buying price and [high-quality] traders' selling price. These people are the merchants. In production these skills are equally necessary—both to be able to identify the quality of inputs and to certify the quality of outputs. And this is one (added) reason why the merchants may logically become the first entrepreneurs.

The problem, of course, is that entrepreneurship may be a scarce resource; no development text leaves entrepreneurship unemphasized. Some treat it as central. Given, then, that entrepreneurship is scarce, there are two ways in which product variations impede development. First, the pay-off to trade is great for would-be entrepreneurs, and hence they are diverted from production; second, the amount of entrepreneurial time per unit output is greater, the greater are the quality variations. . . .

Counteracting Institutions

Numerous institutions arise to counteract the effects of quality uncertainty. One obvious institution is guarantees. Most consumer durables carry guarantees to ensure the buyer of some normal expected quality. One natural result of our model is that the risk is borne by the seller rather than by the buyer.

A second example of an institution which counteracts the effects of quality uncertainty is the brand-name good. Brand names not only indicate quality but also give the consumer a means of retaliation if the quality does not meet expectations. For the consumer will then curtail future purchases. Often too, new products are associated with old brand names. This ensures the prospective consumer of the quality of the product.

Chains—such as hotel chains or restaurant chains—are similar to brand names. One observation consistent with our approach is the chain restaurant. These restaurants, at least in the United States, most often appear on interurban highways. The customers are seldom local. The reason is that these well-known chains offer a better hamburger than the average local restaurant; at the same time, the local customer, who knows his area, can usually choose a place he prefers.

Licensing practices also reduce quality uncertainty. For instance, there is the licensing of doctors, lawyers, and barbers. Most skilled labor carries some certification indicating the attainment of certain levels of proficiency. The high school diploma, the baccalaureate degree, the Ph.D., even the Nobel Prize, to some degree, serve this function of certification. And education and labor markets themselves have their own "brand names."

Conclusion

We have been discussing economic models in which "trust" is important. Informal unwritten guarantees are preconditions for trade and production. Where these guarantees are indefinite, business will suffer. . . . But the difficulty of distinguishing good quality from bad is inherent in the business world; this may indeed explain many economic institutions and may in fact be one of the more important aspects of uncertainty.

6.2 APPLICATIONS

The Contract-Tort Boundary and the Economics of Insurance

WILLIAM BISHOP

Events of Very Low Probability in Contract

In contract cases concerning nonperformance of the promise, the problem of events of very low probability is dealt with mainly under the doctrines of frustration, impossibility, and common mistake. The law of contract also has a doctrine called remoteness of damage, the rule in *Hadley v. Baxendale*. Where nonperformance of the contractual promise is in question, this contract doctrine of remoteness of damage deals most importantly with the adverse selection problem rather than with the problem of low probability events. . . .

Where an event of very low prior probability occurs, should courts hold a promisor to his promise? Performance may have become very expensive or impossible, or the consequences of nonperformance may be very much more expensive than was contemplated when the contract was made. In general a promisor is excused performance where an uncontemplated event of very low prior probability having expensive consequences has occurred, as noted, chiefly under the headings of common mistake and frustration (or impossibility). The main distinction between frustration and common mistake is a formal one. Common mistake deals with cases where both parties make and act on an incorrect assumption about facts existing at the time the contract is made; the usual consequence is that such a contract cannot be enforced. Frustration deals with events arising subsequent to the time of contracting; the usual consequence here is that a contract is discharged from the time of frustration. Unilateral mistake, as we shall see, deals with a different problem, with adverse selection and not with events of low probability.

To illustrate, consider an agreement on April 10 by A to deliver a cargo of wheat to B when the ship carrying it arrives, both parties believing the cargo to be in transit at that date. If the ship had sunk and the cargo been lost before April 10, the case is one of mistake and no rights or liabilities are created by the contract. If the ship sinks after April 10, then the case is one of frustration and such rights and obligations as accrued before sinking are enforceable, but those scheduled to accrue thereafter are not. The sinking has made performance of the promise much more costly for A, and he is excused. Now B bears his own losses, just as he would in tort law if he were the victim of a freak accident caused by A.

However, frustration and common mistake do not deal with all possible events of very low probability. They concern only events that make performance costly to the

William Bishop, The Contract-Tort Boundary and the Economics of Insurance, 12 *Journal of Legal Studies* (1983), pp. 241–266. Copyright © 1983 University of Chicago Law School. Reprinted with permission.

promisor. Where the event increases the victim's loss, the rule in *Hadley v. Baxendale* applies.

The rule in *Hadley v. Baxendale* must deal with two distinct classes of case: first, cases where both victim and "injurer" are equally ignorant of the abnormal risk, and second, cases where the victim is better informed about the possibilities than is the "injurer" (i.e., the performing party). Cases of unusually heavy loss arising from an accident usually fall into the first category. Such cases typically arise in claim under a contractual warranty of quality or a contractual duty of care. Here the contract remoteness doctrine in *Hadley* is and ought to be equivalent to the remoteness doctrine in tort. This is most clearly seen by considering the contractual duty of care. Here the basis of liability is the same in contract as in tort. The duty is almost always implied by the parties' actions rather than explicitly bargained over. The question "What would they have agreed to if . . . ?" is the spectral guest at the feast in contract as well as in tort. This identity of purpose leads to a rule that is identical.

However, the second class of case, where the victim is better informed about the likelihood of unusually large losses, leads to a different analysis. It raises adverse selection possibilities and requires different rules. Once the adverse selection problem has been examined it will become plain that the contract remoteness rules in *Hadley* deal in different cases with two quite distinct problems. In effect, two quite different "rules" are embodied in a single verbal formulation.

Adverse Selection in Contract

The classic formulation of the doctrine of remoteness in contract was that of Alderson, B. in *Hadley v. Baxendale:*

> Where two parties have made a contract which one of them has broken, the damages which the other party ought to receive in respect of such breach of contract should be such as may fairly and reasonably be considered either arising naturally, i.e., according to the usual course of things, from such breach of contract itself, or such as may reasonably be supposed to have been in the contemplation of both parties, at the time they made the contract, as the probable result of the breach of it.

The "first rule" of *Hadley* has been considered in numerous cases. In *Heron II* the court held that damages could be recovered if there were a "serious possibility" or "real danger" of their arising from the breach, but not if (as in tort) they were merely "foreseeable" as one of many possibilities. Lord Reid also used the expression "not unlikely" as a test. Lord Reid gave as reason for the difference between tort and contract the fact that "in contract, if one party wishes to protect himself against a risk which to the other party would appear unusual, he can direct the other party's attention to it before the contract is made. . . . But in tort there is no opportunity for the injured party to protect himself in that way. . . ." Reid's view is widely, and correctly, regarded as the normal explanation for the difference. It has, however, ramifications so far not generally appreciated, which are pursued here.

The line of cases *Hadley v. Baxendale, British Columbia Sawmill Co. v.*

Nettleship, *Victoria Laundry*, and *Heron II* concerns a matter that is usually irrelevant in tort (at least as between strangers): the efficient transfer of information. The law of contract denies recovery to a plaintiff when four conditions are met:

1. The plaintiff possessed information unknown to the defendant.
2. The defendant, had he possessed that information, might have altered his behavior so as to make his breach less likely to occur.
3. The plaintiff could have conveyed the information to the defendant cheaply. (This condition is not mentioned in the cases, though it is clear that it is assumed by the courts to be fulfilled. Of course it is normally not fulfilled in tort.)
4. The plaintiff did not do so.

A good example of these rules in operation is the *Victoria Laundry* case. The defendant manufacturer of boilers contracted to supply them to the plaintiff laundry. In breach of contract the manufacturer delivered late. The laundry sought to recover damages for profit lost on an unusually lucrative dyeing contract. The court limited the plaintiff to such damages as would be normal in a case of this kind.

Less clear is *Heron II*. There the plaintiff charterer under a charterparty to transport sugar to a well known sugar market complained that the shipowner had in breach of contract delayed arrival in port by ten days. In the interim the price of sugar fell. The court held the shipowner liable for the loss. Lord Reid thought that the circumstances (that the charterer might well wish to sell on arrival) ought to have been so clear to the shipowner that the latter ought to have realized the risk without explicit warning. This case is near the line, with everything depending on the circumstances the parties were in. If the circumstances were clear, then to require an explicit warning of the obvious would be wasteful of resources (here labor and time). . . .

The central point here is that where the four conditions above are met the value of the information to the defendant is greater than the cost to the plaintiff of conveying that information to him. To encourage such efficient transfers of information is the purpose of the contract remoteness rule of *Hadley v. Baxendale*.

Note that the first limb of the rule in *Hadley* fits easily into this scheme. The normal case is one in which no information needs to be conveyed, since the defendant, knowing normal business conditions, already knows as much as the plaintiff. To require the plaintiff to inform the defendant of normal conditions would be inefficient, because the cost of transactions here, though low, is not zero. Any expenditure on information transfer is only wasted.

It might seem that there is a *casus omissus*. Consider the case where the consequence of the defendant's breach normally would be a certain loss but in fact is less. Then it seems the plaintiff has no incentive to transfer the information, even though the information would be valuable in that it would allow the defendant to spend less in essential reliance. Such transfer would be efficient: the marginal social value of expenditure on breach avoidance is lower and so less should be spent. But in fact there is no *casus omissus*. The plaintiff has sufficient incentive to inform the defendant, wholly without legal compulsion, if such information transfer is in fact cost justified. The reason is that the plaintiff can obtain a lower price for the defendant's performance if he informs the defendant of the limited damages for breach. The price is lower because such a contract is cheaper to perform than is a "normal" contract. This incentive will induce parties to act efficiently in the case of unusually inexpensive breach as well as in the case of unusually expensive breach.

It should be clear that the function of remoteness in the *Hadley v. Baxendale* line of contract cases is very different from the function of remoteness in tort. The tort measure of foreseeability seeks to define as too remote an event that no one would anticipate at all—one to which the ordinary observer would assign near zero probability. The contract measure of foreseeability will include as too remote many consequences which are merely unusual ones that have quite substantial probabilities of occurring. The defining characteristics of an event that is too remote for the purposes of contract are those set out in conditions 1–4 above. These conditions have nothing to do with unforeseeability in the sense of very low probability. If this analysis is correct, it follows that Lord Scarman was wrong to suggest, as he did in *Parsons v. Uttley,* that the differences between contract and tort remoteness are semantic only. Rather they are distinctions of substance with a coherent purpose behind them.

The promisors or defendants in *Hadley, Victoria Laundry,* and *Heron II* were insurers. Like all insurers they charged a price for that service. Like all insurers they wished to guard against "adverse selection," against high risk promisees obtaining low priced insurance. This is not just a distributional matter between promisor and promisee. If such adverse selection occurs, promisors will make fewer promises and ask higher prices for them. In consequence, planning becomes harder and more costly. So long as planning for the future is of value, this form of adverse selection will be disadvantageous for society.

In sum then, analytically the rule in *Hadley v. Baxendale* is a rule designed to minimize adverse selection. As we have seen, cheap information is the best antidote to adverse selection problems. So in contract the promisor is entitled to assume "usual risks" unless he is notified to the contrary, whereupon he can demand and obtain a high price.

In cases of nonperformance of the promise the contract doctrine of unilateral mistake (that is, a plaintiff cannot hold a defendant to his contractual promise if the plaintiff knew or ought to have known that the defendant entered into the contract under a mistake) is also in a sense addressed to the adverse selection problem. Again, where we label the case as one of mistake the crucial circumstance is one existing at the time the contract was made, and not some subsequent event. But this is only a matter of doctrinal classification, for again the important point concerns information transfer. Where one party to the knowledge of the other is mistaken, information transfer is cheap; where information is valuable to the pricing of the contract, other things equal, it should be transferred. Exceptions to the disclosure requirement can be explained as attempts to overcome the inefficiency arising from the nonappropriability characteristic of information–thus the right to enforce a contract creates a property right in the information, giving people an incentive to invest in generating valuable information.

In cases of a contractual warranty of quality the adverse selection problem is similarly dealt with by the doctrine of fitness for purpose. Under the law of sale in common-law countries the seller can be liable to the buyer for buyer's loss when the goods are unfit for the buyer's purpose, even if the purpose is unusual, provided the buyer notified the seller of that purpose. Again, efficient information transfer is the gist of the rule.

Finally, problems of adverse selection under contractual warranties of quality are catered for by special clauses limiting liability or excluding it in cases of certain specified kinds of loss. These limitations prevent the highest risk (or highest cost) customers from undermining the economic viability of warranties given to the lower risk customers. Such low risk warranties are undermined if adverse selection makes warranties expensive for all. . . .

Adverse Selection in Tort

In strict liability cases the adverse selection problem can be severe and is dealt with in several ways. First, liability is often narrowly defined. As an extreme example, the history of the rule in *Rylands v. Fletcher,* at least in England, is the history of successive circumscription by courts. Second, liability is subject to limitation on recovery. For example, the European Community products liability proposals suggest a cash ceiling on the total claims any one manufacturer must pay the victims of any one defective type of product. Third, it has been suggested that courts ought to adopt a defense of "extrasensitivity." A victim whose loss is much greater than the loss that a "normal" victim would have suffered would be limited to "normal" compensation. For example, there is no good reason why an employer, even if at fault, should have to compensate a young concert violinist with a great future whose left hand is harmed in the employer's steel mill, when the employer has not been informed of the high-risk employee. It would be better ("more efficient") if the violinist were to stay away from that particular summer job. Fourth, where tort rules are used in a consensual setting (for example, products liability cases between producer and purchaser and medical malpractice cases), duty to warn doctrines are concerned with the information transfer problem. Critics of some of the new rules claim that courts have developed the law poorly because of their failure to recognize the importance of contract principles.

The adverse selection problem by definition arises only where there is selection by victims, that is, where, because of the existence of insurance, high-risk persons are more likely to place themselves in positions where they will become victims and so collect that insurance. This can arise only where victims know they are high-risk or high-cost cases. Therefore extrasensitivity should be relevant only in cases where the victims knew of their own riskiness while the injurers did not, and where the victims could either have avoided the risky activity or informed potential injurers so that the latter could have increased their precautions. Given these conditions, an extrasensitivity defense often will not be appropriate in nonconsensual cases. Injurers should face all social costs, even from expensive victims, except in those cases where inexpensive avoidance by high-cost victims is possible or where information can be passed to injurers that improves their cost benefit calculations about avoidance expenditure.

In cases of negligence where the duty is not strict the adverse selection problem

is less severe. But there will be cases where injurers might have modified their behavior had they known of the greater risk. There is really no solution to this problem,
since communication between the parties is usually impossible. The potential injurer
must calculate the cost-justified level of avoidance and must take account of all foreseeable losses, even the very severe ones. Because communication between victim
and injurer is too costly the parties cannot achieve the fine tuning of precaution to risk
that they can achieve in consensual cases.

In cases of negligent misstatement, the problem of adverse selection is solved by
the requirement that the injurer is liable only if the victim acted reasonably in relying
on the statement. Obviously, where large sums are involved, a victim is not behaving
reasonably who relies on mere statement as opposed to a bought promise of guarantee. . . .

The Informational Role of Warranties and Private Disclosure about Product Quality

SANFORD GROSSMAN

A fundamental role of competition is to facilitate the allocation of resources. This is
achieved by prices which, to some extent, reflect and transmit the underlying worth
of resources. The ability of prices to reflect and transmit information derives from the
attempt of economic agents to buy or sell based on their information. Competition
among all those who want to buy wheat because they think that wheat will be scarce
tomorrow drives up the price of wheat. Hence their information can be transmitted
by the price system to those who store wheat but do not have direct access to information about next period's wheat demand. This mechanism works because there is
some future state of nature which will lead to prices which reward those who buy or
sell today.

Unfortunately, there are situations where no such prices exist. An important case
involves information about product quality. Sellers may know the quality of the item
they sell but it may be in their interest to withhold that information. If there is no way
for buyers to learn about the sellers' quality, then this will force all items to sell at the
same price. If there is no way sellers of good-quality items can distinguish themselves
from sellers of low-quality items, then the low-quality sellers will find it in their interest to hide their quality. This has been called the "lemons problem."

In this paper, we will consider cases where good sellers have an incentive to distinguish themselves from bad sellers. Of particular interest is the case where there is
a single seller. Consumer information is often quite poor about those products which
are new. It is just these products where (temporary) monopoly is likely to be found.
Thus, we will be concerned with cases where a monopolist could have an incentive
to reveal his quality even when it is low. In order for this incentive to exist, there must

Sanford Grossman, The Informational Role of Warranties and Private Disclosure about Product Quality,
24 *Journal of Law and Economics* (1981), pp. 461–483. Copyright © 1981 University of Chicago Law
School. Reprinted with permission.

be some event which occurs after the sale which will reward sellers as a function of their true quality. We consider two cases.

In our first case, the seller can make statements about the product's quality which are ex post verifiable. For example, a diamond seller can disclose the weight of a diamond he is selling. He can give the buyer a warranty which states the weight of the diamond. That is, in [the following section] we will consider situations where sellers can make any disclosure about their product's quality and give a complete warranty which guarantees that the disclosure is true (for example, the diamond seller gives the buyer a written statement guaranteeing that the diamond can be returned if an objective party finds that its weight is less than specified).

The second case which we consider is where statements about product quality are too costly either to communicate or verify ex post. For this reason, the statements cannot be guaranteed. However, we assume that there is some characteristic which is observable ex post. For example, the quality of an automobile's construction is difficult to describe or verify ex post. However, it may be easy to verify ex post whether the auto "breaks down." If it is the case that low-quality items have a higher probability of breakdown than do high-quality, then warranties which guarantee against (the ex post cheaply observable) breakdown can substitute for guarantees regarding (the ex post very costly to observe) quality.

This paper is primarily concerned with situations where consumers have had no experience and will have no further experience with the monopolist. This is the case where the monopolist will have the greatest incentive to mislead. Our basic result is, however, that the monopolist will not be able to mislead rational consumers about the quality of his product. . . .

A Model of Private Disclosure

In this section we assume that it is possible for a seller to make ex post verifiable disclosures. For example, a diamond seller can specify the weight of the diamond; a doctor can specify the medical school he graduated from, his class standing, the number of malpractice suits he is engaged in, and so forth. In this section, we will be concerned with disclosures that have negligible ex post verification cost and also negligible communications cost. For example, it would be very costly for a doctor to explain to a patient, in detail, his contribution to the study of ulcers. This might involve imparting four years' worth of medical school training to the patient. Yet, if the patient must choose between two doctors, this information gross of acquisition costs is very valuable (while net of acquisition costs it is of no value). Note further than the doctor may desire to substitute low communications cost information such as, "I am the best ulcer doctor in the world," but such a statement is not easily subject to verification primarily because it is not sufficiently detailed—what does "best" mean?.

In this section we will not model a consumer's decision about verifying ex post the truth of the statement made by the seller. Rather, for simplicity, we will limit attention to situations where all of the seller's statements are costlessly verifiable ex post. A clear example is where the seller is selling boxes of oranges. If the seller states that there are ten oranges in a box, then this becomes verifiable ex post for free. It is important that the information be publicly verifiable for the purpose of this section. In particular, if the seller states that a product will "make the buyer happy," then this fact is not open to easy third-party verification. When the seller states that the diamond weighs one ounce, this is very cheap to verify ex post.

There has been much recent interest in laws which require sellers to make particular disclosures. This is to be distinguished from antifraud laws which make it illegal for a seller to lie. It is sometimes argued that there are disclosures that are of negligible cost but are not made by sellers in an attempt to mislead buyers. Hence a law is needed which requires such disclosures. This section focuses on cases of costless disclosure to derive insight about the issue of what a firm would voluntarily disclose.

We restrict attention to disclosures which are truthful. (For example, a seller who says nothing is making a truthful disclosure. A seller with a diamond which weighs one ounce who states that it weighs at least one-half ounce is making a truthful disclosure, while if the same seller said that his diamond weighs two ounces it would not be truthful.) We consider only truthful disclosures for two reasons. First, we are interested in analyzing the benefits of a positive disclosure law which are above and beyond those provided by a law against lying. Second, if there are zero ex post verification costs, sellers would warranty their disclosures. Any seller who did not warranty his disclosure would immediately be assumed to be lying; that is, saying nothing. As can easily be seen from the analysis to be given below, unwarranteed disclosures could easily be incorporated.

In situations where there is objective, costless ex post verification and no communications costs, we can most clearly elucidate the role of positive disclosure. In situations where sellers do not lie, the only issue is how much of the truth they will decide to tell. In particular, if a seller has a bad product, will he say nothing, leading consumers to believe his product is of average quality? Will adverse selection by the low-quality sellers drive out the high-quality sellers? If the market is competitive, then this will clearly not be the case. That is, if there is free entry into the sellers' activity, then good sellers will make disclosures to distinguish themselves from bad sellers. If any good seller should be lumped with the bad sellers due to nondisclosure, then the good seller could costlessly disclose his quality and be distinguished, getting a higher price.

The case of free entry is reasonably obvious. However, in many important cases involving consumer uncertainty, free entry may be an inappropriate assumption. Consumer information is relatively poor about new products. There may only be one firm selling a new product because of patent protection or because that firm is a particularly rapid innovator. Schumpeter has argued that an extremely important role of the competitive system is in encouraging innovation via the temporary monopoly power won by the fastest innovators. Thus it is important to ask whether a monopolist would find it in his interest to make a full disclosure. It is remarkable that if the monopolist has customers with rational expectations, then it will be in his interest to make a complete disclosure. It will be shown that adverse selection works against a monopolist who makes less than a full disclosure. The idea of the analysis can be seen from a simple example. Suppose the monopolist is selling boxes of apples. He can label the boxes with an exact number of apples, but if he does then this must be the true amount under the above "no lying" assumptions. However, he could also put no label as to the quantity or he could state: "There are at least three apples in the box." Suppose that from the size of the box consumers can tell that the box holds between zero and 100 apples, and they also know that the seller knows how many apples are in the box. Suppose the seller says nothing about the number in the box. Then a consumer could rationally conclude that the box contains no apples, for if there were, say, three apples in the box, the seller could have said, "There are at least three apples in the box." Similarly, suppose a seller makes the statement, "There are at least six apples in the box." This must mean that there are exactly six apples in the box, for if there were really seven, then the seller could have made more profit by saying that there are at least seven apples in the box. This is because the expected number of apples in the box under the latter statement is higher than the former, so consumers will be willing to pay more. Thus there is a kind of adverse selection against a seller who does not make a full disclosure, even though he is the only seller. Consumers rationally expect a seller's quality to be the poorest possible consistent with his disclosure. The seller, knowing that consumers will only offer to pay the lowest amount consistent with his disclosure, finds it optimal to disclose the highest possible quality consistent with the truth; that is, he discloses the truth when he knows it. . . .

Warranties and Indirect Disclosure

. . . As noted earlier, there exist situations where the seller has some information about a product, the disclosure of which would be costly. (For example, a used-car salesman can make statements about the kind of driver who previously owned the car, but the making of verifiable statements might be costly, as would an objective inspection of the car.) In this section, we consider the extreme case where the cost to the seller of making relevant statements or of the buyer determining the quality before purchase is larger than the difference in value between the best and worst possible commodity. We maintain the assumption that the seller knows the quality of the commodity, and this quality is exogenous. As before, we consider a situation where buyers have no experience with sellers and will have no future relationship with them.

Under the above assumptions, all sellers will be judged to be identical. Each seller, whether his commodity is best or worst, will receive the same price. There are two situations to consider. If there are no warranties or any other device other than price to signal quality, then we will have the usual "lemons" problem. There will be adverse selection against high-quality sellers. Each high-quality seller will want to be distinguished from those of average quality, but in this case there is no way for him to do so.

In some situations, sellers can attempt to distinguish their quality even if disclosures are very costly. This can be done with warranties. It seems intuitively plausible that a seller of a high-quality item can offer a better warranty than can be offered by a low-quality seller. Of course, for this to make sense, it is necessary that something be ex post objectively observable by buyers and sellers. For example, a doctor can warranty a patient against the recurrence of an illness, a manufacturer can offer a warranty against breakdown, a lawyer can take a fee contingent on success, etc. In each of these cases, it is observable ex post when the patient gets sick, the car breaks, or the legal battle is lost. In many cases such events will be much easier to verify ex post than will be the ex ante statements a seller makes about his quality, and will have lower communication cost. For example, the lawyer could try to tell the customer about all his previous cases, why losing particular cases was not his fault, compare his record with the records of other lawyers, and so on. Such statements are far more complicated and costly to the buyer and seller than a statement like "Pay me $1,000 if I lose and $10,000 if I win." Sometimes the information is impossible to convey in an ex post verifiable way. A doctor may know that he is the best doctor in existence, but there is no way (at a reasonable cost) that he can prove this to a prospective patient. In situations in which a seller's information cannot be conveyed to a buyer, the seller's warranty can, in effect, transmit that information to the buyer. There is a sense in which the degree of warranty can be a sufficient statistic for the seller's information. . . .

Extensions and Conclusions

. . .

This paper has been concerned with showing that when firms have tools available which they could use to convey information they will do so. It is not in a monopolist's interest to withhold information about product quality. If information transmittal or warranties are costless, then there is no role for government intervention to encourage disclosure. Thus, the argument that there should be a positive disclosure law or government-mandated warranties cannot be justified on the grounds that these tools have negligible social and private cost, and high benefits through giving consumers more information about product quality or less risk about product quality. One might conclude that a positive disclosure law does no harm as well. Unfortunately, disclosure laws are often very broad. Securities law requires the issuer of a new stock to disclose all facts which are material to a purchaser. This requirement may have disadvantages relative to what would arise if there is no positive disclosure law. After the purchase bad events do occur which were not perfectly predictable. Buyers can always bring suit claiming that a material fact was not disclosed regarding the possibility of the bad event. The buyer can then attempt to search the seller's records for evidence. Since this can make the seller bear costs, the seller, anticipating this, discloses an enormous amount of information in the first place. Some of this information may also surely be irrelevant but could be costly to disclose. The seller, by making excessive disclosures, makes the buyer bear more costs in trying to interpret the disclosure. This can convert a situation which involves costless disclosure of truly material facts into a situation where both the buyer and seller must bear costs. An important negative consequence of this is that disclosures may no longer reveal the quality of the seller because they have become so noisy. Thus in the case where disclosure of the (truly) material facts are costless, we are better off without a positive disclosure law.

It would be useful to see how far this policy conclusion can be extended to cases where disclosure or warranties are costly. I disclose that the voluntary disclosure theorem will not be true when disclosure is costly. Further, there are important externalities involved when search or disclosure is costly. Thus the reader should view this policy conclusion with extreme caution.

The Value of Accuracy in Adjudication
LOUIS KAPLOW

Concerns for the accuracy of adjudication permeate analyses of procedural rules and aspects of substantive law. Yet the value of more accurate adjudication is largely taken for granted. When this is done, however, there is no basis for choosing among rules (or for judges to make discretionary judgments when applying them), for it typically is the case that greater accuracy comes at a cost. Even if precise quantification of various benefits of accuracy is impossible, decision-making will be enhanced by understanding why accuracy may be desirable. . . .

Accuracy is a central concern with regard to a wide range of legal rules. One might go so far as to say that a large portion of the rules of civil, criminal, and administrative procedure and rules of evidence involve an effort to strike a balance between accuracy and legal costs. The regulation of lawyers in litigation and some other settings is appropriately viewed, for present purposes, as an aspect of procedural rules concerned with achieving accurate outcomes while not incurring excessive costs. Implicit judgments concerning the value of accuracy are central in assessing major legal reforms (such as substituting an inquisitorial system for an adversarial system), evaluating more modest changes (limiting discovery or the use of expert witnesses), determining how adjudicators should exercise discretion (pretrial orders concerning the conduct of litigation), and designing and using alternative dispute resolution (often specified by contract).

Many aspects of substantive law also are concerned with accuracy. Most obvious are special burdens of proof in particular areas of law (such as burden shifting in employment discrimination cases or *res ipsa loquitur* in tort law) and rules concerning what evidence meets even conventional proof burdens (such as whether a dealer's complaint to a manufacturer is sufficient evidence of an antitrust conspiracy to reach a jury). Also important are components of substantive law that determine which categories of behavior are to be distinguished. For example, when assault with intent to commit murder is made a crime separate from simple assault, the legal system distinguishes two types of behavior, with the result that sanctions are more precisely tailored to individuals' conduct. . . .

This article is primarily concerned with the question of why accuracy is valuable. It is assumed throughout that more accuracy can be obtained only at a higher cost. No attempt is made to determine which legal rules are more accurate. by how much, and at what cost. Such inquiries are best made case by case. Rather, the analysis seeks to illuminate the following sort of inquiry: if a contemplated legal reform would increase

Louis Kaplow, The Value of Accuracy in Adjudication, 23 *Journal of Legal Studies* (1994), pp. 307–402.

accuracy in some specified manner and increase cost by a determined amount, is the reform desirable? . . .

Accuracy in the Assessment of Damages
. . .

Accuracy and Ex Ante Information

How the Value of Accuracy Depends on Individuals' Information at the Time They Act. Consider a scenario in which individuals contemplate committing acts deemed to be harmful. (Such acts may include, for example, torts that involve only a risk of causing harm, breaches of contracts, or violations of intellectual property rights of others.) Acts in a given class are known to cause a particular level of harm, on average; some acts cause more harm than average and others less. (For example, victims' injuries may differ in severity, and given injuries may impose different costs, depending on characteristics of the victim.)

To simplify the discussion, assume that when they cause harm injurers are always liable for damages and the only question is the extent of damages. The adjudicator has two choices: award damages equal to the average harm for the class, or make an inquiry into harm in the particular case, in which event damages will depend on the actual level of the harm. This inquiry entails some cost. . . .

When Individuals Cannot Anticipate the Actual Level of Harm. Assume that individuals, at the time they decide how to act, know only the average level of harm for the type of act they will commit but not the actual harm their act will caused. Then it is apparent that greater precision ex post, in adjudication, is a waste of resources because information learned later cannot improve the earlier decision.

For concreteness, suppose that the contemplated act, using a toxic substance, may cause harm of 5, 10, or 15, each with equal probability, if a storage tank leaks. The average harm is 10. If liability always equals 10, the decision whether to use the substance and about how careful to be in preventing leaks will reflect that, if a leak occurs, liability will be 10. If, instead, damages would equal 5, 10, or 15, depending on the actual harm that results, the analysis is the same, for it is assumed that individuals do not know the actual harm ex ante. Since they will be held liable for damages of 5, 10, or 15 with equal probability, their expected liability is 10, which will induce the same behavior as if the average harm were used as the basis for damages. . . . Thus, greater accuracy has no effect on behavior, while it entails a positive resource cost. As a result, accuracy is of no value and greater accuracy is undesirable.

When Individuals Do Anticipate the Actual Level of Harm. Assume that individuals, at the time they act, do know the actual level of harm their particular acts will cause. If damage payments reflect the level of harm caused, behavior will be improved. For example, if the benefit from using the toxic substance is 8, it will be used if the actual harm and thus the damage award will be 5, but not if the harm and damage award will be 10 or 15. In contrast, if the damage award will be 10, to reflect the average harm, the substance would not be used even when the actual harm is only 5.

Thus, accuracy avoids excessive deterrence. Similarly, if the benefit from using the substance were 12, it would be used even when the actual harm would be 15 if damages were based on the average harm of 10, but not if damages equaled the actual harm of 15. Accuracy thereby provides efficient deterrence that otherwise may be lacking.

The question remains whether this improvement in behavior on account of accuracy is sufficiently desirable to justify the cost of greater accuracy. If most leaks caused harm of 10 and accuracy were expensive, the cost of establishing whether there were exceptions in each case would exceed any benefit in improving behavior. Similarly, accuracy would be undesirable if decisions concerning use of the substance would be unaffected in any event. (This would occur either if most users had benefits less than 5, in which case they would not act regardless of whether damages were estimated accurately, or if most users had benefits exceeding 15, in which case they would act in any event.) In contrast, accuracy will be valuable when it is cheap and the effect on use of the substance would involve substantial benefits.

The Optimal Degree of Accuracy. When individuals are informed about the level of harm that their acts might cause, the optimal level of accuracy will be a matter of degree. It would no doubt be undesirable to treat auto and aircraft collisions as a single group because the average level of harm in each category differs greatly; moreover, the cost of distinguishing these categories is extremely low. Similarly, one would wish to distinguish cases in which cars hit pedestrians from those in which cars crush strands of grass on someone's lawn. It may be very costly, however, to determine whether a victim's loss in future earnings will be 30 percent or 40 percent of his previous potential, while such a difference may have little effect on precautions.

In addition, the more refined the damage inquiry, the less likely it is that any difference in outcome will be anticipated at the time individuals decide how to act. One who drives into a pedestrian in a crosswalk may cause harm of thousands or millions of dollars. But the degree of harm will not be known to the driver in advance. Thus, a practice of using averages—say, for types of accidents or types of injuries—may have little effect on behavior even when the range of potential difference in actual harm is vast.

*The Degree of Ex Post Accuracy and Individuals' Incentives
to Acquire Information Ex Ante*

How Accuracy Ex Post Induces Individuals to Acquire Information Ex Ante. The analysis [in the preceding subsection] assumes that individuals, at the time they decide how to act, either are or are not informed about an aspect of the harm they might cause. Often, however, the extent to which individuals are informed will be a matter of choice. For example, one who contemplates using dynamite may not know very precisely how dangerous it is but may be able to consult experts who are more familiar with the extent of the danger from various uses. The more individuals are willing to spend, the more information they can acquire.

The central point of this subsection is that the extent to which individuals will choose to become more informed ex ante, when contemplating how to act, will depend on the degree of accuracy they expect in adjudication, ex post. An individual will see no value in making an expenditure to learn whether her act will cause harm of 5, 10, or 15 if she knows that a court will award 10 in any event. Only if she anticipates that the court will learn the level of harm more precisely, and make damages reflect the actual harm, will she have an incentive to learn the actual level of harm ex ante. Whether individuals ultimately acquire the information is another matter. Individuals will acquire information when the benefits of being informed exceed the costs of the information. The benefit of being informed is that one can better adjust one's behavior in light of actual legal consequences. For example, if there is a probability that one will learn that actual harm, and thus expected liability, is very high, and one would choose not to act in that instance, then the benefit of information would be the difference between the net value of not acting and that of acting in such an instance, weighted by the probability one expects to learn that actual harm is very high. Similarly, one might learn that harm is lower than anticipated and in that instance commit an act that one otherwise would have refrained from doing.

To illustrate, suppose that the use of a toxic substance will cause harm of 5, 10, or 15, each with equal probability, and that the user does not know the actual harm that would be caused. If the benefit from using the substance is 12, it would be used in the absence of information because the expected liability is 10. If one acquired information on actual harm, the substance would not be used in the event that harm was 15. This produces a gain of 3—liability of 15 is avoided while the benefit of 12 is forgone. Because the probability of this outcome is $\frac{1}{3}$, the expected value of information is 1. Thus, information would be acquired if and only if the cost of the information were less than 1.

Information ex ante is valuable only if what is learned will be reflected in awards ex post. Similarly, as noted [above], accuracy ex post is valuable only if individuals also have accurate information ex ante; that statement can now be interpreted to include both the case in which individuals already have the necessary information and that in which they will be induced to discover it. And individuals will be induced to learn information ex ante only if their benefits from adjusting behavior exceed the cost of the information.

The Social Value of Accuracy Ex Post. Having established when accuracy is valuable, it remains necessary to consider whether the value is worth its cost. In [the discussion above], when individuals simply were assumed to be informed, accuracy ex post was desirable when the benefit of the improvement in behavior was greater than the cost of more accurate adjudication ex post. In the present context, the benefit with regard to behavior must exceed the sum of the cost of accuracy ex post and the cost of individuals' becoming informed ex ante.

Whether Individuals' Incentives to Acquire Information Ex Ante Are Excessive or Inadequate. . . . This subsection explores whether, ex ante, individuals will tend to acquire more or less information than is socially appropriate (taking as given that adjudication will involve the expenditure of resources to reach accurate results ex post). First, consider a benchmark case. Assume that liability is strict, the only uncertainty concerns damages, and individuals properly estimate the value of information to themselves. Then, if the information concerns how a court will properly assess harm, individuals' incentive to acquire information will be socially appropriate. The reason incentives are proper in this simple context relates to familiar properties of a rule of strict liability. When individuals acquire information about the true level of harm, and expected damages equal expected true harm, they will be induced to behave appropriately. Relatedly, the private benefit from their change in behavior will equal the social benefit. (The private benefit will consist of gains or losses from modifying one's activity, which are assumed to be social gains or losses as well. Changes in expected liability payments will equal changes in expected harm, which is a correct measure of the external social cost of one's activity.) Finally, individuals' private cost of information involves real expenditures of resources (usually, time or expenditures to purchase experts' services), which will thus be a proper measure of the social cost of information. Therefore, when individuals compare their private benefits and costs, their calculus will precisely reflect social benefits and costs, and their incentive to become informed will be correct. . . .

Compensatory Objectives and Risk Aversion. When parties are risk-averse, that is, when uncertain losses are more detrimental than a certain loss with the same expected value—the level of actual damage awards will affect their welfare directly, in addition to affecting their behavior. To illustrate this, recall the example in which actual harm may be 5, 10, or 15, each with equal probability.

Consider first the effect with regard to injurers (defendants), who have been the focus of the analysis thus far. In the scenario examined here inaccuracy corresponds to making defendants pay the average harm caused by their type of act (10), while accuracy involves each defendant paying the actual harm caused by her particular act (5, 10, or 15). By construction, the expected payment (the mean) is the same in both instances (10). But the variance differs: there is no variance if the average harm is paid by all injurers, while there is positive variance if each injurer pays the actual harm she causes (the standard deviation is approximately 4.1). As a result, greater accuracy in adjudication increases the risk to which defendants are exposed. On this account, accuracy is less desirable than otherwise. Thus, when individuals are uninformed ex ante, so that there is no behavioral benefit, accuracy would be undesirable even if the greater accuracy were free!

For plaintiffs, the effect would tend to be the opposite. Assume that plaintiffs' damages compensate for pecuniary losses (or nonpecuniary losses, such as physical injuries, that are fully restored by expenditures of money). Then, if they are risk-averse, the optimal amount of compensation equals the actual loss. Hence, if damages equal average harm for classes of plaintiffs rather than actual harm, some plaintiffs (those with harm of 15) will be undercompensated and others (with actual harm of 5) will be overcompensated (as both types receive damages of 10). Therefore, plaintiffs will bear risk when results are inaccurate, but not when they are accurate.

Thus, taking risk aversion into account could make accuracy more or less desirable than otherwise, depending on whether plaintiffs' or defendants' risk aversion were more significant. Of course, all these conclusions are mitigated to the extent parties are insured. . . .

Applications

. . .

Rules versus Standards. Compare a complex—that is, a detailed, precise, accurate—standard to a complex rule with the same content (one that provides for the same level of damages for each particular act that may arise). How, if at all, would their effects on behavior differ? If individuals were, in either case, unaware of the detail of the law, they would act in the same manner under both formulations, in a manner that reflected the expected liability for a class of acts rather than the actual harmfulness of their particular act. Similarly, if they were aware of the precise consequences that would follow from their activity in either case, they would act the same way, in a manner that reflected the particular harm of their act.

But it is plausible to suppose that individuals' information will differ under the two formulations. In particular, individuals may find it easier (cheaper) to become informed under rules. Because the specification of particular consequences is stated in advance, legal consequences are less costly to determine than under a standard, where by definition the statement of damages for various acts is left to an adjudicator, ex post. Thus, even when the content does not differ, individuals may often be more informed about the particulars in rules than in standards at the time they act and thereby conform their behavior more closely to the law's commands.

Moreover, it may sometimes be true that even if one compared a less complex rule to a more complex standard (perhaps one that could be more detailed because hindsight is better than foresight), the former, less complex command may result in behavior that is more precisely in accord with underlying legal norms. The reason is that the standard's added detail may be unknown, and too difficult to predict, ex ante (for the very reason that hindsight has yet to be obtained). Thus, the conventional view that there is an advantage in allowing more room for adjudicators to examine context-specific factors, rather than specifying legal consequences in advance under a simpler scheme, exhibits an error in logic. This view implicitly assumes that the more precise ex post result will be reflected in ex ante behavior; but this will not occur if the ultimate content of the legal command cannot cheaply be predicted in advance (which is often a major premise of those presenting this justification for standards). Moreover, failing to announce (formulate into rules) even simpler aspects of an appropriate outcome may make it more difficult for individuals to take them into account in their behavior, leading to worse results. . . .

Predictability of Legal Outcomes. . . . If one wants the law to be a more precise guide to behavior, individuals need the relevant knowledge. One way to disseminate it is through rules, which state the consequences that attach to various behavior. Another way to disseminate information to guide behavior is through the outcomes of adjudication. Obviously, one benefit of more accurate outcomes is that individuals in the future will have better guidance concerning their behavior.

But adjudication tends to be a costly and ineffective way to collect and disseminate information. Often more will be spent adjudicating a single case than would be required to fund a substantial empirical study of a mass of cases. Moreover, if a case settles, typically few individuals learn of the information produced through the litigation. Even if there is a verdict, it is extremely difficult to interpret a black-box pronouncement by a jury, without statements of reasons, weight given to factors, or basic evidentiary findings.

Thus, while adjudication does, at substantial cost, provide some useful information, governments should consider alternatives that involve directly gathering and disseminating information. For example, if one is concerned about the waste-disposal behavior of dry-cleaning establishments or auto repair shops, study of appropriate methods of disposal and costs of inappropriate methods, followed by distribution of the results, may be an important accompaniment to substantive law regulating the activity. (Of course, such investigation would simultaneously provide the information needed to give appropriate content to the legal commands.) Even if one chose to enforce compliance using a conventional tort regime, there would exist the advantage that, when adjudicators set damages equal to actual harm, individuals would be more likely to know actual harm at the time they decide how to act. (In addition, adjudicators may find that they could assess harm more accurately and more cheaply if such an investigation had been undertaken.)

In contrast, the legal system may function more effectively if its errors and idiosyncrasies are less predictable. Randomly assigning judges and making it difficult to learn how juries actually reach decisions interfere with prediction. . . .

Scheduling Sanctions

Personal Injuries. A conscious decision to reduce accuracy by setting damages equal to average harm rather than particular harm is reflected in proposals to provide a damage schedule for personal injuries. Workers' compensation schemes use such an approach to some extent. But scheduling of damages is not generally employed in the tort context.

It is useful to compare other, similar settings in which schedules are used. First-party insurance is a good example. Policies often provide particular payments for loss of limb, rather than indicating that, in the event of injury, an inquiry will be made to determine actual losses. Presumably, if more precise compensation were thought to be worth the cost of particularized inquiries, policy provisions would differ. Another instance is disability insurance, where policies typically provide for a fixed percentage of one's prior wage for covered disabilities.

In other first-party insurance contexts, such as homeowners' insurance, there is a mix of scheduling and inexpensive forms of alternative dispute resolution. Scheduling arises implicitly from advance appraisals of particular objects, so it is particularized. In other instances, or when property is damaged without losing all its value, a common provision is for binding arbitration. First-party auto insurance for injury by uninsured motorists operates similarly.

Thus, in a wide range of contexts, particularly including contractual settings, one observes damage scheduling or the use of very inexpensive dispute resolution. It is unclear whether these practices should be seen as demonstrating the superiority of employing less accuracy in conventional adjudication. On one hand, the questions posed often are the same. On the other hand, ex ante provision by contractual arrangement offers advantages not available when contracts are silent (rather than providing for liquidated damages) or in contexts such as accidents in which there is no prior contractual arrangement. In addition, the use of alternative dispute resolution may not reflect a differing view about the value of accuracy but, rather, a belief that different procedures can produce as good or better accuracy at lower cost. . . .

Criminal Sentencing. In the past decade, the federal government has adopted criminal sentencing guidelines, involving a highly detailed sanction schedule. For present purposes, two features of these guidelines are notable. First, they often involve substantial differentiation among offenses causing different levels of harm— providing different sentences, for example, for different degrees of offenses against the person and for different amounts of money involved in crimes such as theft or fraud. In most instances, one suspects that individuals contemplating such crimes would know at least approximately how severe the resulting harm would be. (For example, a thief usually knows whether a crime is likely to involve a few thousand dollars or a few million.) Thus, making sanctions depend on harm is desirable with regard to its effect on behavior.

Second, the guidelines do not usually require expensive adjudication of the degree of harm. For example, in determining the amount of the loss for theft and related offenses, the commentary indicates that "[t]he loss need not be determined with precision, and may be inferred from any reasonably reliable information available, including the scope of the operation." And facts are not determined at a trial but, rather, using less formal sentencing procedures. Thus, the system operates in a manner suggesting that approximate accuracy is a reasonable objective, while great precision is not worth the additional costs involved. An important exception, however, is that some of the differentiation provided by the guidelines involves different sentences for formally different crimes rather than for different degrees of harm from a given type of crime. Establishing the crime category requires proof beyond a reasonable doubt at trial or a plea bargain made with the knowledge that the prosecution would otherwise have had to meet such a standard. . . .

NOTES AND QUESTIONS

1. In situations where information is incomplete or asymmetric, the meaning of Pareto efficiency is not entirely clear. There are at least three versions of efficiency that might be thought relevant in evaluating legal and economic institutions: *ex ante efficiency,* which evaluates the parties' expected utility before any private information is learned, *interim efficiency,* which evaluates each party's expected utility from the vantage point of his or her private information, and *ex post efficiency,* which evaluates the parties' utility from the viewpoint of fully revealed information. In the lemons model, for example, a trading regime would be ex ante efficient if it would be preferred initially by parties who did not yet know what kind of car they owned or indeed whether they would be buyer or seller. The regime would be interim efficient if it maximized the total expected gains to be enjoyed by sellers of good cars, sellers

of bad cars, buyers of good cars, and buyers of bad cars. It would be ex post efficient if it allocated good cars to people who most valued having a good car, and bad cars to those who least minded having a bad car. These criteria can diverge; and when they do, it is debatable which is normatively most preferable. The ex post standard seems appealing once all information becomes available, but does not take into account attitudes toward risk or incentives to produce and disclose information. The ex ante standard seems appealing from the vantage point of institutional design, but does not ensure the full exploitation of all exchange opportunities that may arise as new information is revealed. The choice among standards, therefore, is a problem of the second best. See Bengt Holmstrom and Roger Myerson, Efficient and Durable Decision Rules with Incomplete Information, 51 *Econometrica* 1799 (1983).

2. Whatever notion of efficiency we adopt, the private and social value of information are often not the same. In the lemons model, for example, a buyer who invests in learning about the true quality of a car offered for sale (for example, by paying a mechanic to inspect it) may gain from her investment, since it will prevent her from overpaying if the car turns out to be a lemon. Part of this gain, however, is at the expense of low-quality sellers; and there is also a spillover benefit to high-quality sellers, who may now be better able to capture the full value of their wares. In this way, imperfect information can give rise to various externalities. Because of such externalities, it may be possible to improve social welfare either by subsidizing information acquisition or by taxing it, depending on the circumstances. See, for example, Severin Borenstein, The Economics of Costly Risk Sorting in Competitive Insurance Markets, 9 *International Review of Law and Economics* 25 (1989) [arguing that it may improve efficiency to prohibit insurers from spending resources to distinguish high-risk from low-risk insureds].

3. It is worth comparing Stigler's and Akerlof's assessments of the efficiency of the used car market. While Akerlof presents this market as the quintessential illustration of imperfect information, Stigler argues that buyers of large-ticket items such as used cars have efficient incentives to learn about the distribution of prices, and sees car dealers as informational specialists who help to lower the cost of search. Whose argument is more persuasive? Does either author's analysis shed light on the merits of government regulations designed to lower the cost of consumer search, for example, by requiring posted prices or by prohibiting individualized haggling? Would the possibility of price discrimination among consumers affect your answer? See Ian Ayres, Fair Driving: Gender and Race Discrimination in Retail Car Negotiations, 104 *Harvard Law Review* 817 (1991) [presenting empirical evidence suggesting that differential search costs among classes of car buyers results in women and African-American buyers paying significantly higher prices].

4. Evaluate Akerlof's assertion that "dishonesty in business is a serious problem in underdeveloped countries." What is the possible basis for such a statement? Is there any reason to think that dishonesty is any more of a problem in less developed than in more developed economies, or in traditional rather than capitalist economies? For one answer, see Andrzej Rapaczynski, The Roles of the State and the Market in Establishing Property Rights, 10 *Journal of Economic Perspectives* 87 (1996) [arguing, based on the experience of postcommunist countries, that poor societies lacking an established system of market-based incentives find it difficult to provide ef-

fective legal protection for all but the simplest entitlements, and that respect for property and contract rights depend more on economic factors than on moral or cultural ones].

5. Ian Ayres and Robert Gertner have generalized Bishop's analysis of the foreseeability doctrine to explain the more general category of default rules in contract—that is, the background interpretive rules that courts use to fill the gaps in privately negotiated contracts. They have argued, in contrast to the conventional wisdom that courts should set default rules to incorporate terms the average person would want, that optimal default rules should be chosen to penalize parties who fail to disclose their private information. Setting the default contrary to the interest of informed parties, in this view, encourages such parties to reveal themselves and to negotiate terms best suited to their situation. They refer to such a policy as a "penalty default." See Ian Ayres and Robert Gertner, Filling Gaps in Incomplete Contracts: An Economic Theory of Default Rules, 99 *Yale Law Journal* 87 (1989)]; contra, Jason Scott Johnston, Strategic Bargaining and the Economic Theory of Contract Default Rules, 100 *Yale Law Journal* 615 (1990) [arguing that strategic incentives may deter disclosure under a regime of penalty default]; Ian Ayres and Robert Gertner, Strategic Contractual Inefficiency and the Optimal Choice of Legal Rules, 101 *Yale Law Journal* 729 (1992) [responding to Johnston's criticism].

6. Grossman explains why high-quality sellers and buyers often have an incentive to signal their favorable private information. The result is generally efficient if signalling can be done cheaply, but it may be in the individual interest of traders to signal even when it is quite costly to do so. For example, job applicants may find it profitable to invest in otherwise useless educational credentials in order to demonstrate their skills and motivation to potential employers. This is individually rational if there is no cheaper way to signal one's talents credibly, but from the standpoint of employees collectively, it can create a rat race. Conversely, applicants who find it costly to identify their talents to potential employers may underinvest in training and education. The problem is again one of an informational externality; in the former case the employee's training investment injures fellow applicants, while in the latter case it redounds to their benefit. See A. Michael Spence, Job Market Signaling, 87 *Quarterly Journal of Economics* 355 (1973).

7. In order for warranties to serve as a solution to the lemons problem, the buyer must be able to tell after the fact whether the warranty has been breached. This may not be possible at reasonable cost if the underlying good requires experience to evaluate; the difficulty may be especially great for diagnostic services such as medical care or auto repair. Furthermore, even if the buyer can tell whether there has been a breach, it may be difficult or expensive to prove its existence to a third-party tribunal. Do you have any suggestions regarding what arrangements would be efficient in such cases? For one possibility, see George L. Priest, A Theory of the Consumer Product Warranty, 90 *Yale Law Journal* 1297 (1981) [arguing for deference to marketed warranty terms]; for another, see Lisa Bernstein, Opting Out of the Legal System: Extralegal Contractual Relations in the Diamond Industry, 21 *Journal of Legal Studies* 115 (1992) [arguing for private enforcement through reputation within trading community].

8. Kaplow's discussion of optimal accuracy in adjudication leads directly to ap-

plications in a number of substantive legal fields. One example is the tort doctrine of negligence per se: if the defendant's behavior has violated a regulatory statute, courts often take this as sufficient evidence of negligence in and of itself, and do not further attempt to weigh costs and benefits of precaution. Similarly, courts have traditionally looked to custom to give content to the negligence standard; and the debate in antitrust over per se liability versus rules of reason raises analogous issues. In light of Kaplow's discussion, how much tension do you think there is between these rule-based approaches and a more explicit economic balancing test? In which situations is a traditional rule-based approach sufficient to provide efficient incentives for precaution? See Stephen Gilles, Rule-Based Negligence and the Regulation of Activity Levels, 21 *Journal of Legal Studies* 319 (1992) [discussing roles of custom and negligence per se]; Guido Calabresi and Alvin K. Klevorick, Four Tests for Liability in Torts, 14 *Journal of Legal Studies* 585 (1985) [discussing when liability standards should incorporate hindsight]. For an application to property, see Carol M. Rose, Crystals and Mud in Property Law, 40 *Stanford Law Review* 577 (1988).

9. Informational issues are fundamental to the economics of bargaining, as the efficiency of negotiation may depend on the distribution of information among the parties. See, for example, Urs Schweitzer, Litigation and Settlement under Two-sided Incomplete Information, 56 *Review of Economic Studies* 163 (1989). Recently, there has been a proliferation of work in law and economics on the question of how the regime of entitlements affects bargaining under imperfect information. For instance, Ian Ayres and Eric Talley, Solomonic Bargaining: Dividing a Legal Entitlement to Facilitate Coasean Trade, 104 *Yale Law Journal* 1027 (1995), have argued that liability rules are more effective than property rules at promoting informational efficiency in bargaining. In response, Steven Shavell and Louis Kaplow, Do Liability Rules Facilitate Bargaining? A Reply to Ayres and Talley, 105 *Yale Law Journal* 221 (1995), have criticized the Ayres/Talley logic and have suggested that any advantage of liability rules comes from the straightforward internalization of externalities. See also Ian Ayres and Eric Talley, Distinguishing Between Consensual and Nonconsensual Advantages of Liability Rules. 105 *Yale Law Journal* 235 (1995) [responding to Shavell/Kaplow criticism]; Jason Scott Johnston, Bargaining under Rules versus Standards, 11 *Journal of Law, Economics, & Organization* 256 (1995) [suggesting that bargaining is more efficient when entitlements are expressed as general standards than as bright-line rules].

10. While Rosenberg and Shavell (supra, chapter 4) suggested that frivolous lawsuits are profitable because of the sequence in which litigation costs must be incurred, Lucien Bebchuk and Avery Katz have proposed an information-based explanation for the phenomenon. The main reason that frivolous suits are not met with a blanket refusal to negotiate, they contend, is that defendants rarely know the merits of the claim with certainty. Since refusing to take a valid claim seriously can be quite costly, a frivolous plaintiff can take advantage of the defendant's uncertainty over the claim's validity to extract a settlement. Thus, the fraction of lawsuits that are frivolous is directly related to the cost of trial and to the difficulty of evaluating claims on their merits. See Lucien A. Bebchuk, Suing Solely to Extract a Settlement Offer, 17 *Journal of Legal Studies* 437 (1988); Avery Katz, The Effect of Frivolous Lawsuits on the Settlement of Litigation, 10 *International Review of Law and Economics* 3

(1990). On the basis of his informational model, Katz disagrees with Rosenberg and Shavell's claim that a rule shifting litigation costs to the losing party will solve the problem. Instead, he argues that strike suits can be eliminated only by discouraging settlement, by bringing down the cost of litigation generally, or by turning away the meritorious claimants on whose coattails strike suitors ride. Do you agree? Can you think of other policies that would discourage nuisance suits without unduly disadvantaging legitimate claims?

11. There are numerous additional legal applications of information economics, including attorney-client privilege [see Louis Kaplow and Steven Shavell, Legal Advice about Information to Present in Litigation: Its Effects and Social Desirability. 102 *Harvard Law Review* 565 (1989)]; intellectual property [see Symposium on the Law and Economics of Intellectual Property, 78 *Virginia Law Review* 1 (1992)], and civil discovery [see Robert Cooter and Daniel Rubinfeld, Reforming the New Discovery Rules, 84 *Georgetown Law Journal* 61–89 (1995)]. Additionally, Baird, Picker, and Gertner's *Game Theory and the Law* (supra, chapter 4, note 1) contains an excellent survey of legal applications of imperfect information. Good general sources on the economics of information include Eric Rasmusen, *Games and Information*, 2d ed. (Cambridge, Mass.: B. Blackwell, 1994); Jack Hirshleifer and John G. Riley, *The Economics of Uncertainty and Information* (Cambridge: Cambridge University Press, 1992), and Jean-Jacques Laffont, *The Economics of Uncertainty and Information* (Cambridge, Mass.: MIT Press, 1989).

7

Refining the Model IV:
Bounded Rationality

The previous three sections of this book, while introducing important variations on the standard economic model of rational choice, remained grounded firmly within its central paradigm of constrained maximization. In discussing uncertainty and imperfect information, for instance, we focused on how these phenomena influence the ways in which rational individuals attempt to maximize expected utility. As specific examples, we saw how uncertainty over the quality of a good lowers the amount that a risk-averse consumer is willing to pay for it and, similarly, how rational individuals will discount the value of an exchange when they know that the other party has better information than they do.

This section introduces another way of thinking about uncertainty and imperfect information—what has come to be called the *model of bounded rationality*. Under the model of bounded rationality, uncertainty and imperfect information are modeled not as constraints within which rational individuals maximize, but as limits on the reasoning process itself. The first reading in the section, by the economist and cognitive scientist Herbert Simon, compares the models of rational choice and bounded rationality, and argues that while the former is an elegant stylized representation, the latter is a better and truer description of human decisionmaking.

In Simon's view, maximization in the economist's sense—what he calls *substantive rationality*—is an impossible task in the complex and changing world we live in. There are just too many comparisons to make and too much data to consider. Even the standard economic assumptions of consistent preferences and a well-defined utility function are idealized constructs; no existing person or organization actually exhibits them. Instead, in any sufficiently complicated environment, rationality consists of finding and following a reasonable set of procedures for dealing with the tasks of daily life. Such a set of procedures, like any management structure, must be reasonably flexible; it must include arrangements for both monitoring performance and recommending adjustments and corrections in course. A procedurally rational actor, however, does not maximize either utility or profits. Instead, it will *"satisfice"*—that is, follow its ordinary habits and rules of thumb so long as they yield satisfactory results, even if an alternate set of procedures would in theory be superior. Such standard operating procedures are reevaluated only in periods of crisis or unusual opportunity, and even then the choice is not over all possible procedures; such reevaluative episodes have rules and structures of their own.

If Simon is correct, then people using cognitively reasonable procedures will often behave in ways inconsistent with the standard model of rational choice. It should be remembered, however, that the standard model is just that—a model. As such, it should be judged not against an abstract ideal of truth, but rather on its ability to help promote human ends such as prediction, explanation, or the organization of knowledge. In this regard, the model of rational choice has significant virtues: it is parsimonious, has a precise interpretation, and is relatively well understood by its practitioners. It is analytically tractable using common mathematical methods and has an established history of generating testable predictions across a variety of applications. The model of bounded rationality, in contrast, has been less thoroughly studied and has less clear implications for positive and normative economics alike. In order for it to be affirmatively useful as an alternate paradigm—as opposed to remaining a mere critique of the mainstream approach—it needs to be made more systematic. Economists therefore need to model the specific procedures that organizations and individuals actually use and to show that they follow some consistent form, so that predictions and explanations based on them can be generalized across particular institutional settings. Otherwise, the model of rational choice will remain the best practical approximation available, despite its limitations.

The second reading in this section, by the cognitive psychologists Amos Tversky and Daniel Kahneman, illustrates some of the work that has been done toward this end. Tversky and Kahneman show, on the basis of laboratory experiments, that people appear to make economic decisions in systematically different ways depending on how the infomation relevant to those decisions is presented. The evidence, moreover, suggests that experimental subjects follow heuristic rules of thumb rather than maximizing. For instance, subjects asked to evaluate a complex gamble choose differently when the necessary calculation is presented in the form of two steps rather than one. The more salient part of the calculation, or the part that is processed first, seems to carry more weight—as marketing consultants who insist on setting price at $19.99 rather than $20 have long known. Perhaps more significantly, the authors find that experimental subjects react differently, and exhibit different amounts of risk aversion, when choices are measured in terms of potential gains relative to an initial benchmark, rather than in terms of potential losses. This particular finding flies in the face of the fundamental economic concept of *opportunity cost,* which posits that failing to capture a gain is equivalent to suffering a loss. Tversky and Kahneman show that such "framing effects" are apparently widespread, and argue that economists need to take them into account in their analyses of individual consumer choice.

The remaining three readings offer specific applications of the model of bounded rationality to legal rules and institutions. The selection by Alan Schwartz and Louis Wilde draws on work by Tversky, Kahneman, and others to assess the cognitive heuristics likely to be used by individual consumers in markets for complex goods and for secured credit. Schwartz and Wilde argue that, on the whole, there is little reason to conclude that consumers will behave in a systematically overoptimistic fashion when buying goods or when granting security interests. They conclude, accordingly, that the theory of bounded rationality offers relatively weak justification for government intervention in product or credit markets, though they do allow that some kinds of consumer misperceptions are likelier than others.

The selection by Thomas Jackson applies a similar approach to the field of bankruptcy, but reaches substantially different policy conclusions. Jackson focuses on the provisions of the U.S. Bankruptcy Code that govern the discharge of debts. He suggests that these provisions, and the fresh start they award to individual bankrupt debtors, are best understood as a policy response on the part of Congress and the courts to the problem of boundedly rational behavior by debtors. It follows, in this view, that a variety of legal controversies arising under the Code should be resolved in light of this general statutory purpose. Finally, Roger Noll and James Krier discuss the positive and normative implications of bounded rationality for public regulatory policy in the areas of environmental protection, occupational safety, and public health. Noll and Krier suggest that much apparent inconsistency in public risk-management policies (for example, requiring high expenditures for the sake of avoiding trivial risks in some contexts while ignoring much more significant risks in others) can be understood as the rational response of elected and appointed officials to the cognitively biased perceptions of voters and citizens. The officials' interest in maximizing their tenure and job satisfaction, in Noll and Krier's view, leads them to pursue the boundedly rational heuristics employed by their constituents.

7.1 THEORY

Rationality in Psychology and Economics

HERBERT SIMON

The task I shall undertake here is to compare and contrast the concepts of rationality that are prevalent in psychology and economics, respectively. Economics has almost uniformly treated human behavior as rational. Psychology, on the other hand, has always been concerned with both the irrational and the rational aspects of behavior. In this paper, irrationality will be mentioned only obliquely; my concern is with rationality. Economics sometimes uses the term "irrationality" rather broadly and the term "rationality" correspondingly narrowly, so as to exclude from the domain of the rational many phenomena that psychology would include in it. For my purposes of comparison, I will have to use the broader conception of psychology.

One point should be set immediately outside dispute. Everyone agrees that people have reasons for what they do. They have motivations, and they use reason (well or badly) to respond to these motivations and reach their goals. Even much, or most, of the behavior that is called abnormal involves the exercise of thought and reason. Freud was most insistent that there is method in madness, that neuroses and psychoses were patients' solutions—not very satisfactory solutions in the long run—for the problems that troubled them.

I emphasize this point of agreement at the outset—that people have reasons for what they do—because it appears that economics sometimes feels called on to defend the thesis that human beings are rational. Psychology has no quarrel at all with this thesis. If there are differences in viewpoint, they must lie in conceptions of what constitutes rationality, not in the fact of rationality itself. . . .

In its treatment of rationality, neoclassical economics differs from the other social sciences in three main respects: (a) in its silence about the content of goals and values; (b) in its postulating global consistency of behavior; and (c) in its postulating "one world"—that behavior is objectively rational in relation to its total environment, including both present and future environment as the actor moves through time.

In contrast, the other social sciences, in their treatment of rationality, (a) seek to determine empirically the nature and origins of values and their changes with time and experience; (b) seek to determine the processes, individual and social, whereby selected aspects of reality are noticed and postulated as the "givens" (factual bases) for reasoning about action; (c) seek to determine the computational strategies that are used in reasoning, so that very limited information-processing capabilities can cope with complex realities; and (d) seek to describe and explain the ways in which nonrational processes (e.g., motivations, emotions, and sensory stimuli) influence the fo-

Herbert Simon, Rationality in Psychology and Economics, 59 *Journal of Business* (1986) S209–S224.
Copyright © 1986 University of Chicago Press. Reprinted with permission.

cus of attention and the definition of the situation that set the factual givens for the rational processes.

These important differences in the conceptualization of rationality rest on an even more fundamental distinction: in economics, rationality is viewed in terms of the choices it produces; in the other social sciences, it is viewed in terms of the processes it employs. The rationality of economics is substantive rationality, while the rationality of psychology is procedural rationality.

Substantive and Procedural Rationality

If we accept values as given and consistent, if we postulate an objective description of the world as it really is, and if we assume that the decision maker's computational powers are unlimited, then two important consequences follow. First, we do not need to distinguish between the real world and the decision maker's perception of it: he or she perceives the world as it really is. Second, we can predict the choices that will be made by a rational decision maker entirely from our knowledge of the real world and without a knowledge of the decision maker's perceptions or modes of calculation. (We do, of course, have to know his or her utility function.)

If, on the other hand, we accept the proposition that both the knowledge and the computational power of the decision maker are severely limited, then we must distinguish between the real world and the actor's perception of it and reasoning about it. That is to say, we must construct a theory (and test it empirically) of the processes of decision. Our theory must include not only the reasoning processes but also the processes that generate the actor's subjective representation of the decision problem, his or her frame.

The rational person of neoclassical economics always reaches the decision that is objectively, or substantively, best in terms of the given utility function. The rational person of cognitive psychology goes about making his or her decisions in a way that is procedurally reasonable in the light of the available knowledge and means of computation.

Embracing a substantive theory of rationality has had significant consequences for neoclassical economics and especially for its methodology. Until very recently, neoclassical economics has developed no strong empirical methodology for investigating the processes whereby values are formed, for the content of the utility function lies outside its self-defined scope. It has developed no special methodology for investigating how particular aspects of reality, rather than other aspects, come to the decision maker's attention, or for investigating how a representation of the choice situation is formed, or for investigating how reasoning processes are applied to draw out the consequences of such representations. . . .

To move from substantive to procedural rationality requires a major extension of the empirical foundations of economics. It is not enough to add theoretical postulates about the shape of the utility function, or about the way in which actors form expectations about the future, or about their attention or inattention to particular environmental variables. These are assumptions about matters of fact, and the whole ethos of science requires such assumptions to be supported by publicly repeatable observations that are obtained and analyzed objectively. . . .

Attention and Representation

In a substantive theory of rationality there is no place for a variable like focus of attention. But in a procedural theory, it may be very important to know under what circumstances certain aspects of reality will be heeded and others ignored. I wish now to present two examples of situations in which focus of attention is a major determinant of behavior. The first rests on very strong empirical evidence; the second is more speculative, but I will try to make it plausible.

The Purchase of Flood Insurance

Kunreuther et al. (1978) have studied decisions of property owners whether to purchase insurance against flood damage. Neoclassical theory would predict that an owner would buy insurance if the expected reimbursable damage from floods was greater than the premium. The actual data are in egregious conflict with this claim. Instead it appears that insurance is purchased mainly by persons who have experienced damaging floods or who are acquainted with persons who have had such experiences, more or less independently of the cost/benefit ratio of the purchaser.

If we wish to understand the insurance-buying behavior, then we must determine, as Kunreuther and his colleagues did, the circumstances that attract the attention of a property owner to this decision alternative. Utility maximization is neither a necessary nor a sufficient condition for deducing who will buy insurance. The process of deciding—in this case, the process that puts the item on the decision agenda—is the important thing.

Voting Behavior

Voting behavior provides a more complex example of the role of attention in behavior. Both before and since Marx, it has been widely believed that voters respond, at least to an important extent, to their economic interests. Let us assume that is so. A substantial number of empirical studies have shown correlations between economic conditions and votes in American elections. But such studies use a great variety of independent variables as measures of voters' perceptions of the economic consequences of their choices. Some investigators have tried to measure the economic well-being of voters at the time of the election as compared with their well-being at some previous time. Others have measured the state of the economy—the level of the GNP, say, or of employment. Which of these (or what other measure) is the true measure of economic advantage? Quite different predictions can be made if different measures are chosen.

Consider the situation of a voter at the time of the 1984 presidential election who wished to maximize his or her economic well-being. Which of the following facts about the economy should influence the vote? (1) Real incomes of a majority have increased over the past 4 years but at less than the historical rate of a couple of decades earlier. (2) Dispersion of incomes has increased. (3) The rate of inflation has declined dramatically. (4) The rate of interest remains high compared with the historical past. (5) The national debt and deficit have increased dramatically. (6) The balance of trade has "worsened" dramatically. (7) Farm foreclosures have increased substantially. (8) Unemployment has decreased recently but is higher than it was 4 years previously. If we throw noneconomic considerations into the voter's utility function, we may add such facts as, the armament situation has changed in complex ways, et cetera, et cetera, moving into race tensions and equity to minorities, energy, the environment, creationism, abortion, and what not.

To predict how a voter, even a voter motivated solely by concern for his or her economic well-being, will vote requires much more than assuming utility maximization. A voter who attends to the rate of inflation may behave quite differently from a voter who attends to the federal deficit. Moreover, in order to predict where a voter's attention will focus, we may need to know his or her economic beliefs. A monetarist may consider different facts to be salient than the facts to which a Keynesian will attend. In any model of voting behavior that has any prospect of predicting behavior, almost all the action will lie in these auxiliary assumptions about attention and belief that define the decision maker's frame. . . .

Summing Up

Between supporters of substantive and procedural theories of rationality there are fundamental differences about what constitutes a principled, parsimonious, scientific theory. We may put the matter in Bayesian [i.e., probabilistic] terms. Neoclassical economists attach a very large prior probability (.9944?) to the proposition that people have consistent utility functions and in fact maximize utilities in an objective sense. As my examples show, they are prepared to make whatever auxiliary empirical assumptions are necessary in order to preserve the utility maximization postulate, even when the empirical assumptions are unverified. When verification is demanded, they tend to look for evidence that the theory makes correct predictions and resist advice that they should look instead directly at the decision mechanisms [that people actually use. . . .]

Behavioral theories of rationality attach a high prior probability (.9944?) to the assumption that economic actors use the same basic processes in making their decisions as have been observed in other human cognitive activities and that these processes are indeed observable. In situations that are complex and in which information is very incomplete (i.e., virtually all real world situations), the behavioral theories deny that there is any magic for producing behavior even approximating an objective maximization of profits or utilities. They therefore seek to determine what the actual frame of the decision is, how that frame arises from the decision situation, and how, within that frame, reason operates.

In this kind of complexity, there is no single sovereign principle for deductive prediction. The emerging laws of procedural rationality have much more the complexity of molecular biology than the simplicity of classical mechanics. As a consequence, they call for a very high ratio of empirical investigation to theory building. They require painstaking factual study of the decision-making process itself.

What is to be done? What prescription for economic research derives from my analysis?

First, I would recommend that we stop debating whether a theory of substantive rationality and the assumptions of utility maximization provide a sufficient base for explaining and predicting economic behavior. The evidence is overwhelming that they do not.

We already have in psychology a substantial body of empirically tested theory about the processes people actually use to make boundedly rational, or "reasonable," decisions. This body of theory asserts that the processes are sensitive to the complexity of decision-making contexts and to learning processes as well.

The application of this procedural theory of rationality to economics requires extensive empirical research, much of it at micro-micro levels to determine specifically how process is molded to context in actual economic environments and the consequences of this interaction for the economic outcomes of these processes. Economics without psychological and sociological research to determine the givens of the decision-making situation, the focus of attention, the problem representation, and the processes used to identify alternatives, estimate consequences, and choose among possibilities—such economics is a one-bladed scissors. Let us replace it with an instrument capable of cutting through our ignorance about rational human behavior.

The Framing of Decisions and the Psychology of Choice

AMOS TVERSKY AND DANIEL KAHNEMAN

Explanations and predictions of people's choices, in everyday life as well as in the social sciences, are often founded on the assumption of human rationality. The definition of rationality has been much debated, but there is general agreement that rational choices should satisfy some elementary requirements of consistency and coherence. In this article we describe decision problems in which people systematically violate the requirements of consistency and coherence, and we trace these violations to the psychological principles that govern the perception of decision problems and the evaluation of options.

A decision problem is defined by the acts or options among which one must choose, the possible outcomes or consequences of these acts, and the contingencies or conditional probabilities that relate outcomes to acts. We use the term "decision frame" to refer to the decision-maker's conception of the acts, outcomes, and contingencies associated with a particular choice. The frame that a decision-maker adopts

Amos Tversky and Daniel Kahneman, The Framing of Decisions and the Psychology of Choice, 211 *Science* (1981), 453–458. Copyright © 1981 American Association for the Advancement of Science. Reprinted with permission.

is controlled partly by the formulation of the problem and partly by the norms, habits, and personal characteristics of the decision-maker.

It is often possible to frame a given decision problem in more than one way. Alternative frames for a decision problem may be compared to alternative perspectives on a visual scene. Veridical perception requires that the perceived relative height of two neighboring mountains, say, should not reverse with changes of vantage point. Similarly, rational choice requires that the preference between options should not reverse with changes of frame. Because of imperfections of human perception and decision, however, changes of perspective often reverse the relative apparent size of objects and the relative desirability of options.

We have obtained systematic reversals of preference by variations in the framing of acts, contingencies, or outcomes. These effects have been observed in a variety of problems and in the choices of different groups of respondents. Here we present selected illustrations of preference reversals, with data obtained from students at Stanford University and at the University of British Columbia who answered brief questionnaires in a classroom setting. The total number of respondents for each problem is denoted by N, and the percentage who chose each option is indicated in brackets.

The effect of variations in framing is illustrated in problems 1 and 2.

> **Problem 1** [$N = 152$]: Imagine that the U.S. is preparing for the outbreak of an unusual Asian disease, which is expected to kill 600 people. Two alternative programs to combat the disease have been proposed. Assume that the exact scientific estimate of the consequences of the programs are as follows:
>
> If Program A is adopted, 200 people will be saved. [72 percent]
>
> If Program B is adopted, there is $\frac{1}{3}$ probability that 600 people will be saved, and $\frac{2}{3}$ probability that no people will be saved. [28 percent]
>
> Which of the two programs would you favor?

The majority choice in this problem is risk averse: the prospect of certainly saving 200 lives is more attractive than a risky prospect of equal expected value, that is, a one-in-three chance of saving 600 lives.

A second group of respondents was given the cover story of problem 1 with a different formulation of the alternative programs, as follows:

> **Problem 2** [$N = 155$]:
>
> If Program C is adopted 400 people will die. [22 percent]
>
> If Program D is adopted there is $\frac{1}{3}$ probability that nobody will die, and $\frac{2}{3}$ probability that 600 people will die. [78 percent] Which of the two programs would you favor?

The majority choice in problem 2 is risk taking: the certain death of 400 people is less acceptable than the two-in-three chance that 600 will die. The preferences in problems 1 and 2 illustrate a common pattern: choices involving gains are often risk averse and choices involving losses are often risk taking. However, it is easy to see that the two problems are effectively identical. The only difference between them is that the outcomes are described in problem 1 by the number of lives saved and in problem 2 by the number of lives lost. The change is accompanied by a pronounced shift from risk aversion to risk taking. We have observed this reversal in several groups of respondents, including university faculty and physicians. Inconsistent responses to problems 1 and 2 arise from the conjunction of a framing effect with contradictory attitudes toward risks involving gains and losses. We turn now to an analysis of these attitudes.

The Evaluation of Prospects

The major theory of decision-making under risk is the expected utility model. This model is based on a set of axioms, for example, transitivity of preferences, which provide criteria for the rationality of choices. The choices of an individual who conforms to the axioms can be described in terms of the utilities of various outcomes for that individual. The utility of a risky prospect is equal to the expected utility of its outcomes, obtained by weighting the utility of each possible outcome by its probability. When faced with a choice, a rational decision-maker will prefer the prospect that offers the highest expected utility.

As will be illustrated below, people exhibit patterns of preference which appear incompatible with expected utility theory. We have presented elsewhere a descriptive model, called prospect theory, which modifies expected utility theory so as to accommodate these observations. We distinguish two phases in the choice process: an initial phase in which acts, outcomes, and contingencies are framed, and a subsequent phase of evaluation. . . .

In prospect theory, outcomes are expressed as positive or negative deviations (gains or losses) from a neutral reference outcome, which is assigned a value of zero. Although subjective values differ among individuals and attributes, we propose that the value function [v] is commonly S-shaped, concave above the reference point and convex below it.* . . . For example, the difference in subjective value between gains of $10 and $20 is greater than the subjective difference between gains of $110 and $120. The same relation between value differences holds for the corresponding losses. Another property of the value function is that the response to losses is more extreme than the response to gains. The displeasure associated with losing a sum of money is generally greater than the pleasure associated with winning the same amount, as is reflected in people's reluctance to accept fair bets on a toss of a coin. Several studies of decision and judgment have confirmed these properties of the value function.

The second major departure of prospect theory from the expected utility model involves the treatment of probabilities. In expected utility theory the utility of an uncertain outcome is weighted by its probability; in prospect theory the value of an uncertain outcome is multiplied by a decision weight $\pi(p)$, which is a monotonic func-

---

Editor's note on mathematical terminology: In this context, a *concave* function is one that increases at a decreasing rate, while a *convex* function increases at an increasing rate. Thus, a concave utility function would represent diminishing marginal utility. A *linear* function is one that is neither concave nor convex; that is, it increases at a constant rate, like a straight line.

tion of [the probability] p but is not a probability. The weighting function π has the following properties. First, impossible events are discarded, that is, $\pi(0) = 0$, and the scale is normalized so that $\pi(1) = 1$.... [Second,] low probabilities are overweighted, moderate and high probabilities are underweighted, and the latter effect is more pronounced than the former. Third, . . . for any fixed probability ratio q, the ratio of decision weights is closer to unity when the probabilities are low than when they are high, for example, $\pi(.1)/\pi(.2) > \pi(.4)/\pi(.8)$. . . . The major qualitative properties of decision weights can be extended to cases in which the probabilities of outcomes are subjectively assessed rather than explicitly given. In these situations, however, decision weights may also be affected by other characteristics of an event, such as ambiguity or vagueness.

Prospect theory . . . should be viewed as an approximate, incomplete, and simplified description of the evaluation of risky prospects. Although the properties of v and π summarize a common pattern of choice, they are not universal: the preferences of some individuals are not well described by an S-shaped value function and a consistent set of decision weights. The simultaneous measurement of values and decision weights involves serious experimental and statistical difficulties.

If π and v were linear throughout, the preference order between options would be independent of the framing of acts, outcomes, or contingencies. Because of the characteristic nonlinearities of π and v, however, different frames can lead to different choices. The following three sections describe reversals of preference caused by variations in the framing of acts, contingencies, and outcomes.

The Framing of Acts

Problem 3 [$N = 150$]: Imagine that you face the following pair of concurrent decisions. First examine both decisions, then indicate the options you prefer.

Decision (i). Choose between:

A. a sure gain of $240 [84 percent]

B. 25% chance to gain $1,000, and 75% chance to gain nothing [16 percent]

Decision (ii). Choose between:

C. a sure loss of $750 [13 percent]

D. 75% chance to lose $1,000, and 25% chance to lose nothing [87 percent]

The majority choice in decision (i) is risk averse: a riskless prospect is preferred to a risky prospect of equal or greater expected value. In contrast, the majority choice in decision (ii) is risk taking: a risky prospect is preferred to a riskless prospect of equal expected value. This pattern of risk aversion in choices involving gains and risk seeking in choices involving losses is attributable to the properties of v and π. Because the value function is S-shaped, the value associated with a gain of $240 is greater than 24 percent of the value associated with a gain of $1,000, and the (negative) value associated with a loss of $750 is smaller than 75 percent of the value associated with a loss of $1,000. Thus the shape of the value function contributes to risk aversion in decision (i) and to risk seeking in decision (ii). Moreover, the underweighting of moderate and high probabilities contributes to the relative attractiveness of the sure gain in (i) and to the relative aversiveness of the sure loss in (ii). The same analysis applies to problems 1 and 2.

Because (i) and (ii) were presented together, the respondents had in effect to choose one prospect from the set: A and C, B and C, A and D, B and D. The most common pattern (A and D) was chosen by 73 percent of respondents, while the least popular pattern (B and C) was chosen by only 3 percent of respondents. However, the combination of B and C is definitely superior to the combination A and D, as is readily seen in problem 4.

> **Problem 4** [$N = 86$]: Choose between:
>
> A & D. 25% chance to win $240, and 75% chance to lose $760. [0 percent]
>
> B & C. 25% chance to win $250, and 75% chance to lose $750. [100 percent]

When the prospects were combined and the dominance of the second option became obvious, all respondents chose the superior option. The popularity of the inferior option in problem 3 implies that this problem was framed as a pair of separate choices. The respondents apparently failed to entertain the possibility that the conjunction of two seemingly reasonable choices could lead to an untenable result.

The violations of dominance observed in problem 3 do not disappear in the presence of monetary incentives. A different group of respondents who answered a modified version of problem 3, with real payoffs, produced a similar pattern of choices. Other authors have also reported that violations of the rules of rational choice, originally observed in hypothetical questions, were not eliminated by payoffs.

We suspect that many concurrent decisions in the real world are framed independently and that the preference order would often be reversed if the decisions were combined. The respondents in problem 3 failed to combine options, although the integration was relatively simple and was encouraged by instructions. The complexity of practical problems of concurrent decisions, such as portfolio selection, would prevent people from integrating options without computational aids, even if they were inclined to do so.

The Framing of Contingencies

The following triple of problems illustrates the framing of contingencies. Each problem was presented to a different group of respondents. Each group was told that one participant in ten, preselected at random, would actually be playing for money. Chance events were realized, in the respondents' presence, by drawing a single ball from a bag containing a known proportion of balls of the winning color, and the winners were paid immediately.

Problem 5 [N = 77]: Which of the following options do you prefer?

A. a sure win of $30 [78 percent]

B. 80% chance to win $45 [22 percent]

Problem 6 [N = 85]: Consider the following two-stage game. In the first stage, there is a 75% chance to end the game without winning anything, and a 25% chance to move into the second stage. If you reach the second stage you have a choice between:

C. a sure win of $30 [74 percent]

D. 80% chance to win $45 [26 percent]

Your choice must be made before the game starts, i.e., before the outcome of the first stage is known. Please indicate the option you prefer.

Problem 7 [N = 81]: Which of the following options do you prefer?

E. 25% chance to win $30 [42 percent]

F. 20% chance to win $45 [58 percent]

Let us examine the structure of these problems. First, note that problems 6 and 7 are identical in terms of probabilities and outcomes, because prospect C offers a .25 chance to win $30 and prospect D offers a probability of .25 × .80 = .20 to win $45. Consistency therefore requires that the same choice be made in problems 6 and 7. Second, note that problem 6 differs from problem 5 only by the introduction of a preliminary stage. If the second stage of the game is reached, then problem 6 reduces to problem 5; if the game ends at the first stage, the decision does not affect the outcome. Hence there seems to be no reason to make a different choice in problems 5 and 6. By this logical analysis, problem 6 is equivalent to problem 7 on the one hand and problem 5 on the other. The participants, however, responded similarly to problems 5 and 6 but differently to problem 7. This pattern of responses exhibits two phenomena of choice: the certainty effect and the pseudocertainty effect.

The contrast between problems 5 and 7 illustrates a phenomenon discovered by Allais, which we have labeled the certainty effect: a reduction of the probability of an outcome by a constant factor has more impact when the outcome was initially certain than when it was merely probable. Prospect theory attributes this effect to the properties of π. . . .

The first stage of problem 6 yields the same outcome (no gain) for both acts. Consequently, we propose, people evaluate the options conditionally, as if the second stage had been reached. In this framing, of course, problem 6 reduces to problem 5. . . . [Thus, t]he striking discrepancy between the responses to problems 6 and 7, which are identical in outcomes and probabilities, could be described as a pseudo-certainty effect. The prospect yielding $30 is relatively more attractive in problem 6 than in problem 7, as if it had the advantage of certainty. The sense of certainty associated with option C is illusory, however, since the gain is in fact contingent on reaching the second stage of the game. . . .

Many significant decisions concern actions that reduce or eliminate the probability of a hazard, at some cost. The shape of π in the range of low probabilities suggests that a protective action which reduces the probability of a harm from 1 percent to zero, say, will be valued more highly than an action that reduces the probability of the same harm from 2 percent to 1 percent. Indeed, probabilistic insurance, which reduces the probability of loss by half, is judged to be worth less than half the price of regular insurance that eliminates the risk altogether. . . .

The preceding discussion highlights the sharp contrast between lay responses to the reduction and the elimination of risk. Because no form of protective action can cover all risks to human welfare, all insurance is essentially probabilistic: it reduces but does not eliminate risk. The probabilistic nature of insurance is commonly masked by formulations that emphasize the completeness of protection against identified harms, but the sense of security that such formulations provide is an illusion of conditional framing. It appears that insurance is bought as protection against worry, not only against risk, and that worry can be manipulated by the labeling of outcomes and by the framing of contingencies. It is not easy to determine whether people value the elimination of risk too much or the reduction of risk too little. The contrasting attitudes to the two forms of protective action, however, are difficult to justify on normative grounds.

The Framing of Outcomes

Outcomes are commonly perceived as positive or negative in relation to a reference outcome that is judged neutral. Variations of the reference point can therefore determine whether a given outcome is evaluated as a gain or as a loss. Because the value function is generally concave for gains, convex for losses, and steeper for losses than for gains, shifts of reference can change the value difference between outcomes and thereby reverse the preference order between options. Problems 1 and 2 illustrated a preference reversal induced by a shift of reference that transformed gains into losses.

For another example, consider a person who has spent an afternoon at the race track, has already lost $140, and is considering a $10 bet on a 15:1 long shot in the last race. This decision can be framed in two ways, which correspond to two natural reference points. If the status quo is the reference point, the outcomes of the bet are framed as a gain of $140 and a loss of $10. On the other hand, it may be more natural to view the present state as a loss of $140 for the betting day, and accordingly frame the last bet as a chance to return to the reference point [of the morning] or to increase the loss to $150. Prospect theory implies that the latter frame will produce more risk seeking than the former. Hence, people who do not adjust their reference point as they lose are expected to take bets that they would normally find unacceptable. This analysis is supported by the observation that bets on long shots are most popular on the last race of the day.

Because the value function is steeper for losses than for gains, a difference between options will loom larger when it is framed as a disadvantage of one option rather than as an advantage of the other option. An interesting example of such an effect in a riskless context has been noted by Thaler. In a debate on a proposal to pass to the consumer some of the costs associated with the processing of credit-card purchases, representatives of the credit-card industry requested that the price difference be labeled a cash discount rather than a credit-card surcharge. The two labels induce different reference points by implicitly designating as normal reference the higher or the lower of the two prices. Because losses loom larger than gains, consumers are less willing to accept a surcharge than to forego a discount. A similar effect has been observed in experimental studies of insurance: the proportion of respondents who preferred a sure loss to a larger probable loss was significantly greater when the former was called an insurance premium.

These observations highlight the lability of reference outcomes, as well as their role in decision-making. In the examples discussed so far, the neutral reference point was identified by the labeling of outcomes. A diversity of factors determine the reference outcome in everyday life. The reference outcome is usually a state to which one has adapted; it is sometimes set by social norms and expectations; it sometimes corresponds to a level of aspiration, which may or may not be realistic.

We have dealt so far with elementary outcomes, such as gains or losses in a single attribute. In many situations, however, an action gives rise to a compound outcome, which joins a series of changes in a single attribute, such as a sequence of monetary gains and losses, or a set of concurrent changes in several attributes. . . . We propose that people generally evaluate [such] acts in terms of a minimal account, which includes only the direct consequences of the act. The minimal account associated with the decision to accept a gamble, for example, includes the money won or lost in that gamble and excludes other assets or the outcome of previous gambles. People commonly adopt minimal accounts because this mode of framing (i) simplifies evaluation and reduces cognitive strain, (ii) reflects the intuition that consequences should be causally linked to acts, and (iii) matches the properties of hedonic experience, which is more sensitive to desirable and undesirable changes than to steady states.

There are situations, however, in which the outcomes of an act affect the balance in an account that was previously set up by a related act. In these cases, the decision at hand may be evaluated in terms of a more inclusive account, as in the case of the bettor who views the last race in the context of earlier losses. More generally, a sunk-cost effect arises when a decision is referred to an existing account in which the current balance is negative. Because of the nonlinearities of the evaluation process, the minimal account and a more inclusive one often lead to different choices. . . .

The following problem, based on examples by Savage and Thaler, further illustrates the effect of embedding an option in different accounts. Two versions of this problem were presented to different groups of subjects. One group ($N = 93$) was given the values that appear in parentheses. and the other group ($N = 88$) the values shown in brackets.

> **Problem 10:** Imagine that you are about to purchase a jacket for ($125) [$15], and a calculator for ($15) [$125]. The calculator salesman informs you that the calculator you wish to buy is on sale for ($10) [$120] at the other branch of the store, located 20 minutes drive away. Would you make the trip to the other store?

The response to the two versions of problem 10 were markedly different: 68 percent of the respondents were willing to make an extra trip to save $5 on a $15 calculator; only 29 percent were willing to exert the same effort when the price of the calculator was $125. Evidently the respondents do not frame problem 10 in the minimal account, which involves only a benefit of $5 and a cost of some inconvenience. Instead, they evaluate the potential saving in a more inclusive account, which includes the purchase of the calculator but not of the jacket. By the curvature of v, a discount of $5 has a greater impact when the price of the calculator is low than when it is high.

A closely related observation has been reported by Pratt, Wise, and Zeckhauser, who found that the variability of the prices at which a given product is sold by different stores is roughly proportional to the mean price of that product. The same pattern was observed for both frequently and infrequently purchased items. Overall, a ratio of 2:1 in the mean price of two products is associated with a ratio of 1.86:1 in the standard deviation of the respective quoted prices. If the effort that consumers exert to save each dollar on a purchase, for instance by a phone call, were independent of price, the dispersion of quoted prices should be about the same for all products. In contrast, the data of Pratt et al. are consistent with the hypothesis that consumers hardly exert more effort to save $15 on a $150 purchase than to save $5 on a $50 purchase. Many readers will recognize the temporary devaluation of money which facilitates extra spending and reduces the significance of small discounts in the context of a large expenditure, such as buying a house or a car. This paradoxical variation in the value of money is incompatible with the standard analysis of consumer behavior.

Discussion

In this article we have presented a series of demonstrations in which seemingly inconsequential changes in the formulation of choice problems caused significant shifts of preference. The inconsistencies were traced to the interaction of two sets of factors: variations in the framing of acts, contingencies, and outcomes, and the characteristic nonlinearities of values and decision weights. The demonstrated effects are large and systematic, although by no means universal. They occur when the outcomes concern the loss of human lives as well as in choices about money; they are not restricted to hypothetical questions and are not eliminated by monetary incentives.

Earlier we compared the dependence of preferences on frames to the dependence of perceptual appearance on perspective. If while traveling in a mountain range you notice that the apparent relative height of mountain peaks varies with your vantage point, you will conclude that some impressions of relative height must be erroneous, even when you have no access to the correct answer. Similarly, one may discover that the relative attractiveness of options varies when the same decision problem is framed in different ways. Such a discovery will normally lead the decision-maker to reconsider the original preferences, even when there is no simple way to resolve the inconsistency. The susceptibility to perspective effects is of special concern in the domain of decision-making because of the absence of objective standards such as the true height of mountains. . . .

The perspective metaphor highlights the following aspects of the psychology of choice. Individuals who face a decision problem and have a definite preference (i) might have a different preference in a different framing of the same problem, (ii) are normally unaware of alternative frames and of their potential effects on the relative attractiveness of options, (iii) would wish their preferences to be independent of frame, but (iv) are often uncertain how to resolve detected inconsistencies. In some cases (such as problems 3 and 4 . . .) the advantage of one frame becomes evident once the competing frames are compared, but in other cases (problems 1 and 2 and problems 6 and 7) it is not obvious which preferences should be abandoned.

These observations do not imply that preference reversals, or other errors of choice or judgment, are necessarily irrational. Like other intellectual limitations, discussed by Simon under the heading of "bounded rationality," the practice of acting on the most readily available frame can sometimes be justified by reference to the mental effort required to explore alternative frames and avoid potential inconsistencies. However, we propose that the details of the phenomena described in this article are better explained by prospect theory and by an analysis of framing than by ad hoc appeals to the notion of cost of thinking.

The present work has been concerned primarily with the descriptive question of how decisions are made, but the psychology of choice is also relevant to the normative question of how decisions ought to be made. In order to avoid the difficult problem of justifying values, the modern theory of rational choice has adopted the coherence of specific preferences as the sole criterion of rationality. This approach enjoins the decision-maker to resolve inconsistencies but offers no guidance on how to do so. It implicitly assumes that the decision-maker who carefully answers the question "What do I really want?" will eventually achieve coherent preferences. However, the susceptibility of preferences to variations of framing raises doubt about the feasibility and adequacy of the coherence criterion.

Consistency is only one aspect of the lay notion of rational behavior. As noted by March, the common conception of rationality also requires that preferences or utilities for particular outcomes should be predictive of the experiences of satisfaction or displeasure associated with their occurrence. Thus, a man could be judged irrational either because his preferences are contradictory or because his desires and aversions do not reflect his pleasures and pains. The predictive criterion of rationality can be applied to resolve inconsistent preferences and to improve the quality of decisions. A predictive orientation encourages the decision-maker to focus on future experience and to ask "What will I feel then?" rather than "What do I want now?" The former question, when answered with care, can be the more useful guide in difficult decisions. In particular, predictive considerations may be applied to select the decision frame that best represents the hedonic experience of outcomes.

Further complexities arise in the normative analysis because the framing of an action sometimes affects the actual experience of its outcomes. For example, framing outcomes in terms of overall wealth or welfare rather than in terms of specific gains and losses may attenuate one's emotional response to an occasional loss. Similarly, the experience of a change for the worse may vary if the change is framed as an uncompensated loss or as a cost incurred to achieve some benefit. The framing of acts and outcomes can also reflect the acceptance or rejection of responsibility for particular consequences, and the deliberate manipulation of framing is commonly used as an instrument of self-control. When framing influences the experience of consequences, the adoption of a decision frame is an ethically significant act.

7.2 APPLICATIONS

Imperfect Information in Markets for Contract Terms

ALAN SCHWARTZ AND LOUIS WILDE

Contracts between firms and consumers are regulated extensively. Courts and legislatures prohibit the use of certain terms, require the use of others, and, if firms make appropriate disclosures, permit the use of still others. Decisionmakers and commentators often justify this regulation on the ground that "imperfect information" exists in consumer markets. They seldom distinguish, however, the differing forms of imperfect information, nor do they appreciate the various normative implications that attach to each of these forms. This article attempts to clarify the imperfect information justification for regulation as it applies to contract terms. . . .

Imperfect Information in the First Sense: Knowing the Odds

The typical person's estimate of the odds of product failure or of his or her own default will seldom equal true probabilities. Firms are commonly supposed to exploit these errors by imposing unwanted contract terms. Firms, however, respond to consumers as an aggregate, not as individuals; consequently, no firm knows or could know any particular consumer's estimate of the odds. Thus, the question is whether consumers in the aggregate systematically err such that firms have incentives to degrade contract content. We next argue that error of this sort is uncommon.

Market Responses to Consumer Error

We shall begin with consumer beliefs about product reliability. A consumer's subjective estimate of the odds of product failure is related to but is not wholly determined by actual failure probabilities for two reasons. First, a consumer's subjective belief about the odds probably bears some relation to the actual odds. A new car model, for example, is unlikely to be very much more or less reliable than prior models. A consumer often will have owned an earlier model or something similar to it or have talked with friends who have owned one or who own the new model. Moreover,

Alan Schwartz and Louis Wilde, Imperfect Information in Markets for Contract Terms, 69 *Virginia Law Review* (1983), p. 1387–1485. Copyright © 1983 the Virginia Law Review Association and Fred B. Rothman & Co. Reprinted with permission.

magazines and newspapers often discuss the characteristics of many new models. Hence, the actual odds should affect individual consumers' estimates of what those odds are. Second, because consumers lack the expertise and resources to test products and because some product characteristics are only revealed through use, a consumer's estimate of the actual odds will seldom be completely accurate. . . .

Consumer Error Is Random. Suppose that consumers in the aggregate hold subjective beliefs (S) that fluctuate randomly around the true value (A) such that consumer error (e) is "unbiased." An error term is unbiased when positive and negative estimates of the true value cancel out; hence, for consumers in the aggregate the mean estimate $E(S)$ will equal the true value $E(A)$. Because consumers in our model shop randomly, each firm will probably see a representative sample of the market. In this event, firms will respond as if the consumers visiting them knew the odds perfectly. Thus, if consumer estimates of the odds of product defects, or of any other odds, fluctuate randomly around true values, imperfect information about the risks being allocated may exist but will not cause policy problems.

Consumers in the Aggregate Are Pessimistic. Markets also commonly correct for consumer pessimism. . . . Suppose now that consumers would, if perfectly informed, prefer no warranty protection, but pessimism respecting the odds of product breakdown causes consumers to want a warranty. Comparison shopping can ensure that consumers pay competitive prices for warranty coverage, but consumers would be purchasing more coverage than they really want. This problem does not seem serious for two reasons. First, substantial consumer pessimism may be short-lived because firms have an incentive to dissipate it. Pessimistic consumers not only prefer unnecessary warranties when they buy, but also buy fewer products than they would were they well informed. Hence, firms should make efforts to prevent or reduce systematic pessimism. Second, pessimism at worst causes consumers to be overinsured. Consumers seemingly are worse off if they are without protection against product-related losses than if they sometimes have too much protection.

Consumer choices of security terms can be analyzed similarly. If consumers err randomly in their estimate of the odds of default, firms will respond as if consumers knew the actual odds. Respecting pessimism, . . . [t]his problem does not seem serious for the same reasons that the identical warranty problem does not appear bothersome. Firms have an incentive to dissipate pessimism because not only will pessimistic consumers reject security, but they will also be less anxious to incur debt. Moreover, the perceived policy problem in this area is that security interests place consumers at the mercy of firms; pessimism at worst causes consumers to be less at the mercy of firms than they would be if fully informed.

Consumers in the Aggregate Are Optimistic. Markets may correct poorly for consumer optimism. Suppose that consumers would want warranty protection if they knew the odds of product failure, but actual consumers are optimistic about these odds. In this case, optimism is reflected in the consumer's limit price, h_{N}, which is higher than it would be were the true odds known: optimistic consumers have an artificially high willingness to pay for goods without warranties. . . .

If well-informed consumers would want warranties but optimism causes these consumers not to demand them, warranties will probably not appear. Firms lack an incentive to offer broader warranties than consumers demand because warranties are costly: firms must redeem their warranty guarantees. Optimistic consumers might resist the price increases necessary to cover this cost. On the other hand, optimistic consumers who purchase too narrow warranties will often be disappointed; they will experience significant uninsured losses. Firms consequently will lose goodwill. Hence, firms seemingly are better off if consumers would make correct choices, for then firms can preserve good will by making appropriate warranties, yet recover the full costs that these warranties create. Curing consumer optimism, however, could be difficult. Firms would be reluctant to conduct an advertising campaign the theme of which is "Our widgets break a lot." A more promising response is to make correct warranties but bury the cost in the total price of the product. How often this is done is not known. Also, some firms might maximize profits by exploiting consumer optimism in the short run. Thus, systematic consumer optimism respecting product failure rates creates a policy problem, but its seriousness is unknown.

If consumers in security interest markets believe default to be less likely than it is in fact, they will not resist demands for security interests strongly enough, for they will think foreclosure is unlikely. We showed above that firms will not demand security when consumers are willing to pay to avoid it. . . . Optimistic consumers, however, may set h_N too low or h_S (the limit price with a security interest) too high. . . . Thus, if consumers are optimistic regarding the odds of default, they will make too many secured loans. Moreover, lenders will be unlikely to correct consumers' misperceptions by stressing how likely a default may be because such an action will decrease lenders' profits from making loans. Hence, consumer optimism about the odds of default also seems a policy problem. . . .

Cognitive Errors and Optimism

No general theory of how people make inferential judgments exists. In recent years, however, psychologists have extensively studied how these judgments are made. The central theme of this research is that people err in ways that are at once serious, systematic, and predictable. Will these errors cause people in the aggregate to misprocess information and therefore understate the odds of defects or defaults? This section argues that the principal cognitive errors that seemingly plague human inference in most cases will cause people to make random errors, will incline people toward pessimism in the case of products, or will be irrelevant to the question whether people generally are optimistic or pessimistic respecting the odds.

The Odds of Product Failure. Four sources of cognitive error could affect people's assessment of the odds of product defects: cognitive dissonance, misuse of the "availability" and "representativeness" heuristics, and a possible tendency to ignore very low probability events. The cognitive dissonance idea derives from the theory of cognitive consistency. According to this theory, people resist holding in awareness two conflicting ideas simultaneously. Thus, they tend to ignore or distort evidence relevant to the truth of one of these ideas. For example, people are said to believe that they are intelligent and prudent and consequently will make intelligent and prudent choices; hence, the theory predicts that people will devalue evidence that impeaches their choices after these choices have been made. A fair amount of evidence supports the theory. As illustrations, workers taking jobs in unsafe occupations apparently come to believe that the industries are safe—"smart, careful people would not work in dangerous places." Similarly, some buyers may have more affirmative attitudes towards products after purchasing than before.

Cognitive dissonance seemingly could not cause persons to ignore unfavorable information in the case that concerns us, when consumers are deciding whether to buy. Consumer purchases of major items are discrete events that have high salience; people view them as beginnings—"my new car." Dissonance is unlikely to occur when people consciously gather evidence in order to decide.

Even where consumers' self-images are not at stake in their estimation of product risks, the method they use to assess those risks may be flawed. The "availability heuristic" can cause persons to make mistakes about the frequency with which events occur. One making inferential judgments by use of this heuristic tends to ignore statistical data in favor of evidence that seems germane and is "in awareness"—is available. For example, a person may understate the correlation between cigarette smoking and lung cancer because his judgment of this correlation was excessively influenced by his knowledge of two neighbors, each of whom smoked for fifty years and died of stroke. The availability heuristic misleads when the association between cause and effect that is in awareness, or is easily summoned up, correlates poorly with the frequency with which possible causes and effects actually are covariant, as in the cigarette example. Psychologists believe that such mistakes occur frequently because the existence of evidence in awareness is largely a function of its "vividness"—its emotional interest, ability to evoke imagery, spatial and temporal proximity, and concreteness. Vivid evidence is not necessarily the most probative evidence.

If people actually use the availability heuristic to judge product reliability, their errors should in the aggregate either be random (and therefore unbiased) or pessimistic. Respecting the first possibility, suppose potential car buyers assess the reliability of new Saabs not by published repair data but by reference to what they know about cars in general and by what they can recall about Saabs. Evidence of this sort will include rumor and the stories of acquaintances, and it is likely to suffer from the biases of small sample sizes: any one person's sample will have too few data points to reveal the correct odds for a particular model. Yet, although faulty, the method will generate estimates that are influenced by the true odds. Everyone has some knowledge of how cars in general perform, and the performance of Saabs is not excessively dissimilar from the norm. More important, the results of each person's sample will be affected by how reliable Saabs actually are. If Saabs always broke down, no one could have a friend with a good word to say about them. Finally, the errors that this method of assessing data generates are unlikely to lean in one direction. Some people may have had good experiences with cars or know people with good Saabs while others may have had bad experiences or know people with bad Saabs. Hence, if the availability heuristic influences consumer estimates of the odds of product breakdown, those estimates should be unbiased in the aggregate.

If they are not random, consumer errors will likely tend toward pessimism because negative evidence is often more vivid than positive evidence. This fact explains why people tend to draw insufficiently strong inferences from events that fail to happen. If a product performs well most of the time but fails noticeably, people may believe it less reliable than it is because they give too little weight to the absence of failure, and too much weight to its presence. . . .

Another source of cognitive error that might lead to mistaken estimates of the odds is what psychologists have labeled the "representativeness heuristic." A considerable amount of evidence suggests that, when seeking an event's cause, people are strongly influenced by superficial likenesses between some possible causes of the phenomenon under study and the phenomenon itself. The "gambler's fallacy" is one illustration of the error that outcomes "represent" their underlying causes. Each turn of a fair roulette wheel or the toss of a fair coin is uncorrelated with prior turns or tosses. Therefore, the probability that a particular turn will be red is slightly less than .5 (a zero and double zero exist), and the probability of heads is approximately one half. If a long run of blacks or tails has occurred, a victim of the gambler's fallacy will assign a much greater probability than fifty percent to the chance that the next turn will produce a red or the next toss a heads. This mistake is believed to occur because people perceive the process that generates outcomes to be random, and random sequences of reds and blacks or heads and tails seem more representative of such a process than a long run of blacks or tails.

The apparent pervasiveness of the gambler's fallacy suggests that people might make pessimistic assessments of product reliability. Most products, particularly appliances, work reliably. Yet consumers know that appliances are made by people, that human error often exists, and that industrial workers may lack the sense of craft their ancestors had. A consumer whose appliances work well and who sees the manufacturing process in this way could believe his next purchase will be less reliable than his last: consistent success is unrepresentative of a system characterized by human error and a lack of craft sense. In the gambler's fallacy, the consumer errs by overstating the correlation between present and past—a run of heads implies a tails next time. For product purchases, the analogous error would assume that too many product successes imply a forthcoming failure. There is some evidence that consumers actually make this error. A study by the University of Michigan Survey Research Center reported that people perceived a need for repairs in new home appliances that was much greater than their actual need for repairs in the past. Hence, use of the representativeness heuristic may bias people toward pessimism.

In addition to cognitive consistency and the biases in mental processes just discussed, consumers may ignore very low probability risks, even though these risks cause catastrophe when they do materialize. For example, people buy less flood and earthquake insurance than the objective probabilities of those disasters warrant. Similarly, personal injuries are a much less frequent consequence of product defects than ordinary malfunction. People may therefore optimistically ignore the odds that products will physically harm them and demand less warranty protection against personal injuries than they should. . . .

The tendency to ignore low probability events may also reflect use of the availability heuristic. These sorts of events may seldom be in awareness because they occur rarely, so people respond inadequately to them. Some grounds exist for believing that the availability heuristic is partly responsible for the phenomenon. For example, while people seem insufficiently concerned with flood, fire, and earthquake, they express great concern about the risks of nuclear power and recombinant DNA, although the probability that these phenomena will cause harm is quite low. This may be because these latter risks are more "available," as they are much discussed and would cause awful harm if they materialized. . . .

If the availability heuristic is actually at work, it is premature to base policy on the penchant of people sometimes to ignore low probability events. What is needed but does not exist is a way to link the extent to which events may be in awareness with the objective probabilities that persons tend to ignore. For example, if a product carries a .01 risk of causing personal injury, will people act as if that risk is zero? How can a decisionmaker know when a risk of a particular harm is below the threshold of attention? If people are concerned about nuclear power but unconcerned about floods, could they be similarly concerned about cars but not about skateboards? That is, if it is the availability heuristic that is misleading people, are generalizations about odds thresholds warranted? Until cognitive theory develops enough to permit answers to questions of this sort, that people insure insufficiently against certain kinds of low probability events cannot support factual inferences respecting other such events. In the absence of such inferences, it seems unwise to require insurance by mandating warranties.

An exception to this conclusion may exist for frequently purchased, inexpensive items that cause serious personal harm a very small percentage of the time. No case for optimism in the purchase of these products can be derived from the representativeness or availability heuristics, but optimism may be implied by the cognitive dissonance paradigm. This is because consumers often will learn about the possibility of dangerous malfunction after they have made a commitment to the product, for they frequently purchase it. Such negative information could be devalued. The risk that serious personal harm may occur from using such products as soda in bottles, nonprescription drugs, and food thus could be in the class of risks against which insufficient insurance tends to be purchased.

The Odds of Default. Are people optimistic respecting the odds of their own defaults, such that they will resist a creditor's demand for security less than their own better-informed preferences would dictate? This question differs from the one just asked about products, for there the issue was whether consumers could correctly infer an objective frequency—the odds that a product would fail. In this case, people must predict the joint influence of their own abilities and objective circumstances. For example, a person about to take out an auto loan must consider whether he or she is a sufficiently prudent manager to be able to make the payments under stable personal financial circumstances and must also assess the likelihood of unemployment. People generally have as much data about their own abilities as outsiders do; we also have argued that they are likely to have as much data about their objective circumstances as others will have. . . .

This analysis implies that a consumer actor will make at least as good a judgment of how the interaction between his or her traits and circumstances will influence repayment prospects as will a bank/observer unless the consumer uses inferior theories to assess this interaction or uses the same theories as banks do but applies them badly.

Both possibilities are nontrivial, but neither would bias consumers in particular directions.

The clearest example of the first problem is the common "fundamental attribution error." Attribution theory in psychology "is concerned with the attempts of ordinary people to understand the causes and implication of the events they witness." The fundamental attribution error is to place too much weight on characterological factors and too little weight on situational ones when assessing or predicting behavior. For example, people tend to attribute an honest act to an honest disposition rather than to the presence of factors that encourage honesty such as the monitoring of behavior or the need for the approval of others. Attributions are said to be mistaken in life because psychologists have been able in laboratories to induce actors to perform widely divergent behaviors by varying situational factors. Environments may, in short, influence behavior more than many people believe.

Regarding the risk of default, one might suppose that consumers, when assessing this risk, place too much weight on their own traits such as prudence and too little weight on situational factors such as a shaky economy. If people ordinarily think highly of their abilities, they will then be more sanguine about their repayment prospects than their circumstances actually warrant. The fundamental attribution error, however, partly derives from the availability heuristic: people commonly have more salience for observers than situations have; as a result, observers tend to focus more on the influence of actors than on their environments. If this explanation is correct, people should commit the fundamental attribution error less when assessing their own behavior than when evaluating others' actions. The actor, being always present, has relatively less salience for himself than circumstances do. The evidence is consistent with this prediction: actors tend to see their own behavior as situationally determined while observers see the same behavior as dispositionally determined. Hence, at this early stage in the understanding of these issues, there seems an insufficient basis on which to predict that people will be systematically optimistic about the odds because, thinking well of their abilities, they are led by the fundamental attribution error to give those abilities undue weight.

Lay persons also tend to slight statistical data. A bank officer, for instance, will likely use past rates of default among similar consumers to guide lending practices, while individual borrowers may rely on less probative factors such as their own and their friends' histories. The errors that such methods could cause seem random. As an illustration, most people knew that job loss is an important cause of default. If they evaluate this possibility by use of the availability heuristic, they may overstate the likelihood of job loss if they personally know unemployed persons and understate it if they do not. The effect of these errors is presumably random. People may use the representativeness heuristic and ask themselves whether their own traits and circumstances are representative of high- or low-risk debtors, rather than use statistical data on default rates. This inferential process will mislead unless the traits that consumers believe predict default correlate strongly with the traits that actually do predict it. Unfortunately, no one knows whether consumers routinely focus on the wrong traits, nor is it known in which direction their errors run.

In sum, people may sometimes use inferior theories to evaluate the odds of their own default, but there is no reason to believe these theories will routinely lead to optimism. Moreover, people will often use the same theories that banks use: in assessing their fitness to assume debt, people will probably look to their own incomes and job histories just as lending officers do. Potential debtors probably make more mistakes when using these theories than do banks because the debtors have less expertise. But again, there is no reason to think that these mistakes lead to a systematically optimistic bias, nor is there any way to know how serious they are. . . .

The Fresh-Start Policy in Bankruptcy Law
THOMAS JACKSON

The principal advantage bankruptcy offers an individual lies in the benefits associated with discharge. Unless he has violated some norm of behavior specified in the bankruptcy laws, an individual who resorts to bankruptcy can obtain a discharge from most of his existing debts in exchange for surrendering either his existing nonexempt assets or, more recently, a portion of his future earnings. Discharge not only releases the debtor from past financial obligations, but also protects him from some of the adverse consequences that might otherwise result from his release. For these reasons, discharge is viewed as granting the debtor a financial "fresh start." . . .

[This article proposes] two principal hypotheses to account for the nonwaivability of discharge. The first is that most people would choose to retain a nonwaivable right of discharge if they knew of the psychological factors that tempt them to overconsume credit. This hypothesis consists of two distinct, though related, concepts. The first of these—"impulse control"—is the notion that individuals would willingly take steps (such as the creation of a discharge rule) to control impulsive tendencies that tempt them to "mortgage" the future in favor of present consumption. The other concept—"incomplete heuristics"—posits that, in assessing courses of action, people unwittingly use tools for aiding their judgment that systematically induce them to overconsume credit. The second hypothesis is that individuals tend to ignore the full costs that their credit decisions impose on others. Society is accordingly justified, quite apart from any solicitude for individual preferences, in making discharge inalienable in order to reduce such externalities. . . .

The Normative Underpinnings of the Fresh-Start Policy

Not only the American law of contracts, but also much of American society in general, is structured around the premise that individuals should for the most part have the freedom to order their own affairs as they please, because rational, self-interested actors will tend to make decisions that maximize their own utility. In the context of such a presumption, making discharge nonwaivable raises troubling implications:

Thomas Jackson, "The Fresh-Start Policy in Bankruptcy Law," 98 *Harvard Law Review* (1985), pp. 1393–1448. Copyright © 1985 The Harvard Law Review Association. Reprinted with permission.

such a prohibition on contractual waiver both contravenes the principle of contractual freedom and increases the cost of credit.

But this neoclassical model, along with its presumption in favor of contractual freedom, depends on at least two key assumptions: first, that the individual acts rationally out of free will and is capable of discerning his best interests; and second, that no costs are imposed on noncontracting parties. If either of these assumptions fails in a particular case, so might the presumption in favor of contractual freedom. Society has placed a number of restrictions on contractual freedom that do not necessarily undermine the neoclassical model because arguably they are imposed in response to a failure of one of the underlying assumptions. Thus, on the one hand, the doctrines of fraud, duress, undue influence, mistake, mental illness, infancy, the like, and also certain safety-net programs such as social security, respond to a perceived failure of an individual to act in his own best interests. And tort and criminal law, on the other hand, penalize the individual for engaging in actions that adversely affect others. Both types of restrictions appear to conflict with notions of individual autonomy; yet both in fact elaborate the meaning of individual autonomy in a complex and interdependent society.

The question raised by bankruptcy discharge is whether the nonwaivability of the right of discharge, although facially inhibiting a borrower's individual autonomy, does not in fact faithfully protect his own interests and those of noncontracting parties. . . .

Protecting the Individual: Volitional and Cognitive Justifications for the Fresh-Start Policy

Paternalism and the Notion of Regret. Several theories suggest that our decisions about how to allocate our wealth over our lifetimes are systematically biased in favor of present consumption. One theory posits that the bias results from a kind of duress: our will is systematically overborne by the high pressure exerted by those who extend credit. But this justification fails to justify a general right to a fresh start. In contract law, the doctrine of duress operates on a case-by-case basis, not as a blanket rule. Any argument that duress is sufficiently universal in credit cases to warrant a nonwaivable rule must rest on the generalization that the typical individual cannot resist the typical offer of credit. To validate that generalization, one would have to identify the volitional or cognitive weaknesses that lead individuals systematically to ignore or overdiscount the uncertainties of the future. The question is whether we can demonstrate such weaknesses. . . .

Impulse Control: A Volitional Justification. . . . The concept of "impulse" provides at least a partial answer. When presented with an either-or choice, people, like animals, exhibit a tendency to choose current gratification over postponed gratification, even if they know that the latter holds in store a greater measure of benefits. Although, by itself, this predilection might be explained by the rational tendency to discount the value of deferred benefits, such an explanation does not account for this further observation: the same individuals who prefer current to postponed gratification will nevertheless favor a rule that requires them to defer gratification.

This tendency of individuals to desire external restraints on their impulses provides a basis for deciding which of an individual's personalities to favor. One personality is the rational planner; it carefully assesses the relative merits of current versus future consumption. The "impulse" personality, in contrast, approaches life like an addict, unable to consider or plan for the future. The impulse personality does not authentically "choose," because it does not rationally ponder how a given decision will affect the individual's long-term interests. The rational self, to the contrary, suppresses the temptation to act impulsively, resolving instead to act in accordance with the individual's entire set of wants and desires.

The control of impulsive behavior, then, may provide a key to justifying discharge policy. If unrestrained individuals would generally choose to consume today rather than save for tomorrow, and if this tendency stems in part from impulse, they may opt for a way of removing or at least restricting that choice. If individuals cannot control the impulse themselves, they may want the assistance of a socially imposed rule, one that will simply enforce the hypothesized decisions of their fully rational selves.

The question remains what form such a rule would take. In some circumstances the solution can be cast in the form of a "cooling-off" period—an interval during which an individual is permitted to undo the consequences of his impulsive behavior. Yet whereas particular kinds of credit transactions may profitably be subjected to a cooling-off rule, it would be almost impossible to apply such a rule to all credit transactions without substantially undercutting the certainty of expectations essential to the functioning of our credit economy. Consequently, it would be preferable to deal with impulsive credit acquisition by restricting individuals' ability to act impulsively in the first place.

More specifically, a social rule discouraging the extension of credit might be the best means to assist individuals in controlling impulsive credit decisions. A nonwaivable right of discharge controls impulsive credit decisions by encouraging creditors to monitor borrowing. Other, less intrusive rules would not be nearly as effective in controlling the urge to buy or borrow on credit. Consider, for example, a rule that allowed individuals to decide for themselves whether to be subject to a legally enforceable right of discharge. For such a rule to work, the individual's original decision would have to be irrevocable—either it would have to be enforceable by some form of specific performance, or the individual would have to face some penalty for reneging on his initial choice. Otherwise, an individual in the grip of an impulse could revoke his decision to embrace or forgo the right of discharge, just as a smoker can "revoke" his New Year's resolution not to smoke whenever he is seized with the urge to light up. Moreover, even if the decision were made irrevocable, problems would remain in setting limits on when the decision would have to be made. If made too soon, the decision whether to be governed by a nonwaivable discharge right with respect to future credit decisions might unduly constrain one's future self. If made too late, the election might itself be impulsive.

Thus, although the law might respond to the problem of impulsive credit behavior by letting individuals choose whether or not to waive the right of discharge, the problem may be better handled by means of a legal rule that uniformly disallows waiver. . . .

Incomplete Heuristics: A Cognitive Justification. Whereas impulsive behavior is vo-litional, there is a closely related cognitive feature of decisionmaking that makes the need for a legal rule perhaps more evident: because of systematic failures in their cog-nitive processes, individuals appear to make choices in which they consistently un-derestimate future risks. This problem—which I shall call the problem of "incomplete heuristics"—provides a powerful argument that most individuals, whether or not they are prone to impulsive behavior or have undergone personality shifts, would favor a legal rule making discharge nonwaivable. Like impulsiveness, incomplete heuristics may lead the individual to favor present consumption in a way that does not give due regard to his long-term desires and goals. Likewise, incomplete heuristics would jus-tify the decision to adopt a universal, nonwaivable right of discharge on a ground sim-ilar to that cited in the case of impulsive behavior: if individuals in the "original po-sition" had recognized that they would face informational constraints when making credit decisions, they would probably have chosen a system that would make some of the consequences of their borrowing avoidable. . . .

The Justification for a Socially Mandated Rule. The preceeding discussion suggests that what seems initially to be a paternalistic justification for discharge may in fact be consistent with society's preference for individual autonomy, because the non-waivable right of discharge accords with the result of a hypothetical initial delibera-tion behind a Rawlsian "veil of ignorance." If people in the "original position" had known about the problems of incomplete heuristics and impulsive behavior, and about the difficulty of adjusting for these problems in making credit decisions, they pre-sumably would have opted for a legal rule designed to avert those problems in ad-vance. This self-protective course is similar to the one that individuals follow when they take steps to remove their ability to act on later impulses.

A nonwaivable right of discharge may be desirable even if some individuals do not need its protection, as long as (1) a substantial number of people are likely to ex-perience unanticipated regret as a result of impulsive behavior or unwitting reliance on incomplete heuristics and (2) it is either impossible or extremely expensive to dis-tinguish those who will experience such regret from those who will not. To justify a nonwaivable general rule, one need not show that all people require its protection; it is enough to show that the rule promises to be less intrusive, or less costly, than one that attempts to discriminate between people who are likely to experience regret be-cause of impulsive behavior or the use of incomplete heuristics and those who are not.

Development of more discriminating rules does not appear promising. If impul-siveness were the only problem, the necessary sorting might be accomplished by means of a rule providing for a cooling-off period or one that offered each individual an irrevocable choice of whether to embrace the right of discharge. But because the situation is further complicated by the problem of incomplete heuristics, it is much harder to fashion an accurate sorting device. The interval between the time one makes a decision on the basis of incomplete heuristics and the time one comes to regret that decision is likely to be substantially longer than the corresponding interval when re-gret follows an impulsive decision. The time lapse in the former case not only makes it more difficult to restore the status quo, but also makes it harder to determine whether the regret is a product of incomplete heuristics or simply of a calculated gamble that came up a loser.

Our general belief that individuals should be able to set their own priorities suggests a larger problem with sorting devices. We cannot determine whether an individual makes rational credit decisions on his own behalf unless we know how he perceives his present and future wants and needs. When society attempts to distinguish individuals who act impulsively or rely on incomplete heuristics from those who do not, we run the grave risk of substituting an external social judgment for the subjective wants and needs of the individual. Consider, for instance, the broad, prophylactic rules by which the Bankruptcy Code regulates reaffirmation agreements: the cases applying these rules contain little evidence that judges do anything other than impose their own view of the individual's best interests in deciding which reaffirmations to permit. Similarly, we might expect that a case-by-case attempt to single out individuals who need the protection of a right of discharge might well result not in the identification of individuals whose choices do not accurately reflect their personal desires, but rather in the identification of individuals whose choices strike judges as somehow odd or aberrant.

Impulse Control, Incomplete Heuristics, and the Notion of Marketplace Constraints. The notions of impulsive behavior and incomplete heuristics, however powerful, should not be used to prove too much. They are more relevant to an analysis of individual behavior—the sort of behavior that discharge deals with—than to an examination of institutional or market behavior. In the case of market investments by firms, the concepts of impulsive behavior and incomplete heuristics suggest no need for restrictions on contractual freedom because firms that systematically act impulsively or underestimate the risks of investments will, in theory at least, be weeded out and replaced by firms that calculate risks more carefully. This is a desirable, indeed welcome, result. Nor is there reason to think that the market's pricing mechanisms will be systematically distorted by the impulsiveness or incomplete heuristics of investors. If, for example, such behavior leads many individual investors to react overenthusiastically to a biotechnology firm's latest public offering, more skilled investors will be able to capitalize on the impulsiveness or incomplete heuristics of the less skilled. Because of the likely presence of such skilled investors, aggregate (that is, marketplace) price levels should show no sign of any systematic underestimation of risks.

In contrast, individual decisions about investments may fail to reflect accurate estimates of risk because no similar aggregation occurs. As a result of incomplete heuristics and impulsiveness, even individuals with diversified portfolios may systematically underestimate the risks involved in those portfolios; hence they may overindulge in risky investments, consume too much, and save too little. When the relevant issue is the risk-evaluating ability of a particular individual—that is, his ability to decide between spending and saving—there may be no market constraint to produce accurate pricing information. The only meaningful constraint on an individual's credit decisionmaking is the prospect of his individual "failure." But individual financial failure arising from impulsive behavior or incomplete heuristics is exactly what fully informed individuals would presumably contract against in the first instance. . . .

Application of the Normative Theory: Defining the Contours of the Fresh Start in a Market Economy
. . .
The Assets Protected: Human versus Conventional Capital

How should the law accommodate both the need to ensure the availability of discharge and the need to ensure the availability of credit? The accommodation reached in the bankruptcy statutes has been relatively clear and consistent. Discharge, as defined by federal bankruptcy law, focuses on freeing the individual's future income from the claims of pre-bankruptcy creditors while extending no similar protection to the individual's other assets. Yet nonbankruptcy law places certain existing property as well beyond the reach of creditors. The decision whether to make certain assets "exempt" has traditionally been left to the states, and the states' decisions have been incorporated into bankruptcy law with relative indifference. . . . [Yet] recent developments suggest that the line between human capital [i.e, the expected value of a person's future earnings] and other forms of property will become increasingly blurred. . . .

Human Capital. Although no clear conceptual line separates one form of wealth from another, it is not surprising that bankruptcy law has traditionally afforded distinctive protection to human capital. Of the various forms of wealth, human capital is not only the least diversifiable, but also has the most direct bearing on the future well-being of the individual and the people who depend on him. Yet simply because we deem human capital especially deserving of protection through bankruptcy law, we need not conclude that such special protection is equally justifiable in all cases. If an individual holds large amounts of wealth in the form of human capital but has few other existing assets, he might justly be required to have the proceeds of his human capital subject to at least certain debts—in particular, debts such a student loans, which the individual incurred in order to acquire that human capital. In contrast, younger people can make perhaps the most persuasive claim to having their human capital stringently protected, because they are at an age when they are more likely to make decisions that will later induce regret. Whatever the cause of regrettable decisions—be it impulsiveness, incomplete heuristics, or something else—someone who is fifty-five years old is less likely to experience such regret than someone who is twenty. Hence discharge policy, in theory at least, should treat younger debtors more generously than older ones.

   How does this observation translate into policy? It seems to call for a scheme that would protect a decreasing portion of one's wealth as one grows older. Yet this observation would also justify a scheme, such as the existing one, that affords blanket protection to human capital but only selective protection to other forms of wealth. As one advances in age, the portion of one's wealth that consists of human capital is likely to decrease, while the value of one's other assets is likely to increase. Accordingly, limiting discharge to the freeing of human capital may in fact achieve the goal of protecting youth from later regret. . . .

Exempt Property. Many reasons for safeguarding human capital . . . may apply as well to other kinds of assets. Accordingly, such assets may justifiably be protected under a fresh-start policy. Two examples of such protection in current nonbankruptcy law are the exemptions of wage substitutes and certain durable goods from creditor attachment and hence, ultimately, from the bankruptcy estate. The rationale for protecting wage substitutes—such as pension funds and retirement income—may be that they, like human capital itself, are among the least diversifiable assets. The rationale for protecting durable goods may be that they constitute a form of savings, in the sense that their value is consumed over time. Because the main problem caused by impulsive behavior, incomplete heuristics, and externalities is the tendency to overconsume today and undersave for tomorrow, the decision of exemption law to protect savings for future consumption follows reasonably closely from the core notion of the fresh-start policy itself. Individuals may underestimate the extent to which they jeopardize their future consumption of existing goods when they borrow money in order to consume today. This tendency can prove especially costly when the assets have substantially greater value to the individual than to any third party, because the defaulting debtor may have to pay off his debts in assets that are worth more to him than the loan was worth.

As we have seen, however, the fresh-start policy could not protect all existing assets without sharply curtailing the availability of credit. The law therefore needs some mechanism for choosing which assets to shelter. Ideally, individuals would be allowed to protect all those durable assets that they felt were essential to their future well-being or that, if turned over to creditors, would result in a substantial asset-loss cost. In practice, though, it would be necessary to impose some limit on the list of protected assets, or else the debtor seeking discharge would likely deem all of his property essential. To solve this problem, society could formulate a relatively short list of assets considered vital to the typical individual's well-being. Property-exemption systems arguably serve that purpose. . . .

The question whether a debtor should be able to waive his right to exempt certain assets, as he in effect does by granting a security interest in those assets, is generally considered part of the broader question whether society should ever permit an individual to alienate exempt property. Society does not require the individual to buy particular types of exempt property in the first place; nor does it demand that he keep any such property he might acquire. Like the decision to invest in human capital (for instance, by attending college), the decision whether to defer consumption by using durable goods over time is generally left to the individual's discretion.

Notwithstanding the individual's general freedom to alienate exempt property, bankruptcy law might still forbid him to subject such property to general (that is, non-purchase money) security interests enforceable in bankruptcy. An individual who grants a general security interest in an asset retains possession and use of the asset but pledges to relinquish it upon the occurrence of a contingency—default—the probability of which he may underestimate because of incomplete heuristics. Although a ban on the granting of fully enforceable general security interests in exempt property can thus be justified in light of the normative underpinnings of the fresh-start policy, purchase money security interests demand different treatment. Unlike general security interests, purchase money interests are secured by the same asset that the extension of credit helped the debtor to obtain. Because the debtor's ability to acquire—and hence to consume—the asset in the first place derives from his "waiver" of the property's exempt status vis-à-vis the purchase money loan, purchase money security interests should be excepted from any general rule barring secured creditors from reaching exempt assets in bankruptcy. . . .

The Advisability of Allowing Debtors to Choose Which Assets to Protect. One may well ask whether individuals should be able to choose the mix of human and other capital that will constitute their fresh start. We have already seen that it makes sense for discharge policy to focus on protecting human capital. This focus has the arguably desirable result of discouraging exercise of the right of discharge, because protecting human capital to the detriment of other assets may result in asset-loss costs to debtors who resort to discharge. If bankruptcy law allowed the debtor to avoid that cost—by letting him choose to protect current assets and forgo future income—the debtor would more readily take refuge in discharge, and the cost of credit would rise.

If asset-loss cost were the only existing disincentive to invoking the right of discharge, then the question whether to allow debtors to protect current assets rather than human capital would be an easy one. For reasons discussed above, making discharge costless would undermine our entire system of credit. But asset-loss cost, though a significant check on debtors' recourse to discharge, is not the only disincentive to its use. Other exercise costs—such as discrimination by prospective creditors—also provide disincentives. For that reason, the decision whether individuals should be allowed to protect current assets rather than human capital still depends on a balancing of two interests: the need for available credit on the one hand, and the implementation of a fresh-start policy on the other.

None of the three reasons for making discharge nonwaivable—impulsive behavior, incomplete heuristics, and the disregard of externalities—argues against allowing individuals to decide which of their assets should be protected by discharge, assuming that the amount the creditors get is held constant. For two reasons, individuals are unlikely to make systematic errors of judgment in choosing whether to protect human capital or other forms of wealth: first, the measure of what must be handed over to creditors is in either case the market value of the debtor's existing tangible assets; second, human capital and durable goods both represent future consumption items. Although an individual might overestimate the degree to which he will be able to fund payments out of future income, the risk of error seems slight because the debtor's plan will be subject to judicial oversight and his creditors will retain the right to seize tangible assets if the debtor's future income proves inadequate. . . .

Some Implications of Cognitive Psychology for Risk Regulation

ROGER NOLL AND JAMES KRIER

Beginning with a set of books and articles published in the 1950s, cognitive psychologists have developed a new descriptive theory of how people make decisions under conditions of risk and uncertainty. A dominant theme in the theory is that most people do not evaluate risky circumstances in the manner assumed by conventional decision theory—they do not, that is, seek to maximize the expected value of some function when selecting among actions with uncertain outcomes.

The purpose of this article is to consider some implications of the cognitive theory for regulatory policies designed to control risks to life health, and the environment. [We] address, in turn, two central questions about the uses of the theory. First, if people behave in the manner described by the cognitive psychologists, how will this shape the demands that citizens make, through the political system, for risk regulation, and how (if at all) might these demands differ from those that would be expected if citizens behaved, instead, in the manner assumed by conventional decision theorists? Second, if citizens make demands as predicted by the cognitive theory, how (if at all) might their behavior affect the regulatory responses that political actors supply? . . .

Implications Regarding the Demand for Risk Regulation

To consider how the cognitive theory of decisions under risk translates into demands for political action, we first must specify the theory of political processes that maps citizen preferences into incentives to act on the part of government officials. We will begin by exploring the implications of the "positive responsiveness" of majority-rule electoral processes, then amend the treatment to take account of "mobilization bias." Our interest in both cases will focus on two inquiries: first, how citizen demands, given the cognitive view, will depart from those predicted by traditional decision theory; and second, what kinds of decisions are especially susceptible to the possibility of preference reversal with the unfolding of time. The second inquiry has to do with circumstances in which citizens change their minds about the appropriateness of some policy, even though the objective conditions surrounding the policy choice have not changed. These circumstances are especially important to government officials because they create the possibility for intertemporal inconsistencies in citizen demands. A political leader caught in such circumstances has to make a trade-off between dis-

Roger Noll and James Krier, Some Implications of Cognitive Psychology for Risk Regulation, 19 *Journal of Legal Studies* (1990), pp. 747–779. Copyright © 1990 The University of Chicago Law School. Reprinted with permission.

appointing citizens today and disappointing them tomorrow, when their preferences have changed. . . .

Pure Majority-Rule Democracy: Positive Responsiveness

One characteristic of majority-rule decision processes is that the likelihood of a policy being adopted does not decline as the number of citizens who favor it increases, a property called *positive responsiveness.* . . .

If [citizens'] preferences are based on accurate estimates of probabilities and outcome values, then they are unlikely to present intertemporal instability problems for government officials. In essence, those sorts of problems arise because true preferences persistently and systematically diverge from the evaluations arising from the application of traditional decision theory. In this case, the occurrence of a damaging event at a frequency consistent with its objective probability will not cause a citizen's evaluation of policies for dealing with it to change over time.

Circumstances change if . . . the occurrence (or lack of occurrence) of a damaging event can alter the estimates of probabilities, the valuation of outcomes, or both. Moreover, the extent to which preferences are likely to shift over time is predictable.

Consider, for example, the effect of availability, which predicts that people will systematically overestimate the probability of an event if similar events come readily to mind but will systematically underestimate them otherwise. The implication is that, immediately following a widely publicized disaster, citizens will place unusually great demands on their government to take action against recurrence, but as attention subsides, so too will the demand for action. . . .

Political Economy and the Cognitive Theory

Real representative democracies are more complex than the simple majority-rule system hypothesized in the preceding section. Decision-theoretic models of the political process emphasize two key elements of political participation: rational ignorance and mobilization bias. Rational ignorance refers to the lack of incentives on the part of citizens to be fully informed about the policy positions a candidate advocates in an election campaign. The incentive problem is caused, first, by the mismatch between the complexity of policy and the simplicity of the signal a voter can send (yes, no, or abstain) and, second, by the powerlessness of a single vote. To acquire and digest information requires time and effort, yet after exhaustive analysis a voter can send only a very weak signal, one that is not sensitive to the intensity of the voter's preferences, and one that is highly unlikely to influence the outcome of an election. In consequence there is little incentive to compare candidates and their policies carefully. . . .

Mobilization bias refers to the fact that some preferences tend to be more effectively represented in political processes than others. One way to overcome rational ignorance is to rely on someone else to perform issue analyses, then to vote as instructed. Similarly, one way to overcome voter powerlessness is to act in concert with other like-minded individuals. Collectively, a group of voters might have a substantial chance of swinging an election to one candidate or another. Moreover, through campaign contributions, a group can help preferred candidates provide the free entertaining information that might lead others to vote as the group desires. But forming groups, devising strategies, and coordinating efforts also require time, effort, and money. Hence, small, homogeneous groups with high per capita stakes in an issue are more likely to organize successfully than are large, heterogeneous groups with small per capita stakes.

We can now consider how the cognitive theory provides some insights into the bearing of rational ignorance and mobilization bias on risk policies.

Framing . . . suggests that, in a world of rational ignorance, the way political actors describe issues will alter a voter's evaluation of policy options. For example, if an issue is put as one involving alternative ways to obtain gains, voters will exhibit risk aversion; whereas, if the same issue is framed in terms of various ways to experience losses, voters will exhibit loss aversion with risk-taking behavior. Hence, the way an issue is stated can influence outcomes. Consider the case of a proposal to make seat belts mandatory, with a mandatory fine if one is caught unbelted. Outcomes could be expressed variously as follows:

a. a large benefit from driving without a belt, not being caught, and having no accident;
b. a somewhat smaller net benefit from driving belted and having no accident;
c. a smaller net benefit still from driving unbelted, getting caught and fined, and having no accident;
d. a small negative net driving benefit from wearing a belt and having an accident;
e. a more negative net benefit from driving unbelted and having an accident; and
f. a slightly more negative net benefit from driving unbelted, having an accident. and being fined to boot (adding insult to injury).

Facing these possibilities, a voter is likely to exhibit risk aversion because some possibilities are expressed positively.

Alternatively, one could put the seat-belt issue in terms that start with a status quo in which driving safely, unbelted, and unfined is the baseline, so that all other outcomes are negative. This would induce risk-taking behavior and a propensity to search for accident-reduction policies that reduce risks to zero (to avoid probabilistic insurance). In short, adeptness at characterizing issues can go a long way in determining the policy preferences expressed by the electorate.

Issue salience is also related to issue presentation. According to the representativeness heuristic, one can induce preferences concerning policies about one risk by presenting it as being like another risk. Then voters will evaluate the two risks as though they were the same. Therefore, if one environmental risk becomes salient (because, say, a damaging event has recently occurred), political actors may piggyback other environmental risks on the first by presenting them as essentially identical to the salient one.

The availability effect accentuates the importance of salience. A damaging event, if timed appropriately and if widely publicized, induces people to behave as if the likelihood of such events had increased. If the effect makes the issue politically salient—makes it one of the few issues that can animate political participation—then it can serve, for a time, to help overcome mobilization bias. With regard to environmental risk policy, risk producers will normally be the better organized, more effec-

tive participants in the political process. But, a highly publicized environmental catastrophe can cause some citizens to overestimate the degree of risk they face and thus
to be spurred to mobilize for political action. Moreover, other political actors may enhance the success of their own mobilization efforts by clever use of representativeness in the presentation of other issues. Mobilization bias may be temporarily overcome—temporarily because, if the probability of some damaging event is truly low,
then events of that sort are likely gradually to subside from public consciousness and
lose their salience.

The mechanics of how availability and mobilization bias work on a class of risk
policy issues are complex, but a few prototypical examples provide some insight into
the general process. Consider three types of low-probability risk circumstances. One
is a low-probability catastrophic event that, if it occurs, will harm a large number of
people simultaneously, such as the Bhopal or Chernobyl disasters. The second is an
event in which the probability of a disaster at a facility is again low, but where there
are a large number of facilities. Assume here that the number of people harmed per
disaster would be much lower than in the first case, but large enough to make for a
newsworthy event. Examples are airline or bus accidents, where between a few and
a few hundred people are killed or injured. Events of the second type occur more frequently than those of the first type; assume that the lower damage per event is exactly
offset by the greater exposure, so that the average annual number of deaths would be
the same in the two cases. The third type of event is one presenting a risk to only a
single person or a very small group. Events of this sort again have low probability,
but a very large number of exposures causes frequent occurrences—so frequent that,
with such small losses per event, they are not newsworthy.

Availability suggests a different politics for each of the three types of events, even
if they have the same expected losses. Regarding the first, most of the time the issue
is not before the public, so policy questions are not salient. The dominant form of politics here will be mobilization bias, with well-represented groups who have atypically
high stakes in the issue serving as the major drivers of policy. But immediately after
an event actually occurs, it will become very salient because of the large amount of
damage, and this will cause more broad-based political demand for substantial ameliorative action.

In the second case, the higher frequency of events, all of which pass the availability threshold, will tend to keep the issue perpetually salient to people other than
those already well organized into groups. Policy here is less likely to be prone to al-
ternating periods of hectic corrective fire drills and quiescence. Organized interests
will be influential in the quiescent periods, but frequent periods of salience, anticipated by political actors, will tend to work against complete capture even in quiet
times.

In the third case, risks are never salient, so policy will either be nonexistent or
dominated entirely by interests with atypical stakes.

We can think of examples for each of these extremes. Great earthquakes are very
low probability events (even in California) that threaten large disasters. Yet even in
areas that are relatively earthquake prone, enforcement of building standards (especially their retroactive application to old structures) is lax. Airline crashes are relatively frequent, newsworthy events, and airline safety regulation is especially strin-

gent, making it by far the safest mode of transportation in terms of expected annual fatalities per unit of exposure. By contrast, safety standards and their enforcement are quite lax in the case of automobiles and motorcycles, resulting in higher accident rates than those of other modes of transportation in which greater numbers of people are at risk in any given accident.

In the view of some observers, comparisons like these reveal a citizen preference for greater safety when large numbers of people are simultaneously at risk. This may be true, but, even in the absence of that particular preference, the cognitive theory would predict the same pattern of political demands. . . .

Summary of Political Demands for Risk Policy

The preceding discussion leads to several generalizations about the kinds of demands citizens collectively will make on government with regard to risk policies.

1. The intensity of demand for policies to ameliorate risks will tend to be higher for low-probability risks and lower for high-probability risks than is predicted by conventional decision theory.

2. The intensity of demand for policies to ameliorate risks will depend on whether the risk is perceived as a lottery among different levels of improvement, among different levels of losses, or among a mixture of gains and losses. . . .

3. The intensity of demand for risk policies will diverge from the predictions of conventional decision theory depending on the frequency and magnitude of consequential events, holding constant the expected and the actual long-term average losses. . . .

4. Policy preferences will exhibit intertemporal instability, even if underlying information about outcomes and probabilities does not change. . . .

5. The effect of new information on policy preferences depends on how familiar citizens are with the risk—with overreaction to new information in the case of unfamiliarity and underreaction in the case of considerable familiarity. Essentially, if people update expectations as the cognitive theory assumes, one consequence is that expectations are subject to the availability effect when an event is unfamiliar and to overconfidence or anchoring when it is familiar. . . .

Implications regarding the Supply of Risk Regulation

The premise of the following discussion is that, if the cognitive decision theory presents an accurate picture of human behavior, then elected political officials will take into account its effects on the political preferences of their constituents when making policy. We assume that officials seek to secure reelection and that they adopt policies with that end in mind. Political actors may, of course, have their own personal objectives to pursue, but first they have to succeed with the electorate. So we will focus on reelection-maximizing strategies, given the lessons of cognitive theory about electorate behavior.

The first challenge facing elected political officials is to institutionalize systematic differences in the stringency of risk regulation among types of risks so as to reflect the different evaluations of the risks by citizens. (Some risks will be more intensively controlled than others, as measured say, in terms of expenditures per unit of reduction in expected harm.) One way to accomplish this end would be through a statutory specification of differential objectives. This approach can, however, create problems in several respects. First, changes in preferences, knowledge of risks, or technological developments can quickly make the original policy targets obsolete, requiring time-consuming and sometimes unpredictable policy reauthorization. Second, the intertemporal inconsistencies predicted by prospect theory guarantee that a specific statement of a policy objective will be regarded as suboptimal either now or in the future (and perhaps both). Third, if different risks affect different members of the electorate, explicit differences in the commitment to protecting them may provide effective ammunition for future political challengers who can use the issue to appeal to disfavored groups.

A more effective means for producing differential policy outcomes is to specify ambitious targets in all programs but vary resource allocations and the procedural requirements for adopting regulations among various risk-reduction programs. For example, one way to make airline safety standards more rigorous than auto safety standards is simply to give the agency regulating airlines a much larger budget and much more statutory authority, relative to the problem it is asked to solve, than that given to auto regulators. In the case of toxic chemicals, the Food and Drug Administration (FDA) has, in the Delaney amendment, a very strong and easily invoked tool for removing man-made carcinogens from foods, whereas the Toxic Substances Control Act (administered by the Environmental Protection Agency [EPA]) provides only a weak mechanism for keeping off the market new toxic chemicals used in industrial processes. The government bears the burden of proving that a substance is carcinogenic in both cases, but the standard of proof is lower for the FDA than for the EPA. Of course, these differences do not necessarily have their basis in cognitive decision theory; they merely demonstrate how resource allocations and statutory language could lead to policies of quite different stringency among different categories of risk.

Procedural tools are also a potentially useful way to deal with intertemporal instability in preferences. For example, if the demand for policy reflects availability—because it arises from some catastrophic event unlikely to recur soon—strong policy rhetoric, combined with a protracted process and an understaffed agency, assures that the actual standards for ameliorating the risk will be adopted by the implementing agency only long after the event in question has occurred. Detailed regulatory procedures administered by a resource-poor agency thus allow politicians to "lash themselves to the mast" while waiting out the temporary siren calls for immediate overreaction; they allow an agency to "strike when the iron is cold," after the issue has lost its political salience.

The strategies of lashing to the mast and striking when the iron is cold are politically effective but not necessarily normatively correct. The normative problem arises from the fact that, as salience wanes in the course of the agency's developing its regulatory strategy, participation in the policy process will come to be dominated by organized interests, usually the industries subject to regulation. Hence, a publicly interested policy enacted in the midst of public outcry and relatively free of mobilization bias will end up being implemented through a regulatory process hidden from view and subject to special interest favoritism. This may appear to be politically advantageous to elected officials in that it allows them to serve public demands today and special interest desires tomorrow, but even here a problem arises. If an event similar to that in question is likely to recur sometime during the career of politicians responsible for the regulatory policy, then they risk being blamed for unsuccessful regulation. Given this risk, the best political strategy might be to enact rather rigorous and explicitly specified policy targets but set up a process of implementation that provides distributive benefits to participants in the process. An example would be a cumbersome process for setting industry standards, combined with differentially more rigorous standards for new sources of risk than for old ones—so that the costs to industry of the tough standards will be offset (perhaps more than offset) by barriers to competitive entry. A move like this (which happens to reflect regulatory reality) might satisfy political demands in all three periods: there would be an apparently strong response to availability-driven demand for instant overreaction to some incident; there would be a means to cater to the mobilization bias of industrial interests as salience subsides; there would be a reduction in the likelihood that a disastrous event will recur during one's political career.

The choice between strict legislative policy, on the one hand, and delegation of policy objectives to an administrative agency, on the other, turns on the duration of the availability effect as well as the probability of a damaging event recurring. The theory of availability owing to recency does not specify the rate at which overly intense feelings decay as the event recedes into the past, but this rate and the frequency of the sort of event in question determine two critical elements of the delegation/no delegation decision. The first is the proportion of the time that the issue will be salient and thus protected to some degree from mobilization bias. The second is the chance that today's elected officials will later be held accountable for some damaging event. If an event is unlikely to recur within the time horizon of a politician, and the politician (unlike constituents) is free of cognitive pathologies that lead to overestimation of the probability of recurrence, then the politically optimal policy is to do little or nothing of substance with respect to regulation but to cater to the overreaction and temporary mobilization of citizens who seek action. If the duration of the availability effect is short, lashing to the mast through delegation works for the reasons described above. But if the duration of the availability effect is sufficiently long that overreacting citizens remain mobilized throughout the protracted process of developing a policy through an administrative proceeding, there is likely to be excessive regulation. Thereafter, as the availability effect recedes, the politician is left with no political support from previously overreacting citizens (the issue is no longer salient for them) but with a residue of opposition from an industry that remains organized and has had to comply with unnecessary regulations.

One potentially safe escape for a politician facing these circumstances is to seek government subsidies for the risk-abatement policy. Subsidies will soften industry opposition and provide phantom benefits that overreacting citizens might see as larger than the costs. Perhaps the Superfund program for dealing with toxic waste dumps is an example of this strategy. Toxic chemicals are linked to cancer, and cancer and cancer policy seem to be perpetually salient issues. Environmental groups can rely on the representativeness heuristic to link toxic dumps to cancer and to extend the period of salience that follows some newsworthy event involving pollution from a dump. If relatively few dumps are really threats to public health, but large numbers of dumps are politically salient because of availability and representativeness, then the subsidization strategy keeps industry operating profitably, placates citizens, and avoids disaster. (Of course, the factual premises of this example are a matter of controversy, and there are other plausible explanations for a vigorous policy on toxic dumps. We are simply trying to illustrate one of the implications of cognitive theory for policy-making. . . .)

The last case to be considered is where the frequency of recurrence, assuming no regulatory intervention, is shorter than the time horizon of a politician. Here the politician has to trade off demands for excessive regulation against the self-interest of those responsible for creating the risk (they, presumably, want underregulation). If the duration of the availability effect is short, the politician's best available strategy is likely to be a compromise between these interests that is explicitly stated in legislation. Delegation to an agency would be attractive only if the agency could act rather quickly—before salience wanes and before another damaging event occurs. Thus, if delegation is observed, it should be accompanied by strict deadlines for decisions and a commitment to the agency of sufficient resources to make the deadline feasible.

If the period of salience is longer, delegation becomes more attractive, for it can result in an arrangement whereby the representatives of overreacting citizens, and those of self-interested risk producers, are forced to negotiate a compromise under the watchful eye of agencies and the courts. If the duration of the availability effect is long, participation in the policy development process in a regulatory agency will be likely to include both regulated firms and the most intense demanders for risk regulation. Thus, more balanced mobilization will encourage elected political officials to delegate the question of the overall stringency of a policy to an agency.

Our final observation about the optimal political strategy for dealing with risks is the general attractiveness of formulating policy choices in terms of risks of loss rather than as lotteries over gains. In part, loss aversion implies that citizens are easier to motivate when the issue is the threat of loss as opposed to the opportunity for gain. In addition, overreaction by citizens, if it can be orchestrated, is beneficial to political actors. If citizens can be made to attack a problem more intensively than its probability and the gravity of its damage warrant, the ultimate outcome—which is likely to be little or no damage—will cast favorable light on the responsible politicians. In other words, the very nonoccurrence of events will result in greater thanks to the politicians than they deserve. Moreover, by casting risk issues in negative, loss-aversive terms, a political leader is protected against effective challenge by someone adopting the same strategy. This argument is not symmetric: a negative formulation cannot easily be attacked by a positive formulation because the shape of a citizen's valuation function depends on whether outcomes are gains or losses. That is, gains promise smaller changes in welfare than do losses; thus, they are less likely to be salient to the electorate, and less likely to lead to the mobilization of political activity.

Concluding Remarks

The cognitive theory of choice under uncertainty offers two sorts of insights of some relevance to analysts interested in the regulation of risk. First, people take shortcuts (use heuristics) that lead them to make mistakes about variables (such as probability) relevant to evaluating risk. Second, putting the mistakes aside, they approach risk in ways that depart from the norms and assumptions of conventional decision analysis.

The latter category of insights is, of course, much more interesting than the former because the observation that people make mistakes is hardly fresh or startling. Moreover, the cognitive theory seems to contain examples of all kinds of mistakes; for example, while availability may account for overreaction to a catastrophe, anchoring may explain underreaction. As yet, the theory cannot tell us very much about which mistakes are likely to occur in any given circumstance. Even at that, though, the cognitive psychologists' catalog of judgmental errors is a useful contribution, for several reasons. First, knowing something of the types, range, and causes and consequences of misjudgment can guide efforts toward more fruitful communication between the public and policymakers on matters of risk and its regulation, a matter currently (and rightly) of much importance. Second, understanding that judgmental error is fairly regular can help avoid some policy pitfalls. Error is predictable as to kinds of mistakes and their direction, and it is likely to plague any decision maker, even a professional risk assessor. Recognizing these problems guards against such easy answers to problems of risk regulation as faithfully delegating the whole business to "experts." At the same time, an awareness of the problems enables experts to be sensitized to judgmental shortcomings in the course of their training, so as to make them more worthy of their name. . . .

NOTES AND QUESTIONS

1. An alternate account of bounded rationality starts by positing that rational individuals have multiple sets of preferences, possibly inconsistent with each other, that they must somehow reconcile. They wish to run red lights in order to get where they are going more efficiently, but they also wish to obey the law; they wish to quit smoking, but they have a craving for nicotine. In this view, rationality consists of committing to follow one's higher-order preferences—or, more generally, of following the particular preferences appropriate to each role in life that one is called upon to play: consumer, voter, parent, lawyer. See Thomas Schelling, The Intimate Contest for Self-Command, 60 *Public Interest* 94 (1980); Jon Elster, *Ulysses and the Sirens: Studies in Rationality and Irrationality,* rev. ed. (Cambridge: Cambridge University Press, 1984).

2. Simon's concept of procedural rationality has had particular influence on the study of organizational behavior. Indeed, it is unclear what it means for a multiperson organization to engage in substantive rational choice, since organizations, unlike individuals, do not have any preferences to be maximized. Rather, rationally run organizations are those that effectively mediate among the rival preferences of their various constituents. For more systematic discussions of procedural theories of business organizations, see Richard M. Cyert and James G. March, *A Behavioral Theory of the*

Firm (Cambridge, Mass.: Blackwell Business, 1992); Robin Marris, *The Economic Theory of Managerial Capitalism* (New York: Free Press of Glencoe, 1964). For more recent and general discussions of bounded rationality, see Harvey Leibenstein, *Beyond Economic Man,* rev. ed. (Cambridge, Mass.: Harvard University Press, 1980); Richard Thaler, *Quasi-rational Economics* (New York: Russell Sage Foundation, 1991). On the implications of such theories for market equilibrium, see Richard Nelson and Sidney Winter, *An Evolutionary Theory of Economic Change* (Cambridge, Mass.: Harvard University Press, 1982).

3. It is useful to contrast Simon's theory of the business firm with that of Ronald Coase. Recall that Coase puts forward a maximizing theory of the firm, arguing that firms arise in order to conserve on the transaction costs of exchange. Can Simon's and Coase's views be reconciled? Is bounded rationality a type of transaction cost, or does it suggest a decisionmaking process removed from any comparison of costs and benefits? See generally Oliver E. Williamson, *Markets and Hierarchies: Analysis and Antitrust Implications* (New York: The Free Press, 1975); Oliver E. Williamson, *The Mechanisms of Governance* (New York: Oxford University Press, 1996) [both presenting general studies of economics of internal organization].

4. While Tversky and Kahneman argue that people's choices among multiple alternatives can depend on how the decision is framed, they do not address the question of which decision frame is most appropriate. Can economic, political, or legal values provide an answer to this question? In particular, consider their Problems 1 and 2, in which the respondents displayed different attitudes toward risk in the context of a hypothetical decision regarding public health policy, depending upon whether the consequences of the decision were formulated in terms of lives saved or lives lost. Can you think of alternative explanations for this experimental result? Which sets of answers, if any, should be considered more reliable: those obtained in response to the former version of the problem, or the latter?

5. The model of bounded rationality has received increased attention in the legal literature in recent years. Readers interested in applications to other fields of law may wish to consult Melvin Eisenberg, The Limits of Cognition and the Limits of Contract, 47 *Stanford Law Review* 211 (1995) [discussing contract doctrines placing limits on ability to contract]; Edward McCaffery, Cognitive Theory and Tax, 41 *UCLA Law Review* 1861 (1994) [discussing tax policy]; Daniel Kahneman, Edward McCaffery, and Matthew Spitzer, Framing the Jury: Cognitive Perspectives on Pain and Suffering Awards, 81 *Virginia Law Review* 1341 (1995) [discussing jury awards in torts]; and Symposium: Legal Implications of Human Error, 59 *Southern California Law Review* 225 (1986) [discussing variety of doctrinal areas].

6. It is worth considering how the models we have previously studied might be modified to better take account of the phenomenon of bounded rationality. How, if at all, would you evaluate Shavell's conclusions regarding the relative merits of strict liability and negligence (supra, chapter 3) in light of bounded rationality? How would you modify Rosenberg and Shavell's analysis of frivolous lawsuits (chapter 4), Kaplow's analysis of legal transitions (chapter 5), or Bishop's analysis of the doctrine of *Hadley v. Baxendale* (chapter 6)?

7. Schwartz and Wilde conclude that, even though individual consumers are subject to cognitive errors in borrowing and in making purchases, there is no reason to

think that these errors will lead them to be systematically overoptimistic. Do you agree? What sources of systematic bias in consumer decisionmaking might exist? If the authors are correct in their prediction, does it follow that the government should follow a policy of laissez-faire with regard to consumer transactions? See Eric A. Posner, Contract Law in the Welfare State: A Defense of the Unconscionability Doctrine, Usury Laws and Related Limitations on the Freedom to Contract, 24 *Journal of Legal Studies* 283 (1995) [suggesting that government assistance to persons who become insolvent imposes an externality on taxpayers, possibly justifying regulation of individual risk-taking].

8. Are you persuaded by Jackson's defense of the fresh-start policy for personal bankruptcies? Should individual debtors be allowed to bind themselves in advance not to declare bankruptcy, or to waive their statutory right to a discharge of their debts following bankruptcy, in order to obtain credit on better terms? Conversely, does Jackson's argument justify placing any limits on individuals' ability to grant mortgages on homestead property, which under current law do survive bankruptcy? See Michael Schill, An Economic Analysis of Mortgagor Protection Laws, 77 *Virginia Law Review* 489 (1991).

9. What are the incentive effects of government policies that seek to protect individuals from the consequences of their bounded rationality? Does bounded rationality work to mitigate the problem of individual moral hazard, or does it exacerbate it? In this regard, consider the merits of extending government disaster relief to persons who had the opportunity to purchase insurance ex ante, but failed to avail themselves of it.

10. One can assess Noll and Krier's view of regulatory policy from either a positive or normative perspective. From the positive viewpoint, we can ask whether the specific policies they discuss are best explained by their cognitive theory, by a Coasian theory of transaction costs, or by a theory of regulatory capture. Can you think of particular public policies that are boundedly rational, and that cannot be explained by either administrative costs or regulatory capture? Conversely, if the authors are correct that public policy reflects the public's biased heuristics, we can ask whether this outcome is normatively desirable. Should public decisionmakers follow the public's boundedly rational views regarding risk management policy, or should they try to lead public opinion and base their decisions on what may be statistically less biased tools of cost-benefit analysis? In this regard, consider the claim, made by some critics of current public health policy, that the level of government expenditure on AIDS research is too high, and that the money would be better spent on more mundane problems such as heart disease and highway safety.

8

Critiques of the Economic Approach

This next-to-last chapter is devoted to critiques of the economic analysis of law. While it has been more than thirty years now that law and economics has been recognized as a distinct jurisprudential method, the approach continues to engender controversy in the academy and in the legal profession at large. Critics of economic analysis have offered a variety of arguments against it, ranging from its supposed conservative bias to its focus on the efficiency criterion to the limitations of its abstract and mathematical models. The selections reproduced here reflect, in partially overlapping fashion, a survey of such critiques.

The first reading in this section, by Ronald Dworkin, sets forth what might be called a *liberal critique* of law and economics. The essence of the liberal critique is that the economic approach does not take adequate account of individual rights, and is accordingly too ready to sacrifice individuals to the collective interests of society. This criticism has particular force when directed at the Kaldor-Hicks criterion, which holds that a given policy change is desirable if the gains to the winners, measured in monetary terms, exceed the costs to the losers. As Dworkin underscores, the criterion's exclusive focus on the collective sum of gains and losses obscures an important fact—that the individual losers from the change may be very bad off indeed. Dworkin's concern, however, is not just for the poor, or even for a more egalitarian distribution of wealth or income. Economists, after all, differ on the relative importance of allocative efficiency and distributional equity, and many of them agree that it is reasonable to trade off the former for the latter. But even those economists who attach substantial weight to distributional issues tend to follow classical utilitarianism in regarding the social interest as equivalent to an aggregation of individual interests. In Dworkin's view, it is just such an aggregation that is objectionable. In his words, economics, like other utilitarian normative theories, does not "take seriously the difference between people." Instead, it treats individuals as mere instruments of the collective good, rather than ends in themselves.

The second reading, by Duncan Kennedy, also faults economists for not paying sufficient attention to distributional issues. Its main focus, however, is not on distribution, but on what might be called the *paternalist critique* of law and economics. The essence of this critique is that what individuals choose for themselves in voluntary exchange need not correspond to their objective interests. This lack of correspondence, moreover, does not arise from any sort of externality or market failure. At the center of the critique, rather, is the concept of *false consciousness;* the idea that in many circumstances individuals simply are mistaken about what is good for them or about what possible choices are available. In such circumstances, the usual recommendations of laissez-faire economics are inapposite and even misleading. Kennedy argues that despite its obvious tension with liberal principles, paternalism offers a more straightforward and candid justification for public policy than the distributional and efficiency arguments that economists more commonly put forward.

In its emphases on political conflict and on the indeterminacy of efficiency analysis, Kennedy's essay also reflects the distinctively radical perspective of Critical Legal Studies—the contemporary school of jurisprudence most often regarded as being in intellectual competition with the economic approach to law. This *radical critique* is more fully developed here in the selection from Mark Kelman's book *A Guide to Critical Legal Studies.* In contrast to Dworkin, Kelman sees law and economics not as an attack on liberal individualism, but as a conservative attempt to reconstruct it on market terms. From such a perspective, law and economics offers an ostensibly neutral and technical set of principles, based on economic efficiency and cost-benefit analysis, through which individual differences can be mediated. In Kelman's view, however, this attempt is both unsuccessful and deceptive, because no such set of neutral principles is possible. The efficiency criterion is politically biased because its primary standard of value, individual willingness to pay, is determined in large part by wealth; more fundamentally, the criterion is inherently circular, because all individual valuations necessarily depend on the larger social and legal context in which they are situated. Accordingly, Kelman concludes that legal conflicts must ultimately be worked out in the spheres of politics and morality, and not of economics.

Questions about the social determination of value also motivate two of the other critiques presented here. The *sociological critique* challenges economics' account of human behavior as constrained maximization as well as its standard assumption of exogenous preferences. According to this critique, individual persons do not come to society with a preformed set of needs and wants, and more importantly, do not regard the rules of society as mere constraints against which to maximize. Rather, people's preferences are shaped by the social interactions they engage in. Thus, because both law and economic exchange are forms of social interaction, both influence the content of individual desires. This need not mean that social and political conflict are inevitable, as the radical view would have it, but it does mean that individuals do not always react to changes in legal rules in the ways that economic models would predict. Such insights formed the basis for Robert Ellickson's study of rancher-farmer disputes in Shasta County, California—the very setting that Ronald Coase used to develop his famous theorem. Framing his argument in the form of a detailed narrative, Ellickson suggests not just that law and economics has relied too heavily on an individualist methodology, but that it has excessively focused on formal theorizing to the neglect of empirical reality. According to Ellickson, differences in legal entitlements did not change ranchers' and farmers' behavior in the communities he studied, but not for the reasons that Coase had articulated. Rather, the Shasta residents attached more importance to community norms of reciprocity, as embodied in their complex web of social relations, than to either economic efficiency or the formal constraints of the law.

While Ellickson focuses on what he sees as the descriptive and predictive deficiencies of the model of rational choice, the related *communitarian critique,* represented here by the reading by Steven Kelman, is primarily a normative one. The essence of this latter critique is that economics neglects the role that law plays in our collective deliberations over the ultimate ends we choose to pursue as a community. Accordingly, by ignoring the extent to which individual values are socially determined, the economic approach scorns the possibility that legal institutions can influence the formation of private preferences. Kelman develops this critique within the specific context of environmental policy, which in recent years has provided a prominent battleground for public debates over the appropriate scope of market incentives. Why is it, he asks, that so many environmentalist politicians and activists have rejected economists' call to use the price system rather than direct regulation to help protect environmental resources? His answer is that setting a price on environmental resources incorrectly takes as given the public's current valuation of those resources. Many environmentalists, however, do not wish to take people's current views of the environment as given; rather, they see it as a central part of their mission to raise the public's consciousness and to cultivate its appreciation for environmental values. From such a perspective, using the price system to regulate the environment, and the language of the market to talk about it, undercuts this task. It means treating the environment like any other consumer good when it deserves to be seen as something special. In Kelman's view, communitarian concerns such as these carry weight well beyond the environmental sphere; he suggests, for instance, that they also underlie public debates over equality and the obligations of citizenship. For these reasons, he argues, using economic analysis in public discourse may have hidden costs. The rhetoric and culture of economics, by enshrining self-interest and the preferences of the status quo, may inhibit community beliefs and values from developing in more progressive directions.

Finally, the last reading in this section, by Arthur Leff, combines various elements from the foregoing perspectives to present what might be called a *Legal Realist critique*. While Leff concedes the relevance of economics to legal policy, he argues that much of the appeal of law and economics as a jurisprudential movement stems from its relative formalism compared to other fields of social science—specifically, its emphasis on abstract models, quasi-mathematical reasoning, and quantification. In his view, however, social reality is too complex to be adequately captured by any formalistic approach, whether legal, philosophical, or economic. Leff argues that while economic models are useful, it must be remembered that they are overly reductionistic, and that attempting to force legal discussions into the language of economics may distort and impair the resulting analysis. Rather than tie themselves to any single method, therefore, lawyers should follow a more eclectic approach.

8.1   THE LIBERAL CRITIQUE

Is Wealth a Value?

RONALD DWORKIN

. . . Economic analysis holds, on its normative side, that social wealth maximization is a worthy goal so that judicial decisions should try to maximize social wealth, for example, by assigning rights to those who would purchase them but for transaction costs. But it is unclear *why* social wealth is a worthy goal. Who would think that a society that has more wealth, as defined, is either better or better off than a society that has less, except someone who made the mistake of personifying society, and therefore thought that a society is better off with more wealth in just the way any individual is? Why should anyone who has not made this mistake think social wealth maximization a worthy goal?

There are several possible answers to this question. and I shall start by deploying a number of distinctions among them. [First,] social wealth may be thought to be itself a component of social value—that is, something worth having for its own sake. There are two versions of this claim. (a) The immodest version holds that social wealth is the *only* component of social value. . . . (b) The modest version argues that social wealth is one component of social value among others. One society is *pro tanto* better than another if it has more wealth, but it might be worse overall when other components of value, including distributional components, are taken into account. . . .

Another distinction cuts across these. Each of these modes of social wealth claims . . . may be combined with some functional claim of institutional responsibility which argues that it is the special function of courts to pursue social wealth single-mindedly, although it is not necessarily the function of, for instance, legislatures to do so. It might be said, for example, that although wealth maximization is only one among several components of social value, it is nevertheless a component that courts should be asked single-mindedly to pursue, leaving other components to other institutions. . . .

The normative claim of economic analysis, then, admits of many variations. . . . I shall begin by considering whether the claim that social wealth is a component of value, in either the immodest or the modest versions of that claim, is a defensible idea. . . .

Consider this hypothetical example. Derek has a book Amartya wants. Derek would sell the book to Amartya for $2 and Amartya would pay $3 for it. T (the tyrant in charge) takes the book from Derek and gives it to Amartya with less waste in money or its equivalent than would be consumed in transaction costs if the two were to haggle over the distribution of the $1 surplus value. The forced transfer from Derek to Amartya produces a gain in social wealth even though Derek has lost something he values with no compensation. Let us call the situation before the forced transfer takes place "Society 1" and the situation after it takes place "Society 2." Is Society 2 *in any respect* superior to Society 1? I do not mean whether the gain in wealth is overridden by the cost in justice, or in equal treatment, or in anything else, but whether the gain in wealth is, considered in itself, any gain at all. I should say, and I think most people would agree, that Society 2 is not better in any respect.

It may be objected that in practice social wealth would be maximized by rules of law that forbid theft and insist on a market exchange, when it is feasible, as it is in my imaginary case. It is true that Posner and others recommend market transactions except in cases in which the transaction costs (the costs of the parties identifying each other and concluding an agreement) are high. But it is crucial that they recommend market transactions for their *evidentiary* value. If two parties conclude a bargain at a certain price we can be sure that wealth has been increased (setting aside problems of externalities) because each has something he would rather have than what he gave up. If transaction costs are "high" or a transaction is, in the nature of the case, impossible, Posner and others recommend what they call "mimicking" the market, which means imposing the result they believe a market would have reached. They concede, therefore, or rather insist, that information about what parties would have done in a market transaction can be obtained in the absence of the transaction, and that such information can be sufficiently reliable to act on.

I assume, therefore, that we have that information in the book case. We know that there will be a gain in social wealth if we transfer the book from Derek to Amartya. We know there will be less gain (because of what either or both might otherwise produce) if we allow them to "waste" time haggling. We know there can be no more gain in social wealth if we force Amartya to pay anything to Derek in compensation. (Each would pay the same in money for money.) If we think that Society 2 is in no respect superior to Society 1, we cannot think that social wealth is a component of value.

It may now be objected, however, that wealth maximization is best served by a legal system that assigns rights to particular people, and then insists that no one lose what he has a right to have except through a voluntary transaction. Or (if his property has been damaged) in return for appropriate compensation ideally measured by what he would have taken for it in such a transaction. That explains why someone who believes that wealth maximization is a component of value may nevertheless deny that Society 2 is in any way better than Society 1. If we assume that Derek has a right to the book under a system of rights calculated to maximize wealth, then it offends, rather than serves, wealth maximization to take the book with no compensation.

. . . We must notice now, however, that [this argument] justifies only instrumentally rights like Derek's right to the book. The institution of rights, and particular allocations of rights, are justified only insofar as they promote social wealth more effectively than other institutions or allocations. The argument for these rights is formally similar to the familiar rule-utilitarian account of rights.* Sometimes an act that violates what most people think are rights—such as taking Derek's book for Amartya—improves total utility. Some rule utilitarians argue that such rights should nevertheless be respected, as a strategy to gain long-term utility, even though utility is lost in any isolated case considered by itself.

This form of argument is not to the point here. I did not ask whether it is a wise strategy, from the standpoint of maximizing social wealth in the long run, to allow tyrants to take things that belong to one person and give them to others. I asked whether, in the story of Amartya and Derek, Society 2 is in any respect superior to Society 1. . . . If Society 2 is not in any way superior to Society 1—considered in themselves—then social wealth is not even one among several components of social value.

I have assumed so far, however, that you will agree with me that Society 2 is not superior. Perhaps I am wrong. You may wish to say that a situation is better, *pro tanto,* if goods are in the hands of those who would pay more to have them. If you do, I suspect it is because you are making a further assumption, which is this: if Derek would take only $2 for the book and Amartya would pay $3, then the book will provide more satisfaction to Amartya than it does to Derek. You assume, that is, that the transfer will increase overall utility as well as wealth. . . . I must thus make my example more specific. Derek is poor and sick and miserable, and the book is one of his few comforts. He is willing to sell it for $2 only because he needs medicine. Amartya is rich and content. He is willing to spend $3 for the book, which is a very small part of his wealth, on the odd chance that he might someday read it, although he knows that he probably will not. If the tyrant makes the transfer with no compensation, total utility will sharply fall. But wealth, as specifically defined, will improve. I do not ask whether you would approve the tyrant's act. I ask whether, if the tyrant acts, the situation will be in any way an improvement. I believe it will not. In such circumstances, that goods are in the hands of those who would pay more to have them is as morally irrelevant as the book's being in the hands of the alphabetically prior party.

Once social wealth is divorced from utility, at least, it loses all plausibility as a component of value. It loses even the spurious appeal given to utilitarianism by the personification of society. It is sometimes argued by utilitarians that, since an individual is necessarily better off if he has more total happiness in his entire life, even though less on many particular days, so a society must be better off if it has more to-

Editor's note: Rule utilitarianism is a variant of utilitarianism that argues that, because of the practical and institutional difficulties of determining what actions would maximize utility in each individual instance, it is best to devise and follow rules that work to maximize utility over the run of cases. It should be compared with *act utilitarianism,* which argues one should try to choose the utility-maximizing action in every individual case.

tal happiness distributed across its members even though many of these members have less. That is I think, a bad argument in two different ways. First, it is not true that an individual is necessarily better off if he has more total happiness over his life without regard to distribution. Someone might well prefer a life with less total pleasure than a life of misery with one incredibly ecstatic month. . . . Second, society is not related to individual citizens as an individual is related to the days of his life. The analogy is, therefore, one way of committing the ambiguous sin of "not taking seriously the difference between people." . . .

It is important to notice that the Derek-Amartya story shows the failure not only of the immodest but also of the modest version of the theory that social wealth is a component of value. For the story shows not merely that a gain in wealth may be outweighed by losses in utility or fairness or something else. It shows that a gain in social wealth, considered just in itself, and apart from its costs or other good or bad consequences, is no gain at all. That denies the modest as well as the immodest theory. I shall therefore take this opportunity to comment on a familiar idea that, on its most plausible interpretation, presupposes the modest theory, that is, that social wealth is one among other components of social value.

This is the idea that justice and social wealth may sensibly be traded off against each other, making some sacrifice in one to achieve more of the other. Professor Calabresi, for example, begins *The Costs of Accidents* by noticing that accident law has two goals, which he describes as "justice" and "cost reduction," and notices also that these goals may sometimes conflict so that a "political" choice is needed about which goal should be pursued. The same point is meant to be illustrated by the indifference curves I have seen drawn on countless blackboards, on space defined by axes one of which is labeled "justice" (or sometimes "morality") and the other "social wealth" (or sometimes "efficiency").

Whose indifference curves are supposed to be drawn on that space? The usual story speaks of the "political" or "collective" choice in which "we" decide how much justice we are willing to give up for further wealth or vice versa. The suggestion is that the curves represent individual choices (or collective functions of individual choices) over alternative societies defined as displaying different mixes of justice and wealth. But what sort of choice is the individual—whose preferences are thus displayed—supposed to have made? Is it a choice of the society in which he would like to live, or the choice of the society he thinks best from the standpoint of morality or some other normative perspective? . . .

Perhaps the point is that an individual chooses a society that has more rather than less wealth as a whole because the antecedent probability is that he will have more wealth personally in a richer society. . . . [In this view, i]ndividuals choose a mix of justice and efficiency with an eye toward maximizing their individual utility under conditions of dramatic uncertainty; or rather, trading off gains in their prospects, so conceived, against losses in the just character of the society. . . . But surely this is all irrelevant. Calabresi and others contemplate actual political choices—they suppose that the economic analysis of law is useful because it shows how much wealth is lost if some other value is chosen. But in that case we cannot understand the axis of wealth or efficiency, in the indifference curves as generally offered, as a surrogate for judgments about antecedent individual welfare under conditions of uncertainty. We must understand the axis as representing judgments about individual welfare, to be traded off against justice, as things actually stand. *No* particular individual will, then, be concerned about social wealth (or, indeed, about Pareto efficiency). It makes no sense for him to trade off anything, let alone justice, for *that*. He will be concerned with his individual fate, and since, by hypothesis, he now knows his actual position, he can choose amongst societies by trading off justice against increases in his individual welfare in these different societies. *Social* wealth (or Pareto efficiency) simply plays no role in these calculations.

Let us turn to the second interpretation of the supposed trade-off choice. An individual is supposed to be choosing which mix of justice and wealth represents, not the society in which he, as an individual with both moral and self-interested motives, would prefer to live, but the morally best society, all things considered. The very idea of a trade-off between justice and wealth now becomes mysterious. If the individual is to choose the morally best society, why should not its justice alone matter?

We might expect one of two replies to that question. It might be said, first, that justice is not the only virtue of a good society. It surely makes sense, from a normative perspective, to speak of the trade-off between justice and culture, and also to speak of the trade-off between justice and social wealth, as two distinct, sometimes competing social virtues. The second reply is different in form but similar in spirit. It suggests that, when people speak of a trade-off between justice and social wealth, they use "justice" to refer to only part of what that word means in ordinary language and in political philosophy—that is, they use it to refer to the distributional and meritocratic or desert features of justice in the wider sense. They mean the trade-off between those specific aspects of justice and other aspects that are comprehended under "wealth maximization."

These two replies are similar in spirit because they both assume that wealth maximization is a component of social value. In the first, wealth maximization is treated as a component competitive with justice and, in the second, as a component of justice but competitive with other components of that concept. Both replies fail, for that reason. It is absurd to consider wealth maximization to be a component of value, within or without the concept of justice. Remember Derek and Amartya. . . .

But suppose I was wrong to take the trade-off described in the familiar indifference curves, or in texts like Calabresi's, to be a matter of individual preferences, or some collective function of individual preferences. Perhaps the choice is meant to be the choice of society as a whole, conceived as a composite entity. I think that the choice is mentally represented this way, although not reflectively, by many of those who speak of trade-offs between justice and wealth. They have a personified community in mind, as the reference of the "we" in the proposition that "we" want a society of such-and-such sort. Of course, that picture must be disowned when made explicit. It is a silly and malign personification.

Even if society is personified in this silly way, it remains mysterious why society so conceived would want a trade-off between justice and wealth. First, the choice of wealth, taken to be independent of utility information, would make no more sense for society as a composite person than it does for individuals as actual people. Second, and more interesting, the reference of "justice" would be lost. Justice (at least when the trade-off is in question) is a matter of distribution—of the relation among individuals who make up the society, or between the society as a whole and these individuals. Once we personify the society so as to make the social choice an individual choice, there is no longer anything to be considered under the aspect of justice. Society personified can, of course, still be concerned about questions or ordering or distribution among its members. But the dimensions of such orderings do not include that of justice. An individual cares about the distribution of benefits or experiences over the days of his life. But he does not care under the aspect of justice. . . .

8.2 THE PATERNALIST CRITIQUE

Distributive and Paternalist Motives in Contract and Tort Law

DUNCAN KENNEDY

Three Types of Motive in Setting the Groundrules

A decision maker may change (or refuse to change) a rule for many different reasons other than like or dislike of freedom of contract. In particular, he may believe that his act will have desirable distributive, or paternalist, or efficiency consequences. . . .

The decision maker acts out of distributive motives when he changes a rule (or refuses to change a rule) because he wants to increase the success of some group in the struggle for welfare, expecting and intending that this increase will be at the expense of another group (the groups may overlap). The decision maker acts out of paternalist motives when he changes a rule in order to improve someone's welfare by getting them to behave in their "own real interests," rather than in the fashion they would have adopted under the previous legal regime. The decision maker acts out of efficiency motives when he changes a rule so as to induce people to reach agreements that correspond to those they would have reached under the previous legal regime had it not been for the existence of transaction costs.

Distributive Motives

The first hallmark of distributive motive is that the decision maker accepts the beneficiary's definition of what will make the beneficiary better off. The notion is that the decision maker finds two people engaged in a struggle over the distribution of something that each values. They are operating under the previous regime. He changes the regime in a way that helps one. The second hallmark is that the decision maker sees the situation as zero sum: helping one means hurting the other. Some examples of issues that get discussed in distributive terms are: should secondary boycotts by labor unions be tortious? Should there be a minimum wage? . . .

The decision maker operating from distributive motives changes the groundrules

Duncan Kennedy, Distributive and Paternalist Motives in Contract and Tort Law, 41 *Maryland Law Review* (1982), pp. 563–658. Reprinted with permission.

so as to change the balance of power between the various groups in civil society. The change in the rule may operate directly, as in a change in the law of duress that allows one party to do things to the other that were previously illegal (e.g., the legalization of picketing or lockouts), or it may be indirect, as where the prohibition of secondary boycotts changes the balance between a union and its primary target, or a law against monopolistic combinations changes a firm's relations with customers. But the issue is power.

Paternalist Motives

By contrast, where motives are paternalist, the issue is false consciousness. As in the distributive case, the decision maker changes a rule because he believes that under the new regime the objects of his benevolence will end up with a set of experiences that will be "better for them" than those they would have ended up with under the previous regime. What makes this change paternalist rather than distributive is, first, that those who have supposedly benefited do not agree that they are better off, and would return to the previous regime if given a choice in the matter. Second, if there are good or bad consequences for others through the paternalist change, these are seen as side effects, rather than as part and parcel of the decision maker's program.

As I am using the term, all paternalist interventions involve overruling the preferences of the beneficiary in his own best interest, but not all such overrulings are paternalist. One might, for example, refuse to enforce contracts made under duress, thereby overruling the preferences (at the time of contracting) of both parties, but do this in the belief that it would, over the long run, make the weaker party richer at the expense of the stronger. Weak parties, looking at the matter in the abstract rather than at gunpoint, might agree that over the long run a contract defense of duress would be a good thing for them. So long as there is no disagreement as to the values or moral vision on which to act, the decision maker is not acting paternalistically.

Some issues that have often been addressed with paternalist motives are the legality of the possession of prohibited substances; whether there should be required terms in various types of contracts, such as marriage or consumer sales contracts; and the extent to which infants, idiots and seamen are subject to the same contract regime as "normal" people. As in the case of distributive motives, each of these issues can be resolved through the application of tests that do not involve paternalism. One might try to settle each by appeal to distributive considerations, as well as by appeal to fairness, morality, or rights.

Efficiency Motives

A decision maker acting from efficiency motives accepts the rules of the previous regime as legitimate from the point of view of fairness, morality, rights, distribution or whatever. His goal is to modify one of these rules so as to make everyone affected better off, by their own criteria of better-offness, than they would have been under the old dispensation. This will be possible where transaction costs of one kind or another have prevented parties under the previous regime from making an exchange. If the decision maker knows that this exchange would have occurred, he may be able to induce the parties to perform it by the right modification of the background rules.

Some examples of issues that decision makers often approach with efficiency motives are: whether there should be nondisclaimable warranties attached to consumer goods in circumstances where consumers can't cheaply acquire information that would allow them to assess product safety, and sellers have incentives not to provide this information; what should be the rules of damages for breach of contract; whether sports arenas should be liable for damage inflicted by one fan on another. Of course, each of these issues can be approached with quite different motives, or with a set of motives that lead to conflicting resolutions.

Efficiency motives differ from paternalist motives because their premise is that the affected parties will prefer the new situation to the old, so they would not choose to "waive" the benefits the decision maker has attempted to confer on them. The decision maker is not trying to decide what is "really" best for them, without regard to their own views of the matter. On the other hand, an intervention grounded in efficiency concerns will always involve speculation about what the parties "would have done" had they not been prevented by transaction costs. . . .

General Reflections on the Appeal of Efficiency Arguments

Once they have at least somewhat mastered the technical apparatus, people just love to argue for their favorite proposals on efficiency grounds. For years, it was mainly a liberal fad, then it fell into favor with the conservatives, and the liberals are now trying to reappropriate it. Given a choice, almost everyone seems to prefer to cast a difficult rule change proposal in these terms rather than in those of paternalism or redistribution. The paradox is that the standard objection to paternalism and distribution as motives is that they are intrinsically "subjective," "uncertain," and therefore political and controversial. What this means is that they evoke the unresolved conflicts between groups within civil society about who deserves how much and what is the nature of true consciousness. Regimes of compulsory terms are part of that battle, no matter how carefully we refer to efficiency as the only motive for imposing them, and efficiency arguments are, if anything, even more subjective, uncertain, and therefore potentially controversial than the other kinds. Why is it that the patent manipulability of efficiency arguments does not impair their attractiveness, while distributive and paternalist arguments, which are actually easier to grasp and to apply, seem excessively fuzzy?

At least part of the answer, I think, is that the move to efficiency transposes a conflict between groups in civil society from the level of a dispute about justice and truth to a dispute about facts—about probably unknowable social science data that no one will ever actually try to collect but which provides ample room for fanciful hypotheses.

Such a transposition from one level to another makes everyone, just about, feel better about the dispute. The move from a conflict of interests or consciousnesses to a conflict about facts makes it seem—quite falsely—that the whole thing is less intense and less explosive. That it is imaginable that someone could one day actually produce the factual data makes it seem irrelevant that no one is practically engaged in that task, or ever will be. In this sense, the transposition to the cognitive level allows efficiency to act as a mediator of the intensely contradictory feelings aroused by disputes about the shares of groups and the validity of their choices—a mediator that defuses rather than resolves conflict.

It seems obvious to me, but maybe I'm just wrong, that efficiency is also attractive because it legitimates the pretensions to power of a particular subset of the ruling class—the liberal and conservative policy analysts, most of whom are lawyers, economists or "planners" by profession. Efficiency analysis, like many another mode of professional discourse, is an obscure mix of the normative and the merely descriptive; it requires training to master; it provides a basis for an internal hierarchy of the profession that crosscuts political alignments. Its high value in legitimating the outcomes of group conflict in "nonideological" terms is the basis for the professional group's claim to special rewards and a secure niche in the good graces of the ruling class as a whole.

Paternalism

. . .

Varieties of False Consciousness "Cured" by Compulsory Terms

To say that an intervention is paternalist doesn't explain it, beyond identifying the problem as a mistake on the part of the beneficiary about his real interests, or as false consciousness. Decision makers in our society impose compulsory terms because they think buyers suffer from a number of quite specific kinds of false consciousness. For example, buyers underestimate the seriousness of risks of injury from products or situations. The tendency to underestimate risk goes far enough beyond mere misinformation so that when we intervene we can't claim to be achieving an efficient outcome blocked by transaction costs. It amounts to a cognitive bias, a systematic tendency to misinterpret or ignore information, to generate fantasies of safety, to repress unwanted information. It has to do with babyishness, not ignorance. When the decision maker makes the buyer pay for protection against the non-negligent injury, or makes the buyer buy a safety precaution that will prevent injury happening at all, he may be doing so in response to a judgment about this kind of misperception.

But it may be that he is concerned not with a misperception of risk but with willingness to take risks—with recklessness rather than with babyishness about the facts. He may decide that looking at the buyer as a person with a continuous existence in time, as a life rather than as an instant, he can make the buyer better off by forcing him to give up a little now in order to avoid catastrophe later on. That the buyer doesn't think so may be a mistake that seems just a matter of character, or it may be possible to develop an interpretation of the buyer's situation that makes the mistake easy to understand. For example, the buyer may appear to the decision maker to be suffering from addiction, not in the narrower opiate sense, but in the larger sense of needing a continuous flow of commodity fixes in order to keep at bay the pain of being dominated at work or in the family.

Buyers make a third kind of mistake when they fail to obtain guarantees of nonarbitrary treatment. Take the case of consumer remedies and consumer defenses waived by contract, or of extremely favorable creditor remedies written into contracts, or of clauses by which one party determines the venue of any lawsuit favorably to himself, or sets liquidated damages or conditions that create real risks of forfeiture. We have once again the two distinct errors of misperceiving the risk that the terms will be invoked, and of placing too high a discount on the possibility of future loss. But this case also involves a willingness to trust one's present partner to treat one fairly further down the road when what now seem like congruent interests have begun to diverge. The buyer allows the seller to con him—to make it seem unlikely that there will ever be an occasion to which the terms would be relevant and that if there were, the seller can be trusted to act reasonably in the circumstances, rather than standing on a legal right to treat the buyer unfairly.

A fourth type of mistake has to do with the long term consequences of choosing a particular structure for a relationship. . . . [T]hink of the systems of criminal penalties for violating a labor contract that the courts have from time to time struck down as peonage. And why is it that, even before the fourteenth amendment, courts wouldn't enforce a contract of enslavement, though it met the most extreme tests of voluntariness? In all these cases, the objection is not to running a specific risk of loss, or to running a specific risk that the seller will treat the buyer unfairly. The objection is to the whole relationship—it is an objection to *feudalism*, a way of life, or to *slavery*, a way of life. It makes no difference that if we apply to their actions the same tests of voluntariness we apply to our own, some people may sometimes want to be peons. We won't let them be. . . .

Conclusions about Compulsory Terms, and an Example

There are efficiency, distributive and paternalist rationales for regimes of compulsory terms. These are not mutually exclusive. It may be that a given regime will improve efficiency, redistribute income, and "correct" choices all at the same time. But the extent to which compulsory terms accomplish any of the three objectives can be known (to the extent it can be known at all) only on the basis of a study of the particular situation in question. . . . Here is an example: the case of *Steelworkers v. United States Steel* was incorrectly decided. In that case, the workers, represented by Staughton Lynd, argued that the steelworkers union and the Youngstown City Council had acquired an "easement" in the Youngstown plant, which the company had decided to close because it was unprofitable. The U.S. District Court and then the Court of Appeals decided that neither the workers nor the town had acquired, by contract or in any other way, any of the absolute property rights the company had in the plant.

The case was wrongly decided because the court should have implied into every contract of employment between the company and an individual worker the following term: As part payment for the worker's labor, the company promised that in the event it wished to terminate the manufacture of steel in the plant, it would convey the plant to the union in trust for the present workers (along with recently laid-off and retired workers). The company further impliedly promised to condition the conveyance so that if the union as trustee attempted to sell the plant or convert it to a use that would substantially reduce the economic benefit it generated for the town, the town would become the owner in fee simple. I would make this implied promise on the part of the company non-waivable, so as to achieve all three objectives discussed above.

Such terms are not now included in standard collective bargaining agreements because transaction costs, and particularly imperfect information, block the parties from reaching the agreement that would make all as well off as possible. Further, employers like U.S. Steel would find it impossible to pass along the full cost of this term, even if it were possible for them to calculate it accurately. It would therefore probably work a distributive benefit for workers at the expense of employers. Finally, a basic reason why workers have not in the past bargained for and won the kind of property interest in manufacturing enterprises that this term would represent seems to have been that they have miscalculated their true interests. They have underestimated the long-term value of worker control, and also the risks of capital flight and other forms of economic dislocation. They have overestimated the stability of basic arrangements between labor and management, and also overestimated the benefits of a relatively quiet life, with plenty of material goods and no responsibility.

Maybe all these judgments about efficiency, distribution and false consciousness are wrong. I certainly can't claim anything like the knowledge that would allow me to assert them with confidence. The point of my example is not to convince you of their truth. Rather, I am arguing that these are the kinds of judgments you would need to make to decide intelligently. It seems to me on balance likely—more than that I couldn't say—that a compulsory term of the type I've described would do more good than bad. . . .

Paternalism in Public vs. Private Life

When issues of paternalism arise in the context of "private life," the actor is likely to know the other, even to know the other well, and to have a claim to intuitive understanding based on common experience. But the decision maker whose dilemmas we have been examining throughout this essay is a state official deciding cases for people in the abstract, people situated each in his or her particular way in a society divided by class, race, and sex. One might concede that in the private context of intimate knowledge of the other, ad hoc paternalism is unavoidable, but still favor a rule against it in this more complicated situation.

It is no more than a partial response to point out that so long as it is built into free contract through the requirement of capacity, and so long as the concept of capacity has the incoherent quality I sketched just a moment ago, there can be no "rule" against paternalism. Even without a rule, we could be less paternalist in public than in private life. Even if there are no coherent conceptual boundaries we can invoke, it may make sense to move along the continuum in response to the situation of our particular decision maker.

Nor is it an adequate response to reject the public/private distinction as incoherent. . . . State action is not intrinsically more violent than private. But what makes it seem at least conceivable that the state official should be more chary of paternalism than the private actor is that the state official acts on people he doesn't know, which is just a euphemistic way of saying that he acts on people who belong to class, racial and sexual groups different from his own. It is quite likely, moreover, that these groups have a history of oppressive subordination to his own.

The basis of paternalism is intersubjective unity of the actor with the other; it is identification and intimate knowledge. But the most fundamental characteristic of social life in our form of capitalism is social pluralism, which is a euphemism for social segregation and consequent ignorance and fear of one group for another. In the context of segregation, ignorance and fear, the risks of paternalist intervention are multiplied far beyond what they are in private life. . . .

The farther apart they are culturally, the more likely it is that the actor will perceive "mistakes" or false consciousness on the part of the others that they won't recognize as such no matter how much data he lays on them, because they involve basic premises about the world, truth, and the good. Paternalist intervention based on a strong intuition that the supposed beneficiary is wrong on this level—say, in believing that the way to mourn her husband is to throw herself on his funeral pyre—has the unfortunate property of being simultaneously the most imperatively required (when it is required) and the most imperatively forbidden (when it's wrong or officious). In other words, as we move from the personal, private level into the area of relations between large groups that are parts of a single society, the stakes get higher—both for action and for inaction.

It is that the risks escalate on both sides that for me ultimately undermines the case for anti-paternalism in public life. . . . [T]olerance is only half the story. The other half is that life in our form of capitalism conspires to drive its constituent groups so far from each other that they can't communicate very well. It creates a situation in which the working class, the middle class and the welfare class, for example, inhabit incommensurable moral domains. . . .

A decision maker who will not take the risk of imposing housing codes and then enforcing them through tenant remedies—just on the grounds that people are wrong to submit to these conditions—because he doesn't feel confident about what the poor "really want," has let a constituent group slip outside his capacity for intimate intuitive knowledge. Since refusing to act paternalistically involves him in applying state force to execute the law of contracts or torts against those who would have been the beneficiaries of paternalism, he can't claim he's practicing benign neglect. What he's doing, if he tries to be a systematic anti-paternalist in public life, is denying his knowledge of the relative incapacity of groups, of their characteristic mistakes. He is acting to deepen their incapacity by treating them as entitled to their mistakes, and then bringing to bear the apparatus of the state to evict them from their subcode apartments, exclude them under the law of trespass from power over the means of production they created through their labor and leave them to beg for crumbs when accidents they didn't provide for befall them after all.

[Thus, if the decision maker] is concerned about failures of intuition, about the limits of empathy, he has two alternatives he should try before declaring himself a public life anti-paternalist and passing by on the other side. The first is to investigate the consciousness of those he isn't supposed to mess with. This means breaking down the barriers of segregation by knowing others, rather than just making rules for them.

The second is to go beyond exploration to the task of helping mobilize the groups on whose part one may have to act paternalistically. . . . If the others in whose interest you have to act are mobilized, it's more likely that you will have some intuitive knowledge of them because they will have the means of group expression. It's more likely that they'll be able to tell you what to do and correct you when you do it wrong, so that you don't make mistakes on their behalf. And if they're mobilized, there's more chance they will be able to dispense with your services. That is the true paternal goal: that the other should surpass you both in knowledge and in power, and share both.

8.3   THE RADICAL CRITIQUE

Legal Economists and Normative Social Theory

MARK KELMAN

Critical Legal Studies [CLS] is not infrequently paired in observers' minds with Law and Economics, in part because both became prominent as academic movements at the elite law schools in the middle and late 1970s, in part because each represented an attack on the dominant law school stance: centrist, ostensibly pragmatic and antitheoretical, process centered, case-law oriented. Moreover, Law and Economics was frequently thought to represent not just a new method of thinking about legal issues but a substantive attack from the right on the consensus views of the propriety of mildly liberal political policy, while CLS was often seen as the attack from the left on these same policies. Finally, since a fair number of CLS writers attacked Law and Economics writing, either in detail or in passing, CLS was often viewed by outsiders unfamiliar with the range of CLS work as predominantly an anti–Law and Economics group, some kind of negatively charged satellite. In my view, the relationship between CLS and Law and Economics is in fact quite intimate: I believe that to the extent that one can discern general themes within the Law and Economics movement, it is the best worked-out, most consummated liberal legal ideology of the sort that CLS has tried both to understand and to critique. . . .

Law and Economics as Social Theory

To the extent that one can generalize about a movement that is reasonably diverse, Law and Economics has been both an academic school that has advocated, normatively, a certain general vision of state function as well as particular implementing practices and a movement that purports to present a general descriptive theory of existing legal practice.

The descriptive agenda, significant largely to the Chicago school of legal econo-

Mark Kelman, Legal Economists and Normative Social Theory, from *A Guide to Critical Legal Studies* (Cambridge: Harvard University Press, 1987), pp. 114–150. Copyright © 1987 the President and Fellows of Harvard University. Reprinted with permission.

mists, has had two main messages. First, the claim is that judge-made common law has tended to be efficient or wealth maximizing, promoting legal results that "increase the size of the pie"; second, legislatures and administrative agencies are dominated by the more or (frequently) less legitimate distributive demands of groups that tend to capture these bodies. The first point has obviously been the special province of Law and Economics scholars; the themes of illegitimate regulation and legislation have been adopted largely wholesale from conservative economists with little general interest in law. . . . .

At a more general level, the positive theory [that the common law is efficient] is inevitably untestable unless we are sure which decisions are actually efficient, and there is little reason to believe that we can ever identify such decisions. . . . It is surely plausible to say, as Posner does, that the English rule protecting access to light but not view is efficient since it saves the transaction costs of negotiating only those agreements that would generally be negotiated; but the statement that the American rule (ensuring no access to either light or view) represents the hypothetical end point of hypothetical negotiations is surely just as plausible. There is an ad hoc, grab bag character to the bulk of the arguments for the efficiency of actual rules that is hard to miss. Posner actually argues that the English rule protecting light may well have been economically rational in England, while the American rule was rational here, given the fact that we had more room (and were thus less likely to block light than the crowded English). His argument seems wholly unpersuasive: a rule that provided that neighbors not block one another's light would be no less sensible in the United States because land is more plentiful; we would simply expect it to be applied to fewer cases, since neighbors would presumably block light less frequently, given the less crowded conditions. Indeed, the fact that land is plentiful should make a rule against blocking light more efficient, given that the cost of not blocking light would presumably be lower for the blocker where land is more plentiful.

The normative claims have been far more influential, in part, I would claim, because they so assertively deny the omnipresence of those central, painful contradictions that CLS writers have tried to show are inevitably characteristic of liberal discourse.

It is, in my view, in its normative social theoretical mode—the mode in which it most clearly adopts certain stances toward the basic problematics of the formation of the self, the relationship of self to discrete others and social life more generally—that Law and Economics has been a unified and significant movement. One needn't be a Law and Economics adherent at this social theoretical level to do microeconomic analyses of particular legal programs, with more or less market-oriented biases or presumptions, with more or less concern over the distributional impacts of a decision. But I do not think that Law and Economics would, or should, have been discerned as a particularly significant movement had it been no more than a collection of policy analyses of the probable impact of particular decisions. Much of the work would have been, when empirical, indistinguishable from traditional Law and Society impact studies, and when theoretical, simply a better-informed, more sophisticated version of the hypothetical analyses of rules that policy-oriented Legal Realist law teachers had long been doing. It has not, however, simply been a discrete series of micro-studies.

Posner has been the more influential figure in establishing Law and Economics as a school of thought than Guido Calabresi, not because Posner's rather quirky *Economic Analysis of Law* is more informative or better reasoned than Calabresi's breathtaking book *The Costs of Accidents,* but because it is infinitely more imperialistic, complete, catechismic. It may offer mind-numbingly off-target answers to many, many questions, but it does have an answer for every legal issue, and, perhaps more important, the answers can be derived from a very short list of normative and descriptive propositions about individuals, markets, and the political process. Posner's massive outpouring of work has occasionally been economically unsophisticated (as where he confuses many-seller markets with fully competitive markets without any regard for variations in buyer behavior); occasionally unfathomable (as in his efforts to define a meaningful concept of wealth maximization, or worse, to claim that this wealth-maximization criterion avoids many of the horrors of classic utilitarianism, though it is quite clearly subject to all the standard deontological rights theorists' critiques of a system that can sacrifice the individual to the whim of others); occasionally downright comic in its effort to explain all court-centered legal practice as derivative of the search for efficiency (as in his claim that premeditated murder is punished more severely than unpremeditated murder because we must equalize the expected punishment of all killings, and premeditated killers are more likely to get away with their crimes). This really doesn't matter, though; Posner gets the underlying world view completely right, probably even more right by being so obstinate in his drive for completeness, in his powerful urge to flatten human experience and deny complexity in the service of a desperate Panglossian optimism about the "straight" world of middle-class barter and an utter cynicism about the worlds of the outsiders and those who at least claim to care for them. It is, I think, because Posner would never have written a book called *Tragic Choices* (as Calabresi did, along with Philip Bobbitt) that he is the culturally, social theoretically central figure in the Law and Economics movement.

But it is not solely my claim that the Chicago school is the truly significant one because it brought its simple message to an academic and social world in search of simplicity while both the old economically influenced Realists (like Calabresi, Bruce Ackerman, and Richard Markovits) or the prominent neoinstitutionalists (like Oliver Williamson, Charles Goetz, and Robert Scott), who focused on transaction-cost reduction in organizational and legal form, gave too few simplistic answers. It is my claim that even the more sophisticated thinkers (like Calabresi) in fact must adopt, more or less consciously, the social theory of the Chicago school adherents, even if they genuinely bristle at any summary statement of the theory, because it really is the social theory of economics, of a coherent liberal individualism that sees society as fundamentally successful when it responds to the will of individuals, and mediates the conflicts between individuals simply by making everyone pay his way.

As social theory, Law and Economics starts with the supposition that values and desires are the arbitrary assertions of individuals. In fact, legal economics much more wholeheartedly embraces value skepticism than either of the main antecedent liberal movements, libertarianism and utilitarianism. Unlike the libertarians, legal economists try to assign initial entitlements without any regard to the preferred status of particular activities. Battery, for instance, is tortious not, as for the libertarians, because the taste to be free from battery is morally preferable to the taste to batter, but because we seek either to eliminate needless transactions (it is an empirical proposition that those who wish to be free from battery would end up having to pay off those who would batter them) or to assign entitlements as they would have been assigned in an auction among people with arbitrary desires to gain the entitlement. . . .

The sense of contradiction that dominates CLS writing is utterly absent in legal economic literature. At the technical level of the choice of form of legal pronouncement, the rules-standards "dilemma" is reduced to an argument over the relative costs of substantively misgoverned conduct and administratively costly case-by-case fact-finding. At the philosophical level, the conflict between individualism and altruism simply disappears, because the very notion of altruism is unfathomable. It is simply reduced to (even perhaps defined as) an individual's arbitrary taste to incorporate the interests of others in making his own selfish calculations, and like other tastes it is neither to be condemned nor encouraged. To the extent that (apparent) altruism is ever to be applauded, it is simply in situations in which one party (for example a helpless child) is transactionally disabled from negotiating with others to purchase help; thus, a parent's emotional concern for a child is needed solely because we cannot readily substitute enforceable contracts in which a child agrees to trade some of the future wages she will earn if cared for in exchange for better care in early life. Finally, as I have indicated, the fear of disorder or excessive individualism simply disappears: proper background rules can restrain harm, ensure that everyone's interests are properly accounted for in the incentive structures each of us faces.

The legal economists maintain a strong commitment both to the general adequacy of intentionalist discourse and to drawing the traditional line between the private domain of intentionality and the public domain of coercion. Actors are treated as sufficiently self-determined to respond to price signals, to alter and control their behavior. Law is generally treated entirely as an objective price constraint that tells actors what it will cost them to obtain a certain good (for example, a battery "costs" X dollars in fine or Y days of imprisonment, an in-kind fine substitute). Thus, punishment is a guide or restraint on presumptively self-determined action, not a person-creating, determinist force affecting, say, whether people will want to batter or not.

Moreover, implicit contracts are assumed to dominate private life, expanding the assumed domain of self-determination: people hypothetically consent to nearly all conditions in private life because the exit option (working elsewhere, buying a different product, leaving a battering husband) is invariably formally available, legally privileged, not precluded by sovereign command. Law and Economics, as a social theoretical movement, has a fundamentally complete or gapless vision of what law should aspire to as well as a remarkably inclusive short list of rules of thumb for realizing the ideal in practice. . . .

Welfare Economics: The Normative Critique

. . .

Is the Kaldor-Hicks Position Coherent?

. . .

The CLS claim, quite simply, is that there is absolutely no politically neutral, coherent way to talk about whether a decision is potentially Pareto efficient, wealth maximizing, or whether its benefits outweigh its costs. Essentially, Critics cite two distinct reasons for indeterminacy, each of which makes it impossible to ascertain which rule one ought to adopt to be efficient without already knowing what rules are in place, including the rule that is at issue when one undertakes an efficiency analysis. The first concerns the problem of wealth effects; the second involves the disparity between offer and asking prices.

Law and Economics commentators invariably acknowledge that wealth effects exist, but they trivialize the phenomenon. Surely, they say, there will be some impact on the downstream user's demand for clean water if she is made richer by assigning her a right to clean water, but since the increase in aggregate wealth represented by granting her this entitlement is minor, it will undoubtedly have trivial effects on her demand for each particular good, including clean water. While one can imagine cases where the initial rights assignment would always prove efficient owing to wealth effects (for example, where the only bottle of water is assigned to one of two parties desperate to survive in the desert and neither could purchase the water from the other once the other received the entitlement to it) such cases have little to do with our usual concerns.

There are three sorts of problems with this dismissive, minimizing response, only the last of which has particularly preoccupied CLS commentators. The first is that there are significant legal disputes that are far more like the water-in-the-desert dispute than the marginal-shift-in-pollution-rules dispute. For instance, assume that one is trying to decide whether Native American hunters and grazers are making the sort of "efficient" or "best and highest" use of the land that economically justifies their ownership claims. Ordinarily, we assume that land is put to the highest use as long as it is freely alienable, so that the party who desires (values) it most will purchase it. It would surely be far easier to decide that Native Americans are not efficient users if one were to imagine their effective demand for land, given that we have already made a decision that they don't own any land because they use it inefficiently. Clearly, Native Americans stripped of their land rights could not purchase back the land assigned to European settlers, but it is at least a great deal more ambiguous whether the settlers could have purchased it from the Native Americans had the Native Americans been considered its owners.

The second problem is that it is simply unclear why we treat any rule as marginal, rather than recognizing that we must simultaneously assign all entitlements if the regime as a whole is to be efficient. If it is not the case that there are only trivial wealth effects from the assignment of all entitlements, it is unclear why we can ignore the general wealth effects of entitlements by assuming that we have correctly set all other entitlements except the one we are discussing.

Finally, and most interesting, the Critics have emphasized that serious wealth-effects problems must occur if we picture people being granted entitlements to manifest their moralistic attachment to particular end states—that is, given the power either to be bought off or to refuse to waive a right to maintain a state they approve of. Here is an example: In one entitlement scheme, mine workers have no right to safety that I, a concerned moralist, can enforce. If I want to make the mines safer for workers, I must pay the owners to do so. If, however, I must be paid by the owners to waive my right to insist on a particular safety level, I might not be bought out at any imaginable price. Even if I have no such enforceable rights the cost-benefit analysis we may reasonably use might demand that we measure the value of my attachment to the end state (the benefit of the rule) by reference to the dollar value I would place on waiving the benefit. Thus, just as the bottle of water in the desert seems to be efficiently assigned no matter who starts out with it, because the party without the right cannot induce voluntary transfer of the right, so either a legal regime with a right to safe mines or one with no right to safe mines would appear efficient or wealth maximizing, since the sum of the hypothetical bids of workers and moralists to improve the mines cannot induce the change from the unsafe situation, while the hypothetical bids from those who find it cheaper to operate unsafe mines cannot shake the moralists' desire to exercise their right to maintain safe mines. . . .

The issue of valuing the desires of the third-party moralist ties into the second of the major CLS critiques of the potential Pareto efficiency criterion: the critique that there is a disparity between offer and asking price separate from that caused by wealth effects. The descriptive claim is that the amount a person might pay to attain an end state might frequently be lower than the amount he would have to be paid to forgo the end state. The implication is that, once more, if both a legal rule and its opposite can be efficient because neither affected party can induce the other to waive the initially assigned rights, then either legal rule seems to be wealth maximizing. . . .

The difference in valuation, though, may well not be the result of a general wealth effect. . . . Instead, people may simply cling to existing states rather than seek parallel substantive ones; they may value more highly the things that they are declared to be entitled to because these things are sanctified by the entitlement; they may be averse to valuing things except when comparing items they are choosing to purchase, so that they typically are able to do no more than imagine a selling price valuation for items they already own that bears no relation to any offer they might conceivably receive ("Sure, I'd sell my house for a million dollars").

We see many instances of this in daily life and observable markets: for instance, people may not sell goods they own when they wouldn't buy the same goods for the selling price net of the costs they must bear in selling the goods . . . ; people often refuse fairly high offers to give up airline seats on overbooked flights although they would almost surely pay far less to retain their seats; baseball owners may have refused to pay as much to sign players after the advent of free agency as they implicitly used to spend on them when they refused to sell them to other owners in the regime in which players were bound to sign only with the team that "owned" them.

The problem of evaluating third-party moralisms is difficult because both wealth effects and nonwealth-based disparities between offer and asking price undoubtedly exist. A third-party moralist might pay less to save the miners than he would ask to be paid to waive his right to insist on safety both because he is being granted enormous illiquid change-blocking wealth if he is given the right to enforce his moralistic concerns and because we may systematically pay out less to manifest our moral concerns than we would have to be paid not to, in part because we might feel that we had *caused* the victim harm, rather than simply failed to prevent it, if we waived a protective right. It also implicates, once more, the profound problem raised in discussing the ambiguity of desire or defining "wants." In the absence of asymmetry between offer and asking price, it would be easier to convince ourselves that people's moral tastes can be understood as privately held individual preferences for particular end states that they define. Collective decision on social organization could be seen as potentially responsive to these tastes. But what seems to be demonstrated by the existence of the price disparity is the impossibility of defining desires for end states without recognizing that the end states can scarcely be understood without reference to the collective and legal background. There is no single-valued "desire" as such for miners to be safe; there may be a discrete desire to purchase mine safety and a desire not to waive a right to insist on a safe mine, but neither is an abstract, presocial, authentic representation of the "real" desire; each is simply the contextually influenced understanding of both a want and an end state that cannot be abstracted from the legal setting in which one understands precisely what it is that one is seeking.

At any rate, the existence of an offer-asking problem not only calls the empirical validity of the Coase Theorem into question but also poses the same definitional problems for the potential Pareto efficiency test that a widespread wealth-effects problem would. When judging a proposed legal rule, it is simply unclear if it meets the test if the "loser" could neither buy out the "winner" nor the "winner" buy out the "loser." When both a rule and its opposite seem efficient, depending on who has the hypothetical burden of paying compensation or the benefit of resisting bribes to change, a judgment that one of the rules is efficient will almost surely be made by covertly privileging one party's interests, for unstated political reasons. . . .

The political point of all these technical attacks on the Kaldor-Hicks criterion is really quite simple, although it is undoubtedly obscure to many CLS sympathizers with a deep aversion to the jargon of economics. The point is to revive the political person's instinct that there really is no technical, prepolitical way to determine that one society is materially better off than another, or better off under one regime than another. When we get more widgets at the cost of more widget workers' lives, we can use different, technically coherent procedures to value this change that will make us think, alternatively, that it is either perfectly dreadful or completely unexceptionable. Choosing between the procedures will so often reflect underlying substantive debates about whether materialist or life-preserving value structures are preferable that the notion that there even exists a separate procedural technique for evaluating the polity, grounded in simple responsive preference aggregation, is, while not invariably wrong or self-deluding, so frequently wrong that it is almost certainly a pernicious myth.

8.4 THE SOCIOLOGICAL CRITIQUE

Of Coase and Cattle: Dispute Resolution among Neighbors in Shasta County

ROBERT ELLICKSON

. . . In his landmark article, "The Problem of Social Cost," economist Ronald Coase invoked as his fundamental example a conflict between two neighbors—a rancher running cattle and a farmer raising crops. Coase used the Parable of the Farmer and the Rancher to illustrate what has come to be known as the Coase Theorem. This unintuitive proposition asserts, in its strongest form, that when transaction costs are zero, a change in the rule of liability will have no effect on the allocation of resources. For example, the Theorem predicts that as long as its admittedly heroic assumptions are met, the imposition of liability for cattle trespass would not cause ranchers to reduce the size of their herds, erect more fencing, or keep closer watch on their livestock. . . .

To explore the realism of the assumptions underlying the Farmer-Rancher Parable, I searched for a jurisdiction that had imposed varying rules of liability in cattle trespass situations and had changed those rules with some frequency. After briefly surveying a half-dozen candidates in California, I settled on Shasta County. Since 1945, a specific California statute has authorized the Shasta County Board of Supervisors, the county's elected governing body, to determine whether in the county an owner of cattle is liable for damages stemming from unintentional cattle trespass on unfenced land. Although most of Shasta County is "open range"—territory where a cattleman is not liable for trespass damages of that sort—the Board has the authority to "close the range" in subareas of the county. A closed-range ordinance makes a cattleman strictly liable for any damage his livestock might cause while trespassing within the area affected by the ordinance. The Shasta County Board of Supervisors has exercised this power to close the range on dozens of occasions since 1945, thus changing the exact rule of liability that Coase used in his famous example. I traveled to Shasta County to determine whether these legal changes had had any impact.

This study presents findings that cast doubt on many of the assumptions undergirding the Coasean Parable. It also strives to help bridge the chasm lying between the law and economics and law and society movements, perhaps the two most significant social-scientific schools of legal research. On the whole, the law and society scholars have gathered the better field data on dispute resolution practices, and the law and economics scholars have developed the more explicit, rigorous, and testable theories of human behavior. Although one might think that members of these two schools would perceive irresistible benefits from collaboration, these two groups have worked largely in isolation from one another. They have separate journals. They

Robert C. Ellickson, Of Coase and Cattle: Dispute Resolution among Neighbors in Shasta County, 38 *Stanford Law Review* 623 (1986), pp. 623–687. Reprinted with permission.

gather at separate conferences. They rarely read, much less cite, work by scholars in the other camp. This absence of cross-fertilization stems not only from lack of familiarity with the working language of the other group, but also from a mutual lack of respect, even a contempt, for the kind of work that the other group does. To exaggerate only a little, the law and economics scholars believe that the law and society group is deficient in both sophistication and rigor, and the law and society scholars believe that the law and economics group is not only out of touch with reality but also short on humanity. . . .

The Shasta County evidence indicates that Coase's Farmer-Rancher Parable correctly anticipates that a change in the rule of liability for cattle trespass does not affect, for example, the quality of fences that separate ranches from farms. The Parable's explanation for the allocative toothlessness of law is, however, exactly backward. The Parable's explanation is that transaction costs are low and that parties respond to a new rule by agreeing to an exchange of property rights that perpetuates the prior (efficient) allocation of resources. The field evidence I gathered suggests that a change in animal trespass law indeed fails to affect resource allocation, not because transaction costs are low, but because transaction costs are high. Legal rules are costly to learn and enforce. Trespass incidents are minor irritations between parties who typically have complex continuing relationships that enable them readily to enforce informal norms. The Shasta County evidence indicates that under these conditions, potential disputants ignore the formal law. . . .

The Resolution of Animal Trespass Disputes in Shasta County

. . .

Animal Trespass Incidents

Every landowner interviewed, including all thirteen ranchette owners, reported at least one instance in which his lands had been invaded by someone else's livestock. Hay farmers grow what cattle especially like to eat, and thus expect frequent trespasses. Owners of large ranches are also common victims because they cannot keep their many miles of aging perimeter fence cattle-tight. Thus, when a rancher gathers his animals on his fenced pastures each spring, he is not startled to find a few head carrying a neighbor's brand.

Because cattle eat almost incessantly, a trespass victim's vegetation is always at risk. Nevertheless, a victim usually regards the loss of grass as trivial so long as the owner removes the animals with reasonable promptness—that is, within a day or two if the animals are easy to corral. Trespassing livestock occasionally do cause more than nominal damage. Several ranchette owners reported incidents in which wayward cattle had damaged their fences and vegetable gardens; one farmer told of the ravaging of some of his ornamental trees. . . .

Rural residents especially fear trespasses by bulls. In a modern beef cattle herd, roughly one animal in twenty-five is a bull, whose principal function is to impregnate cows during their brief periods in heat. Bulls are twice as heavy as the other herd animals, and tend to be much more ornery. Several respondents had vivid memories of bull trespasses. A farmer who owned irrigated pasture was amazed at the depth of the hoof marks that an entering bull had made. A ranchette owner and a rancher told of barely escaping goring while attempting to corral invading bulls. Because an alien bull often enters in pursuit of cows in heat, owners of female animals fear illicit couplings that might produce offspring of an undesired pedigree. Although no cow owner reported actual damages from misbreeding, several mentioned that this risk especially worried them.

Animal Trespass Law

One of the most venerable English common law rules of strict liability in tort is the rule that an owner of domestic livestock is liable, even in the absence of negligence, for property damage that his animals cause while trespassing. This traditional English rule also applies in the closed-range areas of Shasta County.

In the open-range areas of the county—that is, in the great bulk of its rural territory—the English rule has been rejected in favor of the pro-cattleman "fencing out" rule that many grazing states adopted during the nineteenth century. . . . [Under this rule] even a livestock owner who has negligently managed his animals is generally not liable for trespass damage to the lands of a neighbor.

Even in open range, however, there remain three important pockets of trespass liability. First, owners of goats, swine, and vicious dogs are subjected to the English rule throughout Shasta County. Second, when a cattleman's livestock have trespassed through, or over, a "lawful fence" that entirely encloses the victim's premises, the cattleman remains strictly liable for trespass damages. (A California statute, unamended since 1915, defines the technological standard that a fence must meet to be "lawful.") Third, common law decisions make a livestock owner in open range liable for trespass when he intentionally causes his animals to enter the unfenced lands of another. . . .

When the law of either open- or closed-range entitles a trespass victim to relief, the remedy is usually an award of compensatory damages. A plaintiff who has suffered from continuing wrongful trespasses may also be entitled to an injunction against future incursions. Moreover, California's Estray Act entitles a landowner whose premises have been wrongly invaded by cattle to seize them as security for a claim to recover boarding costs and other damages. . . .

The formal law provides trespass victims with only limited self-help remedies. A victim can use reasonable force to drive the animals off his land. In addition, a trespass victim willing to give the animals proper care can seize estrays and bill the costs of their care to their owner. But a victim is generally not entitled to kill or wound the offending animals. For example, a fruitgrower in Mendocino County (a closed-range county) was recently convicted for malicious maiming of animals when, without prior warning to the livestock owner, he shot and killed livestock trespassing in his unfenced orchard. We shall see that in this respect the formal law diverges from Shasta County mores.

The distinction between open-range and closed-range has formal relevance in public as well as private trespass law. Shasta County's law enforcement officials are entitled to impound cattle found running at large in closed range, but not those found in open range. Brad Bogue, the county Animal Control Officer, relies primarily on mediation and warnings when responding to reports of loose animals. Regardless of whether the trespass has occurred in open or closed range, Bogue's first priority is to locate the owner of the offending animals, and to ask him to retrieve the livestock promptly. For example, if the owner of a ranchette situated in open range were to complain of trespassing mountain cattle, Bogue would inform the complainant of the cattleman's rights in open range, but he would also find the owner of the animals and explain why it would be in the owner's interest to take better care of the livestock. Bogue asserts that little else is required in the usual case. In most years, Bogue's office does not impound a single head of cattle or issue a single criminal citation for failure to prevent cattle trespass.

Knowledge of Animal Trespass Law

. . .

Laymen's Knowledge of Trespass Law. . . . I found no one in Shasta County—layman or professional—with a complete working knowledge of the formal trespass rules just described. The persons best informed are, interestingly enough, two public officials without legal training: Brad Bogue, the Animal Control Officer, and Bruce Jordan, the Brand Inspector. Their jobs require them to deal with stray livestock on almost a daily basis. Both have striven to learn applicable legal rules, and both sometimes invoke formal law when mediating disputes between county residents.Both Bogue and Jordan possess copies of the closed-range map and relevant provisions of the California Code. What they do not know is the case law; for example, neither is aware of the rule that an intentional trespass is always tortious, even in open-range. Nevertheless, Bogue and Jordan, both familiar figures to the cattlemen, and (to a lesser extent) the ranchette owners of rural Shasta County, have done more than anyone else to educate the populace about formal trespass law.

What do ordinary rural residents know of that law? To a remarkable degree the landowners I interviewed did know whether their own lands were within open- or closed-range. . . . This level of knowledge is probably atypically high. Most of the landowner interviews were conducted in the Round Mountain and Oak Run areas. The former was the site in 1973 of the Caton's Folly closed-range battle. More importantly, Frank Ellis' aggressive herding had provoked a furious closed-range battle in the Oak Run area just six months before I conducted the interviews. Two well-placed sources—the Oak Run postmaster and the proprietress of the Oak Run general store—estimated that this political storm had caught the attention of perhaps eighty percent of the area's adult residents. In the summer of 1982, probably no populace in the United States was more alert to the legal distinction between open- and closed-range than the inhabitants of Oak Run.

What do laymen know of the substance of trespass law? In particular, what do they know of how the rules vary from open- to closed-range? Laymen tend to conceive of these legal rules in black-and-white terms: Either the livestock owners or the trespass victims "have the rights." We have seen that the law of animal trespass in open-range is quite esoteric. Even there, an animal owner is liable, for example, for intentional trespass, trespass through a lawful fence, or trespass by a goat. Only a few rural residents of Shasta County know anything of these subtleties. "Estray" and "lawful fence," central terms in the law of animal trespass, are not words in the cattlemen's everyday vocabulary. Neither of the two most sophisticated open-range ranchers that I interviewed were aware that enclosure by a lawful fence elevates a farmer's rights to recover for trespass. A traditionalist, whose cattle had often caused mischief in the Northeastern Foothills, thought estrays could never be seized in open-range, although a lawful fence gives a trespass victim exactly that entitlement. . . .

As most laymen in rural Shasta County see it, trespass law is clear and simple. In closed-range, an animal owner is strictly liable for trespass damages. (They of course never used, and would not recognize, the phrase "strict liability.") In open-range, their basic premise is that an animal owner is never liable. When I posed hypothetical fact situations designed to put their simple rules under stress, the lay respondents sometimes backpedaled a bit, but they ultimately stuck to the notion that cattlemen have the rights in open-range and trespass victims the rights in closed-range.

Legal Specialists' Knowledge of Trespass Law. The laymen's penchant for simplicity enabled them to identify correctly the substance of the English rule on cattle trespass that formally applies in closed-range. In that regard, the laymen outperformed the "legal specialists"—the judges, attorneys, and insurance adjusters. Although I sought out specialists who I had reason to believe would be knowledgeable about rural legal problems, I found that in two important respects the legal specialists had a worse working knowledge of trespass and estray rules than did the lay landowners. First, in contrast to the landowners, the legal specialists immediately invoked negligence rules when asked to analyze rights in trespass cases. . . . Second, unlike the lay rural residents, the legal specialists knew almost nothing about the location of the closed-range districts in the county. For example, two lawyers who lived in rural Shasta County and raised livestock as a sideline, were ignorant of these boundaries; one incorrectly identified the kind of range in which he lived, and the other admitted he did not know what areas were open or closed. The latter added that this did not concern him because he would fence his lands under either legal regime.

I interviewed four insurance adjusters who settle trespass-damage claims in Shasta County. These adjusters had little working knowledge of the location of closed-range and open-range areas or of the legal significance of those designations. One incorrectly identified Shasta County as an entirely closed-range jurisdiction. Another stated that he did not keep up with the closed-range situation because "closed-range" just signifies places where there are fences, and the fence situation changes too rapidly to be worth following. The other two adjusters knew a bit more about the legal situation. Although neither possessed a closed-range map, they were able to guess how to locate one. On the other hand, both implied that they would not bother to find out whether a trespass incident had occurred in open- or closed-range before settling a claim. The liability rules that these adjusters apply to routine trespass claims seemed largely independent of formal law.

The Settlement of Trespass Disputes

If Shasta County residents were to act like the farmer and the rancher in Coase's Parable, they would settle their trespass problems in the following way. First, they would look to the formal law to determine who had what entitlements. They would then regard those substantive rules as beyond their influence ("exogenous" to use the economists' adjective). When they faced a potentially costly interaction, such as a trespass risk to crops, they would resolve it "in the shadow of" the formal legal rules. Because transactions would be costless, enforcement would be complete: No violation of an entitlement would be ignored. For the same reason, two neighbors who interacted on a number of fronts would resolve their disputes front-by-front, rather than globally. My findings cast doubt on the realism of each of these implications of the Parable. . . .

Norms, Not Legal Rules, are the Basic Sources of Entitlements. In rural Shasta County, trespass conflicts are generally resolved not "*in* the shadow of the law" but, rather, *beyond* that shadow. Most rural residents are consciously committed to an overarching norm of cooperation among neighbors. In trespass situations, their salient lower-level norm, adhered to by all but a few deviants, is that an owner of livestock is responsible for the acts of his animals. Allegiance to this norm seems wholly independent of formal legal entitlements. Most cattlemen believe that a rancher should keep his animals from eating a neighbor's grass, regardless of whether the range is open or closed. Cattlemen typically couch their justifications for the norm in moral terms:

> Marty Fancher: "Suppose I sat down [uninvited] to a dinner your wife had cooked." Dick Coombs: It "isn't right" to get free pasturage at the expense of one's neighbors. Owen Shellworth: "[My cattle] don't belong [in my neighbor's field]." Attorney-rancher Pete Schultz: A cattleman is "morally obligated to fence" to protect his neighbor's crops, even in open range. . . .

Incomplete Enforcement: The Live-and-let-live Philosophy. The norm that an animal owner should control his stock is modified by another norm that holds that a rural resident should "lump" minor damage stemming from isolated trespass incidents. The neighborly response to an isolated infraction is an exchange of civilities. A trespass victim should notify the animal owner that the trespass has occurred and assist the owner in retrieving the stray stock. Virtually all residents have telephones, the standard means of communication. A telephone report is regarded not as a form of complaint, but rather as a service to the animal owner, who, after all, has a valuable asset on the loose. Upon receiving a telephone report, a cattleman who is a good neighbor will quickly retrieve the animals (by truck if necessary), apologize for the occurrence, and thank the caller. . . .

Several realities of rural life in Shasta County help explain why residents are expected to lump trespass losses. First, it is commonplace for a country landowner to lose a bit of forage or to suffer minor fence damage. The area northeast of Redding lies on a deer migration route. During the late winter and early spring thousands of deer and elk move through the area, easily jumping the barbed wire fences. Because wild animals trespass so often, most rural residents come to regard minor damage from alien animals not as an injurious event, but as an inevitable part of life.

Second, most residents expect to be on both the giving and receiving ends of trespass incidents. Even the ranchette owners have, if not a few hobby livestock, at least several dogs, which they keep for companionship, security, and pest control. Unlike cattle, dogs that trespass may harass, or even kill, other farm animals. If trespass risks are symmetrical, and if residents lump all trespass losses, accounts balance in the long run. Under these conditions, the advantage of reciprocal lumping is that each person is made whole without having to expend time or money to settle disputes.

The norm of reciprocal restraint that underlies "live-and-let-live" also calls for ranchers to lump the costs of boarding another person's animal, even for months at a time. A cattleman often finds in his herd an animal wearing someone else's brand. If he recognizes the brand he will customarily inform its owner, but the two will often agree that the simplest solution is for the animal to stay put until the trespass victim next gathers his animals, an event that may be weeks or months away. The cost of "cutting" a single animal from a larger herd seems to underlie this custom. Thus, ranchers often consciously provide other people's cattle with feed worth perhaps as much as $10 to $100 per animal. Although Shasta County ranchers tend to regard themselves as financially pinched, even ranchers who know that they are legally entitled to recover feeding costs virtually never seek monetary compensation for boarding estrays. . . .

The Complexity of Interneighbor Relations: Comprehensive Mental Accounts of Who Owes Whom. Residents with few animals may of course not perceive any average reciprocity of advantage in a live-and-let-live approach to animal trespass incidents. What if, for example, a particular rancher's livestock repeatedly caused minor mischief in a particular farmer's fields? In that situation, Shasta County norms call for the farmer to keep track of those minor losses in a mental account. Eventually, the norms entitle him to act to remedy any imbalance.

A fundamental feature of rural society makes this enforcement system feasible: Rural residents deal with one another on a large number of fronts, and most residents expect those interactions to continue far into the future. In sociological terms, their relationships are "multiplex," not "simplex." They interact on water supply, controlled burns, fence repairs, social events, staffing the volunteer fire department, and so on. Where population densities are low, each neighbor looms larger. Thus any trespass dispute with a neighbor is almost certain to be but one thread in the rich fabric of a continuing relationship. . . .

The live-and-let-live norm also suggests that neighbors should put up with minor imbalances in their aggregate accounts, especially when they perceive that their future interactions will provide adequate opportunities for settling old scores. Creditors may prefer having others in their debt. For example, when Larry Brennan lost six to seven tons of baled hay to Frank Ellis' cattle in open range, Brennan (although he did not know it) had a strong legal claim against Ellis for intentional trespass. Brennan estimated his loss at between $300 and $500, hardly a trivial amount. When Ellis learned of Brennan's loss he told Brennan to "come down and take some hay" from Ellis' barn. Brennan declined this offer of compensation, partly because he thought he should not have piled the bales in an unfenced area, but also because, to paraphrase his words, he would rather have Ellis in debt to him than be in debt to Ellis. Brennan was willing to let Ellis run up a deficit in their aggregate interpersonal accounts because he thought that as a creditor he would have more leverage over Ellis' future behavior.

The Control of Deviants: The Key Role of Self-help. The rural Shasta County population includes deviants who do not adequately control their livestock and do not adequately balance their informal accounts with their neighbors. . . . To discipline deviants, the residents of rural Shasta County use the following four types of countermeasures, listed in escalating order of seriousness: (1) self-help retaliation; (2) reports to county authorities; (3) claims for compensation informally submitted without the help of attorneys; and (4) formal legal claims to recover damages. The law starts to gain bite as one moves down this list.

Self-help. Because most trespass disputes in Shasta County are resolved according to extralegal rules, the fundamental enforcement device is also extralegal. A measured amount of self-help—just enough to "get even," to invoke a marvelously apt phrase—is the predominant and ethically preferred response to someone who has not taken adequate steps to prevent his animals from trespassing.

The mildest form of self-help is negative gossip. This usually works because only the extreme deviants are immune from the general obsession with neighborliness. Although the Oak Run–Round Mountain area is undergoing a rapid increase in population, it remains distinctly rural in atmosphere. People tend to know one another, and they value their reputations in the community. Some ranching families have lived in the area for several generations and plan to stay indefinitely. Members of these families seem particularly intent on maintaining their reputations as good neighbors. Should one of them not promptly and courteously retrieve an estray, he might fear that any resulting gossip would permanently besmirch the family name. . . .

When milder measures such as gossip fail, a person is regarded as being justified in threatening to use, and perhaps even actually using, tougher self-help sanctions. Particularly in unfenced country, a victim may respond to repeated cattle trespasses by herding the offending animals to a location extremely inconvenient for their owner. Another common response to repeated trespasses is to threaten to kill a responsible animal should it ever enter again. Although the killing of trespassing livestock is a crime in California, six landowners—not noticeably less civilized than the others— unhesitatingly volunteered that they had issued death threats of this sort. . . . Another landowner told of running the steer of an uncooperative neighbor into a fence. The most intriguing report came from a rancher who had had recurrent problems with a trespassing bull many years ago. This rancher told a key law enforcement official that he wanted to castrate the bull—"to turn it into a steer." The official replied that he "would have deaf ears" if that were to occur. The rancher asserted that he then carried out his threat.

It is difficult to estimate how frequently rural residents actually resort to violent self-help. Nevertheless, fear of physical retaliation is undoubtedly one of the major incentives for order in rural Shasta County. Ranchers who run herds at large freely admit that they worry that their trespassing cattle might meet with violence. One traditionalist reported that he is responsive to complaints from ranchette owners because he fears they will poison or shoot his stock. A judge for a rural district of the county asserted that a vicious animal is likely to "disappear" if its owner does not control it. . . .

Complaints to Public Officials. The long-time ranchers of Shasta County pride themselves on being able to resolve their problems on their own. Except when they lose animals to rustlers, they do not seek help from public officials. Although ranchette owners also use the self-help remedies of gossip and violence, they, unlike the cattlemen, sometimes respond to a trespass incident by contacting a county official who they think will remedy the problem. These calls are usually funneled to the Animal Control Officer or Brand Inspector, who report that most of their callers are ranchette owners with limited rural experience. As already discussed, these calls do produce results. The county officials typically contact the owner of the animal, who then arranges for its removal. Brad Bogue, the Animal Control Officer, reported that in half the cases the caller knows whose animal it is. This suggests that callers often think that requests for removal have more effect when issued by someone in authority.

Mere removal of an animal may provide only temporary relief if its owner is a mountain lessee whose cattle repeatedly descend upon the ranchettes. County officials therefore use mild threats to caution repeat offenders. In closed-range, they may mention both their power to impound the estrays and the risk of criminal prosecution. These threats appear to be bluffs; the County never impounds stray cattle when it can locate an owner, and it rarely prosecutes cattlemen (and then only when their animals have posed risks to motorists). In open-range, county officials may deliver a more subtle threat: not that they will initiate a prosecution, but that, if the owner does not mend his ways, the Board of Supervisors may face insuperable pressure to close the range in the relevant area. Because cattlemen perceive that a closure significantly diminishes their legal entitlements in situations where motorists have collided with their livestock, this threat can catch their attention.

A trespass victim's most effective official protest is one delivered directly to his elected county supervisor—the person best situated to change stray-cattle liability rules. Many Shasta County residents are aware that traditionalist cattlemen fear the supervisors more than they fear law enforcement authorities. . . . When a supervisor receives many calls from trespass victims, his first instinct is to mediate the crisis. . . . If a supervisor is not responsive to a constituent's complaint, the constituent may respond by circulating a closure petition.

The Rarity of Claims for Monetary Relief. Because Shasta County residents tend to settle their trespass disputes beyond the shadow of the law, one might expect that the norms of neighborliness would include a norm against the invocation of formal legal rights. And this norm is indeed strongly established. Owen Shellworth: "I don't believe in lawyers [because there are] always hard feelings [when you litigate]." Tony Morton: "[I never press a monetary claim because] I try to be a good neighbor."

Norman Wagoner: "Being good neighbors means no lawsuits." Although trespasses are frequent, Shasta County's rural residents virtually never file formal trespass actions against one another. . . . [T]hey are also strongly disinclined to submit informal monetary claims to an owner of trespassing animals or that owner's insurance company.

The landowners who were interviewed clearly regard their restraint in seeking monetary relief as a mark of virtue. When asked why they did not pursue meritorious legal claims arising from trespass or fence-finance disputes, various landowners replied: "I'm not that kind of guy"; "I don't believe in it"; "I don't like to create a stink"; "I try to get along." The landowners who attempted to provide a rationale for this forbearance all implied the same one, a long-term reciprocity of advantage. Ann Kershaw: "The only one that makes money [when you litigate] is the lawyer." Al Levy: "I figure it will balance out in the long run." Pete Schultz: "I hope they'll do the same for me." Phil Ritchie: "My family believes in 'live and let live.'" . . .

Shasta County landowners regard a monetary settlement as an arms-length transaction that symbolizes an unneighborly relationship. Should your goat happen to eat your neighbor's tomatoes, the neighborly thing for you to do would be to help replant the tomatoes; a transfer of money would be too cold and too impersonal. When Kevin O'Hara's cattle went through a break in a fence and destroyed his neighbor's corn corp (a loss of less than $100), O'Hara had to work hard to persuade his neighbor to accept O'Hara's offer of money damages. O'Hara insisted on making this payment because he "felt responsible" for his neighbor's loss. . . . There can also be social pressure against offering money settlements. Bob Bosworth's father agreed many decades ago to pay damages to a trespass victim in a closed-range area just south of Shasta County; other cattlemen then rebuked him for setting an unfortunate precedent. . . .

The landowners, particularly the ranchers, express a strong aversion to hiring an attorney to fight one's battles. To hire an attorney is to escalate a conflict. A good neighbor does not do such a thing because the "natural working order" calls for two neighbors to work out their problems between themselves. The files in the Shasta County courthouses reveal that the ranchers who honor norms of neighborliness—the vast majority—are simply not involved in cattle-related litigation of any kind.

My field research uncovered two instances in which animal trespass victims in the Oak Run–Round Mountain area had turned to attorneys. In one of these cases the victim actually filed a formal complaint. Because attorney-backed claims are so unusual, these two disputes deserve elaboration. . . . Although both arose in open range, in each instance legal authority favored the trespass victim. . . . In both instances, the victim, before consulting an attorney, had attempted to obtain informal satisfaction but had been rebuffed. Each victim perceived that the animal owner had not been honest. Each dispute was ultimately settled in the victim's favor. In both instances, neither the trespass victim nor the cattle owner was well-socialized in rural Shasta County norms. Thus other respondents tended to refer to the four individuals involved in these two claims as "bad apples," "odd ducks," or otherwise as people not aware of the natural working order. Ordinary people, it seems, do not often turn to attorneys to help resolve disputes.

Summary and Implications

Coase's Parable of the Farmer and the Rancher, like most writing in law and economics, implies that disputants look solely to formal legal rules to determine their entitlements. In rural Shasta County, California, residents instead typically look to informal norms to determine their entitlements in animal trespass situations. . . .

In Shasta County, the law of trespass had no apparent feedback effects on trespass norms. In no instance did the legal designation of an area as open (or closed) range affect how residents resolved a trespass or estray dispute. Thus Rancher Kevin O'Hara paid a neighbor for the loss of a corn crop because he "felt responsible," a feeling he said would not have been influenced by formal trespass law. Being located in closed range did not appear to make a trespass victim more likely to perceive a grievance or to exercise self-help. Insurance adjusters paid virtually no attention to the distinction between open-range and closed-range when settling trespass claims.

Other findings suggest the unreality of other literal features of the Coasean Parable. Victims of stray cattle did not treat the formal legal rules as exogenous; they were aware that one way to use limited resources is to lobby for legal change. Victims' enforcement of their norm-based entitlements was far from complete; they ignored some trespasses altogether and used others to offset outstanding informal debts. Victims tended to shun monetary settlements and instead preferred in-kind transfers, including ones effected through self-help. Although these findings are at odds with the literal features of the Coasean Parable, they are fully consistent with Coase's central idea that, regardless of the specific content of law, people tend to structure their affairs to their mutual advantage.

The Shasta County evidence suggests that law and economics scholars need to pay more heed to how transaction costs influence the resolution of disputes. Because it is costly to carry out legal research and to engage in legal proceedings, a rational actor often has good reason to apply informal norms, not law, to evaluate the propriety of human behavior. Contracts scholars have long known that norms are likely to be especially influential when disputants share a continuing relationship. A farmer and a rancher who own adjoining lands are enduringly intertwined, and therefore readily able to employ nonlegal methods of dispute resolution. Law-and-economics scholars misdirect their readers and students when they invoke examples—such as the Parable of the Farmer and the Rancher—that greatly exaggerate the domain of human activity upon which the law casts a shadow. . . .

8.5   THE COMMUNITARIAN CRITIQUE

Ethical Theory and the Case for Concern about Charges

STEVEN KELMAN

I believe that it is justified to be hesitant about the use of economic incentives in environmental policy out of concerns about the kind of society we help create when we choose to change over to doing so. These concerns are of four types:

1. If a society uses economic incentives in environmental policy, it makes a social statement of indifference towards the motives of polluters in reducing pollution. If people may justifiably care about the motives others have for behaving as they do, and if one further believes that using economic incentives endorses self-interested behavior that one may not in the circumstances wish to endorse, then one has reason to be concerned with using economic incentives in environmental policy.

2. If a society uses economic incentives in environmental policy, it fails to make a statement stigmatizing polluting behavior. If one believes that people may justifiably wish the societies they live in on occasion to make approbatory or stigmatory statements about certain behaviors, and if one further believes that polluting behavior should be stigmatized, then one has reason to be concerned with using economic incentives in environmental policy.

3. Using economic incentives in environmental policy means bringing environmental quality into a system of markets and prices of which it previously has not been a part. If one believes that people may justifiably oppose incorporating some additional previously unpriced but valued things into a system of prices and markets, and if one further believes that environmental quality is something that should not be so incorporated, then one has reason to be concerned with using economic incentives in environmental policy.

4. Frequently, although not specifically in the special case of the use of charges in environmental policy, economic incentive approaches produce a situation where wealthier people choose to pay the charge and continue behaving as before, while poorer people, to avoid the charge, are the ones to change their behavior. If one believes that equity considerations may justifiably play a role in the choice of policy instruments in specific policy areas, then one has reason to be concerned with economic incentive approaches that change the behavior of those who are worse off, while those who are better off continue to behave as before.

. . . [I]t should be noted that common to these concerns is an importance attributed to thoughts going on inside people's heads. A person who cares about the motives of others in behaving the way they do cares not only about how much pollution potential polluters decide to create, but also about what is going on inside their heads when they make such decisions. A person who cares about whether society stigmatizes polluting behavior cares not only about how much pollution is created, but also about having the opportunity to experience such stigmatization and about what is going on inside the heads of citizens when they decide how to judge polluters. A person who cares about whether environmental quality, previously unpriced and untraded in the market, becomes assimilated into a system of prices and markets, also cares about the diminution of certain positively valued feelings associated with nonmarket relationships. . . . A decision to fail to allow equity concerns to be reflected in the design of individual programs has an impact on the ideas people develop about the importance of equity in social life. . . .

This concern about what goes on inside people's heads contrasts with the behavior[al] and results-oriented approach of economic theory. Yet certainly such a concern corresponds to the everyday experience of most people. Those parts of life that involve simple feelings, without any material goods associated with them, constitute an important part of what makes people happy or unhappy—whether we are praised or condemned, respected or ignored, loved or unloved, laughed with or laughed at.

First, caring about what goes on inside people's heads reflects an interest in what preferences people have and a concomitant interest in the process of preference formation (since, if we care what preferences people have, we are likely to care about the processes that produce preferences). Second, caring about what goes on inside people's heads reflects a concern about the existence and continued production of certain valued feelings. The explicit refusal of economists to worry about what preferences people have and about the process of preference formation makes them insensitive to arguments that hinge on preferences or influences on the process of preference formation. The greater ease with which economists analyze material goods makes them less receptive to arguments based on concerns over the production of simple feelings, unrelated to any material goods. . . .

It should also be noted that common to the four concerns expressed above is the view that "society" is a reality. "Society" is a reality not in some mystical sense, but, first, because people's preferences and behaviors are influenced by what they experience around them. It is more likely that a person will believe x or behave in y way if everybody around him does so than if only one percent of everybody around him does so. . . .

Society is also a reality because it is one source of the production of feelings that an individual values or disvalues and one unit of analysis for certain moral judgments. Other specific individuals clearly have an impact on whether we experience various valued or disvalued feelings such as love or rejection. But experiencing these feelings can also depend on how widespread a certain attitude is in society as a whole. Hence the development of "preferences about society." For a person who wishes to live in a society with a high level of patriotism or altruism, society is the relevant unit of analysis. Furthermore, for certain moral judgments about relationships among people, such as judgments about the distribution of wealth, the respect accorded various human rights, or the method for reaching social decisions (democratic or dictatorial), society is the appropriate unit of analysis. . . .

Caring about Motives

Polluters may have differing motives for decisions to reduce pollution. They may do so out of a belief that they ought to abide by the law. They may do so out of a belief that they should not harm others. They may do so out of a desire to save money by reducing pollution so as to avoid a larger tax payment. Through a decision that endorses self-interest in achieving environmental policy goals, society makes a statement, both to polluters and to citizens in general, that people's motivations in behaving how they do are a matter of indifference. One may be concerned about the use of economic incentives in environmental policy because one does not wish to see such a statement made. . . .

The view that we ought to care only about results seems sensible at first blush. Isn't how clean the air and water are, and how much it costs to get them that way, what we care about? In fact, it appears that to suggest otherwise is either to engage in naive sentimentalism or to indulge the professional bias lawyers have for legal edicts that command duties and prohibitions.

I would suggest, however, that what initially may appear to be sensible in fact may be less so. Certainly, the view that we shouldn't care about what motivates people to act as they do is inconsistent both with elements of our legal system and with commonly held views that ordinary people hold. We frequently care about motivations for two reasons. First, bad motivations often tend to produce bad results, when a series of cases is observed. Second, good motivations tend to produce positively valued feelings in those who experience the good motivations. . . .

But it is possible to take the argument even further. Doing so involves what philosophers would regard as the intrinsic value of good motives and/or the intrinsic moral rightness of acts with good motivations. . . . This was Kant's position when he wrote of "the absolute value of good will," which, even if it "should be wholly lacking in power to accomplish its purpose" would still "sparkle like a jewel in its own right, as something that had its full worth in itself." A person who believes that good motivation is intrinsically good would argue that it was better for an employer to act sincerely towards employees even if it didn't make the employees more satisfied than they otherwise would have been and even if nobody else, who might feel differently about the well-motivated and badly motivated acts, knew about it. . . .

Wishing to Stigmatize Polluting Behavior

One of the most frequently expressed concerns of critics about the use of economic incentives in environmental policy is that they would grant an unacceptable "license to pollute"; to use economic incentives would express society's indifference about whether or not polluting behavior occurred as long as the charge was paid. Economists who discuss objections made to charges have had trouble suppressing contempt for the "license to pollute" contention, which they interpret merely as an argument that, faced with a pollution charge, polluters simply will pay the charge and continue to pollute as much as before. Such an argument, in their view, demonstrates ignorance of the insight, fundamental to economics, that if something becomes more expensive to do, less of it will be done. To the extent that they are willing to accept the "license to pollute" vocabulary at all, economists would give the counterargument that license fees are low charges, too low to diminish by much the behavior in question. If the pollution charge is set at an appropriately high level, it is continued, it certainly will decrease polluting behavior.

The thinly disguised contempt that most economists feel about the "license to pollute" worry displays, I believe, a failure to appreciate the importance many people attribute to feelings inside people's heads. And the counterarguments fail to address the worry at that level. Licenses are normally given out to authorize behavior toward which society takes a positive, or at least neutral, attitude. Practicing medicine, driving a car, or getting married are not regarded as undesirable activities. A license is needed to perform these behaviors in order to assure that those performing them are competent enough to meet certain legal requirements. In other cases, as with a license to own a dog, the behavior is also approved, and the main purpose of the license requirement is to raise revenue. What many of those who worry about charges being a "license to pollute" react against is the implied authorization of polluting behavior that charges represent. The fear is that by replacing standards, which are social statements of acceptable and unacceptable behavior, with charges, society would be saying, in effect, that "it's OK to pollute as long as you pay a fee." . . .

One may note in this context the writings of some economists regarding the "optimal" amount of fraud and the "optimal" level of the enforcement of laws. Most people would react, probably with some vehemence, that the "optimal" level of fraud is zero and the "optimal" level of enforcement is complete. When economists suggest this is not the case, the point they are making is basically the simple one that at some point the fraud or crime detected by marginal efforts to prevent them do not justify additional marginal resources devoted to prevention. If explained this way, most people would probably agree. However, they would likely add something like, "But 'optimal' is normally a moral term, something that expresses a positive attitude towards a situation. You are using the word in a different way." And thus the heart of the disagreement is reached: to most people, a moral condemnation of fraud or other crime is important, and the statement that some level other than zero is optimal counteracts that necessary moral condemnation. . . .

The first reason why environmentalists would wish to avoid an environmental policy that did not include as part of it a social statement of condemnation is that they wish society to express a judgment stigmatizing such behavior. In so wishing they are, I think, expressing a judgmentalism that is very common among people and applying it to something they feel strongly about. . . . The second reason for wishing to see a stigmatization of polluting behavior take place is to give citizens a signal encouraging them to develop preferences that give the environment a strong weight. There is an unresolved debate over to what extent laws change people's values: did civil rights laws tend to erode racial prejudice in the South, for instance? But environmentalists are deeply involved in the business of battling for the minds of citizens, trying to encourage people to develop preferences that give achievement of good environmental quality a high weight. Statements by environmentalists frequently emphasize the importance of developing an "environmental ethic," which is the way they express the importance of what they would call the "consciousness-raising" process they are involved in. Hence they can be expected to welcome a social statement condemning polluters as something that will help them in their battle. And, since decisions about what levels of environmental cleanup to seek are social decisions, environmentalists have obvious reason to be concerned about what preferences other citizens have, because these preferences will influence the outcome of the social decisions. Whether or not polluting behavior is stigmatized will tend, therefore, to influence what level of cleanup demands society ends up making. . . .

An important distinction should, however, be noted. It is one thing to recognize that advocates of various values seek to get others to value highly the things they value highly themselves. It is another thing to suggest that the power of the state be used to favor one set of values over another. It is a strongly embedded aspect of our traditions of liberal tolerance that we accept the legitimacy of people with different values jostling for the support of citizens, while government remains neutral among the contenders. By this view, it may be argued that it is an illegitimate grounds for favoring standards over economic incentives in environmental policy to state that the former policy expresses an attitude of social stigmatization towards polluting behavior. Such stigmatization, in this view, is not a legitimate role for government to play.

The problem with this argument is that any time government adopts laws in an area, these laws may well turn out to influence the values citizens hold. This is the case even for the most minimal laws that libertarians would favor, such as laws against murder. That murder is illegal almost certainly plays a role in affecting people's attitude towards it, for making murder criminal says more about society's attitude towards it than that society will punish those who murder. It also stigmatizes murder and encourages the development of preferences that give respect for life a strong weight. In any society with laws, complete government neutrality about values cannot exist. Once government adopts a policy for dealing with environmental pollution—a policy decision that advocates of charges question no more than advocates of standards—the choice between standards and charges is not a choice between government partiality and government neutrality. Either policy will reflect a different social statement about polluting behavior and will tend thus to influence people's preferences. . . .

Reasons for Opposing Establishment of Markets or Prices for Things Not Previously Traded in Markets or Priced

. . .

Anxiety about the spread of the market has been one of the main currents in the writings of sociologists about market-based societies ever since Ferdinand Toennies wrote of the distinction between *Gemeinschaft* and *Gesellschaft* and Max Weber worried about the consequences of the "demystification" of all aspects of life under the influence of the spread of rational calculation. The unease is by no means confined to scholars who write about society. It is given expression—above all, one is tempted to add—in the metaphors of everyday language and the images of everyday life. We attack people for "prostituting" themselves by selling some things for money. The expression "businesslike" may be fine to describe the relationship among businessmen, but hardly between parents and children or between friends. We tell ourselves that love of money is the root of all evil. Such unease is a common theme in literature as well, even dominating one of the important literary movements (Romanticism) in the period around the beginnings of industrialization. Blake decried the "dark satanic mills," and Ruskin praised the Gothic architecture of the Middle Ages as a representation of a better view of relationships among people. Dickens wrote of Ebenezer Scrooge in *A Christmas Carol,* "squeezing, wrenching, grasping, scraping, clutching," and of Gradgrind in *Hard Times,* "a man of facts and calculations" who believed that "everything, was to be paid for" and "nobody was ever on any account to give anybody anything, or render anybody help without purchase"—and who ended up destroying the lives of the people around him. The novels of Flaubert presented fathers who sold their children for money and businessmen who rushed away from the funeral of a partner, glad that it didn't take too long.

That themes so common, not only in the writings of scholars in other fields but in literature and in everyday life as well, are largely ignored by economists when they praise markets is striking. One reason for this may be that the unease tends to be presented in intuitive form. People, including scholars and novelists, aren't forced to analyze their intuitions unless faced with challenges to them. Few people ever get challenged unless they have dealt on a sustained basis with economists. . . .

The essential argument is that the very use of the market for steering production and allocation of a thing imposes costs. This occurs in two ways.

First, using the market may decrease production of certain behaviors that induce valued feelings. It need hardly be argued that there exist many feelings to which most people ascribe a positive value. One feeling, pleasure (broadly defined), is recognized as an important intrinsic good by almost everyone who has thought about the subject, and the only intrinsic good by some. Other positive feelings—such as exhilaration, gaiety, a sense of security, pride in accomplishment, the feeling of love, the feeling of being loved, and so forth—may then be valued, just as other things (such as television sets or yacht trips) are valued, as means to achieving greater feelings of pleasure. Other feelings have a negative value for most people. One could cite loneliness, sadness, or the feeling of being disliked, but, again, the list could be extended considerably. If an action produces decreased production of valued feelings or increased production of disvalued feelings, it therefore imposes costs.

Second, exchanging something on the market and placing a price on it may itself reduce the perceived value of the thing. What is taking place is not a cost in terms of decreased production of positively valued feelings due to market exchange in good *x*. It is, instead, a direct decrease in the perceived value of *x* itself. This is not a mere shift in tastes as would occur if people value hula-hoops one year and skateboards another. The strength of the preference for *x* has decreased without the strength of preferences for any other things increasing.

These two processes may be dubbed the "feeling-falloff effect" and the "downvaluation effect." Together they constitute psychological costs of using the market. . . .

The Feeling-Falloff Effect

. . .

Achieving the full efficiency benefits of markets requires that interactions in market exchange be impersonal. This is because if personal interactions were necessary, the scope for market production and allocation would be reduced to a small circle of people who could have personal relationships with each other. An efficient market requires large numbers of producers, buyers, and sellers. It thus must be impersonal. Firms seeking to expand their markets and consumers seeking a "better deal" for the things they buy are all impelled to widen the network of people they deal with. If I stay with a certain supplier of a product, not because he provides the best value for the money, but because I am a personal friend, market exchange becomes more personal, but at the cost of hurting competitive price formation. . . . Many of the people I deal with at the marketplace must be people I have never seen before and will never see again. Similarly, efficient markets impel a division of labor where most things people buy are produced by others they never met and who will never meet them. . . . Finally, the use of money in market exchange depersonalizes interactions as well, because money is an abstract, "dead" means of exchange that, unlike any specific good which incorporates the human effort put into its production, has no direct connection with human effort. The Marxist notion of alienation may be understood as involving the separation of the producer from any opportunity to receive recognition from the user for the product of his labor. The impersonal relationships of market exchange would appear to be what Marx had in mind when he wrote that in the marketplace "there is a definite social relation between men, that assumes . . . the fantastic form of a relation between things."

In impersonal relationships, the value of the things being exchanged must itself be sufficient to make the exchange worthwhile, since the relationship imparts nothing of value. By contrast, in many nonmarket relationships, the opportunity the interaction generates to display feeling-inducing behavior is a major value. A classic example is the gift exchanges in primitive societies investigated by such anthropologists as Marcel Mauss and Bronislaw Malinowski. Reciprocal gift-giving was frequently used to initiate friendship ties (for example, among groups living on different islands) or cement an emotional bond among the gift exchangers (for example, between families being united by marriage). Goods were transferred through the gift exchanges described by Mauss and Malinowski, but what was important in the interaction was not so much the specific identity of the things transferred as the interaction itself. Similarly, when one neighbor bakes a cake for another, the value of the interaction comes in significant measure from the display of feeling-inducing behavior.

It is in this context that many people fear that introducing elements of market exchange into human interactions will reduce them to impersonal market relationships and thus destroy a significant part of their value. . . .

The Downvaluation Effect

. . .

The very act of putting behavior on the market or placing a price on something may make us value it less. . . . Examples of the perceived cheapening of a thing's value by the very act of buying and selling it abound both in everyday life and in everyday language. The horror and disgust that accompany the idea of buying and selling human beings is based on the sense that this would dramatically diminish human worth. Praise that is bought is worth little, even to the person buying it. "He prostituted himself" and "He's a whore" said of people who have sold something reflect the view that certain things shouldn't be sold because doing so diminishes their value. Thus condemnation is heaped on those responsible for such diminution. . . .

This [diminution occurs] for another reason. If one values the existence of a non-market sector because of its connection with production of certain valued feelings, then one ascribes added value to any nonmarketed thing simply as a representative for, and part repository of, values represented by the nonmarket sector one wishes to preserve. This status removed, the thing loses its repository character and, hence, part of its perceived value. This seems certainly to be the case for things in nature such as pristine streams or undisturbed forests: for many people who value them, part of their value comes from their position as repositories of values the nonmarket sector embodies.

There is a second way that placing something in the market may decrease its perceived value. The market is a realm of inequality, where rewards are not equal and goods are not shared equally. For someone who values equality, a thing may be valued more highly simply because it is such that it can be shared equally among all citizens. Thus, part of the perceived value of things as varied as our national parks and our right to vote would appear to be their very character as part of a common treasure that we all share equally. That is, part of their value is as a repository for the value ascribed to equality. Such equal sharing would not occur if those things were placed in the market; it is possible only through nonmarket allocation. Placing such things in the market would thus tend to decrease their perceived value.

The other ways that placing a price on something may decrease its perceived value involve ways that one is able to proclaim the special value of something simply by keeping it outside the system of markets and prices of which most valued things form a part. Emile Durkheim noted in *The Elementary Forms of Religious Life* that a key element of religion was the separation of a small subset of things adjudged "sacred" from the larger category of things that were "profane." That separation was signaled, Durkheim noted, by treating sacred things in ways that the run of things were not normally treated. Animals are normally eaten, but sacred animals are not. Ground can normally be used for farming or for building shelter, but sacred ground is not. The very act of keeping something outside the realm of the market, then, is a way of proclaiming its special value: most things are bought and sold, but special things are not. . . .

The Appropriate Scope for Markets

So much for the psychological costs of markets. It might be noted that there also may exist psychological benefits of using the market. A person who feels that spontaneity or present-orientation are excessively prevalent in people might favor markets precisely because they encourage calculation and future-orientation. Early proponents of capitalism argued that capitalism promoted thrift, something they regarded as virtue in itself, independent of any positive consequences it had, because they regarded the very ability to abstain from present enjoyment as a virtue. Furthermore, it was also argued that, in comparison with alternative forms of allocation such as plunder, the market represented a benign form of human interaction. Similarly, impersonal relationships might be welcomed by those who fear they will be excluded from any means of getting things otherwise—say, minority groups, or anyone with personal characteristics unpleasing to others. (In the marketplace a black is able to say, "My money is as good as yours.") And the division of labor in society that is part of the efficiency benefits of markets also, in the view of Emile Durkheim, is a force that makes people more dependent on each other and thus increases ties among people. Furthermore, a firm searching for things to produce for the market needs to have a good insight into what the desires of others, his potential customers, are—as any textbook in marketing emphasizes. Such a search can increase production of feelings, if not of altruism, at least of empathy towards others. Finally, the market may lead to placing a higher valuation on something than would otherwise be the case. The attachment of a market price to something clearly signals its value; it is easier to ignore the value of something to which no price is attached. . . .

These potential psychological benefits of markets should be taken into consideration, of course, but it is important to note that they, just like the psychological costs discussed, tend to be left out of economists' analyses of the subject, which analyze market exchange simply in terms of its efficient properties and not as a form of human interaction. . . .

The full-blown psychological costs of using the market occur in instances where prices are established and where market exchange (with the attendant decrease in production of positively valued feelings) occurs as well. These would be relevant in discussions of proposals by economists for greater reliance on the market in areas such as health care or education. They are not, at least conceptually, fully relevant to proposals for using charges in environmental policy (although they would be for marketable rights proposals). The costs in terms of decreased perceived value for things to which a price has been attached do, however, apply. . . .

In conclusion, then, there are reasons for being concerned about the use of economic incentives in environmental policy out of a fear that "using the market" or establishing a price for pollution—and hence environmental quality itself—imposes costs. . . .

8.6   THE LEGAL REALIST CRITIQUE

Economic Analysis of Law: Some Realism about Nominalism

ARTHUR ALAN LEFF

With the publication of Richard A. Posner's *Economic Analysis of Law,* that field of learning known as "Law and Economics" has reached a stage of extended explicitness that requires and permits extended and explicit comment. But one of the dangers of reviewing a work "mainly designed for use either as a textbook in a law school course . . . or as supplementary reading for law students . . . " and getting down to the job later than one ought, is that one perforce reads it straight through, as if it were a real book. Having done that one cannot help being nagged throughout by what may be the literary critic's most pernicious and unavoidable naggerie: Where have I seen this before? At any rate, from my first glance at the table of contents, with its relentless item by item march through all of law—property, contracts, crimes and torts, labor law, corporations, taxation, racial discrimination, civil procedure . . . all the way through to a final "Note on Jurisprudence"—I smelled a familiar genre. But for the longest time, I couldn't place it. A manual of possible uses, the kind that comes with a new chain saw? A text on herbal healing? Not quite. But what? I was more than half way through the book before it came to me: as a matter of literary genre (though most likely not as a matter of literary influence) the closest analogue to *Economic Analysis of Law* is the picaresque novel.

Think of the great ones, *Tom Jones,* for instance, or *Huckleberry Finn,* or *Don Quixote.* In each case the eponymous hero sets out into a world of complexity and brings to bear on successive segments of it the power of his own particular personal vision. The world presents itself as a series of problems; to each problem that vision acts as a form of solution; and the problem having been dispatched, our hero passes on to the next adventure. The particular interactions are essentially invariant because the central vision is single. No matter what comes up or comes by, Tom's sensual vigor, Huck's cynical innocence, or the Don's aggressive romanticism is brought into play, forever to transform the picture of the pictured world (without, by the way, except in extremis, transforming the hero).

Richard Posner's hero is also eponymous. He is Economic Analysis. In the book we watch him ride out into the world of law, encountering one after another almost all of the ambiguous villains of legal thought from the fire-spewing choo-choo dragon to the multi-headed ogre who imprisons fair Efficiency in his castle keep for stupid and selfish reasons. In each case Economic (I suppose we can be so familiar) brings

Arthur Allan Leff, Economic Analysis of Law: Some Realism about Nominalism, 60 *Virginia Law Review* (1974), pp. 451–482. Copyright © 1974 Virginia Law Review Association and Fred B. Rothman & Co. Reprinted with permission.

to bear his single-minded self, and the Evil Ones (who like most in the literature are in reality mere chimeras of some mad or wrong-headed magician) dissolve, one after another.

One should not knock the genre. To hold the mind-set constant while the world is played in manageable chunks before its searching single light is a powerful analytic idea, the literary equivalent of dropping a hundred metals successively into the same acid to see what happens. The analytic move, just as a strategy, has its uses, no matter which mind-set is chosen, be it ethics or psychology or economics or even law. In each case, of course, the approach has its limitations, some of which arise from the single-approach strategy itself, and some of which arise from the particular mode of apprehension chosen to be single-minded with. I expect, in this review, to deal with some of those costs and benefits.

But the peculiarly relentless tone of this book moves me also to another inquiry: what pressures in contemporary legal scholarship might be responsible for the appearance, now, of four hundred pages of tunnel vision and, assuming one could answer that, why this particular tunnel? For Posner's book is, after all, just a fatly reified symbol of a currently important trend. It is a matter of common knowledge that economic analysis of the type Posner's book exemplifies is growing ever more popular among legal scholars. Not only have major and important analytical works recently been written in that style, . . . not only do almost all recent "law and" law review articles and law school courses turn out to be "law and economics," but even people like me, with no formal background in and little natural taste or aptitude for economic analysis, seem to be drawn to reading it, learning it and, in primitive fashion, even writing it. What is our problem?

In order to approach that question at all, . . . we will have to take rather a bleak architectural tour of a modern intellectual box. Only then will we be able to comment usefully upon the most recently designed and most elegantly constructed of the presently purported doors.

The Way We Live Today

Let us start with a couple of vicious intellectual parodies. Once upon a time there was Formalism. The law itself was a deductive system, with unquestionable premises leading to ineluctable conclusions. It was, potentially at least, all consistent and pervasive. Oh, individual judges messed up, and even individual professors, and their misperceptions and mispronouncements needed rationalization, connection, and correction. But that was the proper job of one of the giants we had in the earth in those days. The job of legal commentators, and a fortiori of treatise writers, was to find the consistent thread in the inconsistent statements of others and pull it all together along the seam of what was implicit in "the logic of the system." When you found enough threads and pulled them just hard enough, you made a very neat bag—say, Beale's *Conflict of Laws*.

Then, out of the hills, came the Realists. What their messianic message was has never been totally clear. But it is generally accepted that, at least in comparison to the picture of their predecessors which they drew for themselves, they were much more interested in the way law actually functioned in society. There were *men* in law, and the law created by men had an effect on other men in society. The critical questions were henceforward no longer to be those of systematic consistency, but of existential reality. You could no longer criticize law in terms of logical operations, but only in terms of operational logic.

Now such a move, while liberating, was also ultimately terrifying. For if you were interested in a society, and with law as an operative variable within that society, you would have to find out something about that subject matter and those operations. You would, it seems, have to become an empiricist. That, as we shall see, is no picnic when the facts you are searching out are social facts. But there is a worse worry yet. If you no longer are allowed to believe in a deductive system, if criticism is no longer solely logical, you no longer can avoid the question of *premises*. . . . If "good" were seen solely in terms of effects, the only good premises were those that came up with good effects. Thus, by dropping formalism we (quite rightly) fell into the responsibility of good and evil.

But not, alas, the knowledge thereof. While all this was going on, most likely conditioning it in fact, the knowledge of good and evil, as an intellectual subject, was being systematically and effectively destroyed. The historical fen through which ethical wanderings led was abolished in the early years of this century (not for the first time, but very clearly this time); normative thought crawled out of the swamp and died in the desert. There arose a great number of schools of ethics—axiological, materialistic, evolutionary, intuitionist, situational, existentialist, and so on—but they all suffered the same fate: either they were seen to be ultimately premised on some intuition (buttressed or not by nosecounts of those seemingly having the same intuitions), or they were even more arbitrary than that, based solely on some "for the sake of argument" premises. I will put the current situation as sharply and nastily as possible: there is today no way of "proving" that napalming babies is bad except by asserting it (in a louder and louder voice), or by defining it as so, early in one's game, and then later slipping it through, in a whisper, as a conclusion.

Now this is a fact of modern intellectual life so well and painfully known as to be one of the few which is simultaneously horrifying and banal. As I said, I raise it here only because it seems so very important both in explaining and understanding Posner's book, and the impulse in current legal scholarship it exemplifies.

Let us say you found yourself facing a universe normatively empty and empirically overflowing. What I suppose you would want most to do, if you wanted to talk at all, would be to find some grid you could place over this buzzing data to generate a language which would at the same time provide a critical terminology ("*X* is bad because . . . ") and something in terms of which the criticism could be made (that is, something to follow the "because . . . "). Now "because it is" is a bit naked as a satisfactory explanation. "Because you won't get to *Y* that way" is better, but when you make "good" teleological, you rather promptly run into "what's so great about *Y*?" It is hardly convincing if you explain the goodness of *X* in terms of the desirability of *Y* if you can't say anything more about *Y* than that it is desirable. You might just as well skip the intervening step and stick with *X*, saying all the pretty things about it itself, rather than about its product.

But what if you said *X* wasn't "good" or anything like that, that is, wasn't normative at all? What if you described X solely in empirical terms, for instance, X is what people, as a matter of fact, want. That way you can get to the well-known neo-Panglossian position of classic utilitarianism: while all is not for the best (because the best is what people want and they don't have it yet), the best is still nothing more than what they do want. Admittedly, this is just an example of one of the now-classic normative copouts—essentially, "good" becomes just a function of nosecounting—but it does have the advantage of providing a ready-made critical vocabulary: because there is now a clear area between what people want and what they have, while you can no longer say that doing anything is bad, you *can* say of some things that they are being done badly.

Of course, you still haven't solved all your intellectual (or practical) problems. The world may no longer be normatively empty (you've filled it by definitional fiat), but it is still full of all sorts of puzzling things. You have, that is, solved only one difficult problem. True, you need no longer ask if people *ought* to desire other things than they do desire, for those desires are the measure of all things. But what do people (some or all and is that relevant?) desire? If you don't know that, then you can't criticize what they presently have, or what they are right now doing, with reference to their failure to reach that desire. That is, while you are now working with *is*-terms only (you have escaped the dreaded *ought*), they are, as a matter of fact, very difficult matters of fact: what indeed *is* of "value" must be known before one rates the "efficiency" in getting there. Thus it is possible that all you have ended up doing is substituting for the arbitrariness of ethics the impossibilities of epistemology.

Now all of the above is but by way of introduction to Posner's solution to these scarifying problems. He does indeed solve the normative "oughtness" problems by the neo-Panglossian move: good is defined as that which is in fact desired. But then he makes a very pretty move one that renders his work, and work like it, so initially attractive to the dwellers of the box: in place of what one might have expected (and feared)—a complex regimen for an empirical investigation of human wants and values—he puts a single-element touchstone, so narrow a view of the critical empirical question as to be, essentially, a definition. "What people want" is presented in such a way that while it is in form empirical, it is almost wholly non-falsifiable by anything so crude as fact.

To follow this initially attractive development in legal criticism (for purposes both of admiration and scorn), one will have to master the critical early moves. The first and most basic is "the assumption that man is a rational maximizer of his ends in life. . . ." As Posner points out, this assumption "is no stronger than that most people in most affairs of life are guided by what they conceive to be their self-interest and that they choose means reasonably (not perfectly) designed to promote it." In connection with this assumption, several "fundamental economic concepts" emerge. "The first is that of the inverse relation between price charged and quantity demanded." The second is the economist's definition of cost, "the price that the resources consumed in making (and selling) the seller's product would command in their next best use—the alternative price."

> The third basic concept, which is also derived from reflection on how self-interested people react to a change in their surroundings, is the tendency of resources to gravitate toward their highest valued uses if exchange is permitted. . . . By a process of vol-

untary exchange, resources are shifted to those uses in which the value to the consumer, as measured by the consumer's willingness to pay, is highest. When resources are being used where their value is greatest, we may say that they are being employed efficiently.

Now it must immediately be noted, and never forgotten, that these basic propositions are really not empirical propositions at all. They are all generated by "reflection" on an "assumption" about choice under scarcity and rational maximization. While Posner states that "there is abundant evidence that theories derived from those assumptions have considerable power in predicting how people in fact behave," he cites none. And it is in fact unnecessary to cite any, for the propositions are not empirically falsifiable at all.

> Efficiency is a technical term: it means exploiting economic resources in such a way that human satisfaction *as measured by aggregate consumer willingness to pay* for goods and services is maximized. Value too is defined by willingness to pay.

In other words, since people are rationally self-interested, what they *do* shows what they value, and their willingness to pay for what they value is proof of their rational self-interest. Nothing merely empirical could get in the way of such a structure because it is definitional. That is why the assumptions can predict how people behave: in *these* terms there is no other way they can behave. If, for instance, a society dentist raises his prices and thereby increases his gross volume of business, it is no violation of the principle of inverse relation between price and quantity. It only proves that the buyers now perceive that they are buying something else which they now value more highly, "society dentistry," say, rather than "mere" dentistry. And if circularity isn't sufficient, the weak version of the rational maximization formula ("most people in most affairs of life . . . choose means reasonably (not perfectly) designed . . . ") has the effect of chewing up and spitting out any discordant empirical data anyway. Any puzzling observation fed into that kind of definition will always be able to find a "most," or a "reasonably," way out.

Thus what people do is good, and its goodness can be determined by looking at what it is they do. In place of the more arbitrary normative "goods" of Formalism, and in place of the more complicated empirical "goods" of Realism, stands the simple definitional circular "value" of Posner's book. If human desire itself becomes normative (in the sense that it cannot be criticized), and if human desire is made definitionally identical with certain human acts, then those human acts are also beyond criticism in normative or efficiency terms; everyone is doing as best he can exactly what he set out to do which, by definition, is "good" for him. In those terms, it is not at all surprising that economic analyses have "considerable power in predicting how people in fact behave."

I shall argue that lovely as all of this is, it is still unsatisfactory as anything approaching an adequate picture of human activity, even as expressed in that subcategory of living loosely called "law." But one can still admire the intelligence with which it is tried, and the genuine, though limited, illuminations the effort provides. More than that, one can now understand the forces that shaped the attempt. All of us are unable to tell (or at least to tell about) the difference between right and wrong. All of us want to go on talking. If we could find a way to slip in our normatives in the

form of descriptives, within a discipline offering narrow and apparently usable epis-
temological categories, we would all be pathetically grateful for such a new and more
respectable formalism in legal analysis. We would leap to embrace it. Since that is the
promise of economic analysis of law, to an increasing (and not wholly delusive or pernicious) extent, many of us are leaping.

To summarize, the move to economic analysis in law schools seems an attempt
to get over, or at least get by, the complexity thrust upon us by the Realists. There was
nonsense in Beale, but, ah, there was a feeling of elegance and power too. It was
lovely to be able to say things like, "law . . . is not a mere collection of arbitrary rules,
but a body of scientific principle." It was marvelous not to be embarrassed to say that
"Law . . . in great part . . . consists in a homogeneous, scientific and all-embracing
body of principle. . . ." Now we have a book in which it is apparently plausible to declare, "it may be possible to deduce the basic formal characteristics of law itself from
economic theory . . . " *and then do it in a two-page chapter*. What bliss.

We are, I think, beginning to see in the speedy spread of economic analysis of law
the development of a new basic academic theory of law. Since its basic intellectual
technique is the substitution of definitions for both normative and empirical propositions, I would call it American Legal Nominalism. . . .

Avoiding Complexity

As I have suggested several times, this aura of repetitive relentlessness that Posner's
book gives off is not solely the product of Posner's personal literary style. It is dic-
tated, I think, by a basic analytic strategy consciously chosen for the book: first, vigorously to exclude from consideration any normative statements of any kind, and second, to allow in empirical data only of particular kinds and only under the most
restrictive of conditions. To put it another way, it was Posner's conscious choice in
writing the book to deal only with what is, and then to exclude any description of that
is-ness in any uncongenial vocabularies, say, those employed in sociology, anthropology, and psychology. This decision has the natural collateral effect of excluding
any *data* relevant to the categories used by those disciplines but not by Posner's.

Such a choice has obvious advantages, as I have earlier pointed out. But let me
just sketch, in connection with some of Posner's particular analyses, what the possible effects of a somewhat less narrowly tunnelled vision might have been. Take sociology. Now I don't pretend to know what sociology is, but I can mention one thing
that seems to interest or at least to bother some sociologists—social groups and
classes. The following kind of question is frequently posed: Can I say anything interesting, or predictive, or even amusing about a number of people more than one and
less than all? Posner considers questions of that form throughout his book, for it is
implicated every time individual demand schedules are talked about as joined into a
general demand curve. Much of the time he addresses the problem of class formation
explicitly, sometimes indeed to criticize the alleged grossness and insensitivity of
other observers of human activity. But note the effect a greater sensitivity to classes
formed on criteria other than the particular ones Posner admits as relevant might have.

Consider, for instance, Posner's discussion of the role of judges. For reasons I shall not go into here, it is of some importance to Posner to establish that common-law adjudication is superior, at least in efficiency terms, to legislative decision-making. One pillar of this "proof" is the freedom of the judge from large dollops of allocative bias:

> [L]aw resembles the market in its impersonality, its subordination of distribution considerations. The invisible hand of the market has its counterpart in the aloof disinterest of the judge. The method by which judges are compensated and the rules of judicial ethics are designed to assure that the judge will have no financial or other interest in the outcome of a case before him, no responsibility with respect to the case other than to decide issues tendered by the parties, and no knowledge of the case other than what the competition of the parties conveys to him about it. Jurors are similarly constrained. . . . Judicial impersonality is reinforced by the rules of evidence. . . .

Well that's fine; the judge doesn't personally give a damn how the case comes out. At this point, of course, to go along with the game we must overlook the fundamental confusion in this passage between formal law and law in action, that is we must disregard one of the central lessons of legal sociology. We will just pretend that the legal realists had never lived, and that no one is allowed to look through the rules of pleading and evidence to see what judges and juries actually do know, despite these restrictions, about the parties and cases before them. We will even shut our eyes to the fact that it is in the service of getting before judges and jurors these "irrelevant" facts, like the parties' wealth and class, that many lawyers spend most of their time and skill.

Let us assume, then, that the judge is so exquisitely shielded from the world. Why then should he come up with decisions that favor efficiency? "This is a difficult question," Posner has the good grace to note. The "tentative" answer, surprisingly enough, is that judges are not without personal interest in how the case comes out. Especially (but not only) "where the judges do not have lifetime tenure" they frequently aspire to higher office, judicial or political." Indeed "[i]t seems appropriate to view these judges as the agents of the executive or legislative organs of the state." Will that skew the allocative disinterest of the judge? Not at all. Efficiency is also valuable to society, and judges will opt for that. Why? It must be because the efficiency-oriented decisions will help them in *their* quest for higher office. But if that is so, and legislators who also aspire to higher office apparently respond by taking distributive matters very seriously, why do we assume that judges won't? What is it (a sociologist might wonder) about the class "ambitious politician" that changes the predicted behavior of the two subclasses "legislator" and "judge" with respect to favoring certain interests, notably their own and those of their own class?

This is a particularly piquant question when asked in the context of Posner's extraordinarily shallow definition of the term "interest." Consider the following:

> It has sometimes been argued . . . that a judge's decisions can be explained in terms of the interests of the group or class in society to which he belongs—that the judge who owns land will decide in favor of landowners, the judge who walks to work in favor of pedestrians, the judge who used to be a corporate lawyer in favor of corporations.

Note first what Posner means by "class;" his examples are landowner, pedestrian, and corporate lawyer. Those are funny "classes" to choose; one might have expected other classes, for instance "bourgeoisie," or "upper middle class," or "elitist education group," or "Caucasian" or "male." If classes of that kind were chosen, it would have an interesting effect on the remainder of Posner's point:

> There are two points to be made here. First, where a particular outcome would promote the interests of a group to which the judge no longer belongs (our last example) [i.e., corporate lawyer], it is difficult to see how the judge's self-interest is advanced by adopting that outcome. *The judge's previous experience may, however, lead him to evaluate the merits of the case differently from judges of different backgrounds.* [emphasis supplied] Second, the increase in a judge's income from a ruling in favor of a broad group, such as pedestrians or homeowners, to which he belongs will usually be so trivial as to be easily outweighed by the penalties . . . for deciding a case in a way perceived to be unsound or biased.

Had Posner chosen other classes it would, first, have been clear that there were many *important* classes (that is, classes with more significant permanence than pedestrianism) in which judges can be placed which cannot *be* left, thereby making his first point inapplicable. It would so have made him reconsider the meaning of the word "income" in the modified sentence "the increase in a judge's income from a ruling in favor of a broad group, such as the upper middle class or Caucasians . . . will usually be so trivial. . . ." Most important, it would have pointed toward a considerably more sophisticated treatment of the cognitive and emotional effects of class membership than the single sentence italicized above; it is arguable at least that being a Caucasian male member of the bourgeoisie with a professional school education (which describes almost all judges) has a somewhat more striking effect on one's very perceptual and cognitive apparatus than being, say, a corporate lawyer or a pedestrian. This class-conditioned status might even affect the judge's evaluation of his "income" from a particular ruling, and of the "cost" to him of *his* class's perception of what "unsoundness" or "bias" might be.

Again, the question is not whether a sociologist's approach or an economist's would, on this issue, be superior. They are obviously complementary over a wide range. Even a lawyer's views, say Cardozo's, or Llewellyn's, might be useful. But that too is not the point. It is no more than this: class analysis is beggared by limiting the classes analyzed solely to those defined by narrow explicitly economic characteristics. To do otherwise, of course, would require new data, even empirical information rather hard to come by, certainly harder to come by than assumed responses to changes in price or supply of goods. But it might still help us to "understand" more than we now do about judicial behavior and its springs.

Or take anthropology. What if Posner knew some of that (or let on that he did)? One needn't get into the minutiae of disputes among anthropologists (which, at least in the area of "structuralism," is like falling into a tub of still-warm taffy) to suggest a lesson upon which they would all agree: one cannot say anything conclusive, or even particularly assertive, about any aspect of a culture without trying to place it (as much as one ever can) within all its other aspects. One need not, for instance, believe with Lévi-Strauss that all cultural artifacts are in some useful sense to be seen as isomers and polymers of each other to recognize, as I have suggested elsewhere, that a culture's "political" system and its "economic" system together form another system in which the "contradictions" within each subpart may turn out more transcended than one would otherwise suspect.

What about psychology? Oh sure, it has its problems. But various psychologists have said some shrewd things from time to time. Posner concedes that "the assumptions of economic theory are to some extent, certainly, oversimplified and unrealistic as descriptions of human behavior" but that "there is abundant evidence that theories derived from those assumptions have considerable power in predicting how people in fact behave." I am sure there is such evidence (though at this point none is cited) but there is evidence of other kinds too. What happens, for instance, to questions of "utility" if one accepts, even as a hypothesis, the idea of unconscious desires? In speaking about the maximization of utility, does one rate success in achieving what people "conceive to be their self-interest" in terms of their conscious or unconscious aims? If a man kills himself out of incandescent rage at his wife (or vice versa) has he "succeeded"? What is the social utility function when, in misery, one substitutes architectural erection for sexual and finds himself equally miserable but with an extra house? Nothing normative, mind you, just a question about the usefulness of defining value solely in terms of people's objective acts, and then generating a social utility function out of their aggregation. Can one actually, now, write four hundred pages about human desire without adverting to Freud, his followers, or even his enemies?

It must once again be emphatically stated that all of these considerations do not destroy Posner's contribution to the understanding of law. Suggesting that other intelligent men, also honestly groping for understanding, have used other matrices to place against society, thereby coming up with other assumptions, definitions, and expectations about the relevance of data, is not to suggest the primacy, or greater virtue or power, of any of the approaches. But do allow me to say that again: *any* of the approaches.

Of course, giving in to the temptation to eschew recognition of any approach but one's own is hardly merely self-indulgent. It may well be an absolute precondition to getting anything thought out or written down at all. Especially in the analysis of social states which are the product of many variables, and especially when one's interest is in a predicted state some substantial time in the future, *even if one restricts [one's] vocabulary of inquiry very sharply,* the problems are enormous. . . .

Indeed, one of Posner's key points is that because of the knottiness of such decisions, it is best if, as much as possible, everyone is given the opportunity to untie them for himself. For many of the decisions of life, if anyone can come out more or less where he wants to, it is he himself for himself. Another way to put this, I suppose, is to say that "the market" looks dreadfully inadequate as a way of maximizing human satisfactions, but only until one asks "compared to what?" If, however, one is going to criticize a society, one has in effect to do it for others as well as oneself. Once again, one need not do anything as vulgar as use ethical categories. One can say, after all, "if you are trying to get to *q* you ought [in the non-normative sense] to do *p* (or stop doing *p*, or do ~*p*, or do *something*)." Posner would, I think agree, especially since he spends much of the book doing just that. And after all, it is frequently possible to tell what a person is trying to do even if he seems to be going about it in rather a peculiar way. That is, one ought to be able, sometimes, to point out that someone is making a big mistake, and even tell him sometimes what it is.

But even while one is giving that kind of advice, it cannot be overemphasized how on the edge of arbitrariness the game is, especially in long-term multi-variable contexts. . . .

Smuggling Normatives: How to Win for Friends and Influential People

I have saved for last the question that is really most basic, for me, and I think for Posner. For all of his claims to non-normativity, it is obvious that there is at least one value qua value that directs and informs Posner's whole analysis. God (and history) knows it's one that does him credit: individual human freedom. . . . As normatives go, freedom is a good good, and there's no reason for anyone to be embarrassed by its espousal. For Posner, freedom—individual freedom—is a merit good, and why not; it's certainly no worse than, say, equality.

But having said that human freedom is the subterranean value upon which *Economic Analysis of Law* stands, and having praised that foundation, I cannot bring myself to stop. I know I should. Normative premises are just that; they don't get any more proved by being talked about. But I am just not up to resisting the modern moralist's temptation: even if I cannot say anything sensible about the choice of an intuitionist good, I shall nonetheless run on a while about the logical consistency and intellectual elegance of its deployment.

All right, let us consider, one last time, Posner's key definitional paragraph:

> Despite the use of terms like "value" and "efficiency," economics cannot tell us how society should be managed. Efficiency is a technical term: it means exploiting economic resources in such a way that human satisfaction as measured by aggregate consumer willingness to pay for goods and services is maximized. Value too is defined by willingness to pay. Willingness to pay is in turn a function of the existing distribution of income and wealth in the society. Were income and wealth distributed in a different pattern, the pattern of demands might also be different and efficiency would require a different deployment of our economic resources. The economist cannot tell us whether the existing distribution of income and wealth is just, although he may be able to tell us something about the costs of altering it as well as about the distributive consequences of various policies. Nor can he tell us whether, assuming the existing distribution is just, consumer satisfaction should be the dominant value of society. The economist's competence in a discussion of the legal system is limited to predicting the effect of legal rules and arrangements on value and efficiency, in their strict technical senses, and on the existing distribution of income and wealth.

In such a system whatever is, is. If you do not "buy" something, you are *unwilling* to do so. There is no place for the word or concept "unable." Thus, in this system, there is nothing which is coerced. For instance, let us say that a starving man approaches a loaf of bread held by an armed baker. Another potential buyer is there. The baker institutes an auction; he wants cash only (having too great doubts about the starveling's health to be interested in granting credit). The poor man gropes in his pockets and comes up with a dollar. The other bidder immediately takes out $1.01 and makes off with the bread. Now under Posner's definitional system we must say that the "value" of the bread was no more than a dollar to the poor man because he was "unwilling" to pay more than that. An observer not bound within that particular definitional structure might find it somehow more illuminating to characterize the poor man's failure as being the result of being unable to pay more than a dollar. But one cannot, consistent with Posner's system, say any such thing. One's actual power is irrelevant. . . .

If, however, one thinks intellectual consistency is worth talking about, it is worth pointing out that a similar argument can, perhaps must, be made if one applies Posner's definitional structure to political decisions. There are two ways to put the case. One is that if the poor man is forever to be deemed "unwilling" to buy, then the individual (rich or not) must be deemed "unwilling" to change or leave the political system, and so we will not hear his complaints about being coerced. That is tempting, but maybe it would be more instructive to say no more than that in both cases, he is "unwilling" to pay the price charged. The poor man could grab the bread (and risk being shot); the political man, unsatisfied with his lot, could revolt and seek to form his own polity (and risk getting squashed). In each case, all that stands in his way is a serious worry about his likelihood of success, given the inequality of power between him and the others.

What this all means is that Posner has not played fair with the question of power, or inequalities thereof. He has made a very common move: *after* something of value has been distributed he has defined *taking* as illicit and *keeping* (except when paid) as in tune with the expressed wishes of the universe. It is not as if force is never to be used; Posner assumes, indeed commands, its use against theft. One of the purposes of the state is to detect the terrible inefficiencies of nonconsensual transfers by having the government really smash those who persist in such behavior. But by and large the government is to have no role in even annoying those who choose to exclude others from what they already have. Keepers keepers, so to speak. . . .

But why is that? . . . If force, organized or not, were admissible as a method of acquisition there is no reason to assume that eventual equilibrium would not be reached, albeit in different hands than it presently rests. After all, as Posner would be the first to tell you, "force" is just an expenditure. If a man is "willing" to pay that price, and the other party is "unwilling" to pay the price of successful counterforce, we have an "efficient" solution. That is, we are "exploiting economic resources in such a way that human satisfaction as measured by aggregate consumer willingness to pay for goods and services is maximized."

In brief, there seems to be some normative content in Posner's neo-Panglossianism after all. Only some kinds of inequality are to be accepted as an unquestionable *grundnorm* upon which to base efficiency analyses. The transfers that come about against a background of wealth inequality are fine; any that come about against a background of inequality in strength, or the power to organize and apply strength, are unjustifiable. Some inequalities are apparently more equal than others—and all without reference to any apparent normative criterion at all.

Conclusion

There is none, and that's the point. We all know that all value is not a sole function of willingness to pay, and that it's a grievous mistake to use a tone which implies (while the words deny) that it is. Man may be the measure of all things, but he is not beyond measurement himself. I don't know how one talks about it, but napalming babies *is* bad, and so is letting them or even their culpable parents starve, freeze, or merely suffer plain miserable discomfort while other people, more "valuable" than they are or not, freely choose snowmobiles and whipped cream. . . . *And "the law" has always known it; that is the source of its tension and complexity.* If economic efficiency is part of the common law (and it is), so is *fiat justitia, ruat coelum.*

Thus, though one *can* graph (non-interpersonally comparable) marginal utilities for money which are the very picture of geometric nymphomania, we still preserve our right to say to those whose personalities generate such curves, "You swine," or "When did you first notice this anal compulsion overwhelming you?" or even "Beware the masses." And indeed "the law," even "the common law," has on impulses like those often said, even against efficiency—"Sorry buddy, you lose."

I admit that it is not easy these days to be a moralist manqué, when what it is that one lacks is any rational and coherent way to express one's intuitions. That's why it is, today, so very hard to be a thinking lawyer. But I will tell you this: substituting definitions for both facts and values is not notably likely to fill the echoing void. Much as I admire the many genuine insights of American Legal Nominalism, I think we shall have to continue wrestling with a universe filled with too many things about which we understand too little and then evaluate them against standards we don't even have. That doesn't mean that any of us—especially bright, talented and sensitive people like Richard Posner—should stop what they are doing and gaze silently into the buzz. What he is doing and has done (including this book) enriches us all. But (to get back to where we started) he (and all of us) should keep in mind what I think is the most lovely moment in *Don Quixote*. When asked by a mocking Duke if he actually believes in the real existence of his lady Dulcinea, the Don replies:

> This is not one of those cases where you can prove a thing conclusively. I have not begotten or given birth to my lady, although I contemplate her as she needs must be. . . .

One can understand the impulse, and be touched by the attempt, but the world is never as it needs must be. If it ever so seems, it is not the thing illuminated one is seeing, but the light.

NOTES AND QUESTIONS

1. Many objections to the particular policy recommendations of the economic approach can be understood in terms of the liberal critique. For instance, consider the concept of efficient breach of contract, which suggests that it is economically desirable for a promisor to breach a contract whenever the costs of performance exceed the benefits of breach. This concept has been widely controversial among lawyers and traditional legal scholars; for many critics, breaking a promise on such grounds flies in the face of what it means to make a promise in the first place. One way to interpret such objections is to observe that the gains from an efficient breach are achieved only by depriving the promisee of his or her due. From this viewpoint, such gains are the promisee's property; and efficient breach constitutes an unjust taking. A fuller development of this argument can be found in Daniel Friedmann, The Efficient Breach

Fallacy, 18 *Journal of Legal Studies* 1 (1989), who, like Dworkin, argues that the pursuit of economic efficiency can never justify the violation of individual rights. The economic account of negligence embodied in the Hand formula is open to similar attack; for such a critique, see George Fletcher, Fairness and Utility in Tort Theory, 85 *Harvard Law Review* 537 (1972) [also presenting an excellent comparison of liberal and utilitarian accounts of modern tort law]. In your view, does the Hand formula fail to respect the individual rights of accident victims, injurers, or anyone else? If so, to what extent should such rights take priority over the legitimate interests of the community in reasonably allocating its resources between accident prevention and the pursuit of other social objectives?

2. The economist and philosopher Amartya Sen (whom Dworkin obliquely honors in his central hypothetical example) has argued that properly understood, liberalism is in tension not only with the Kaldor-Hicks criterion, but with the weaker and ostensibly uncontroversial principle of Pareto superiority. Sen claims that even a policy change that unanimously benefits every individual in society can violate liberal principles. His argument appears in Amartya Sen, The Impossibility of a Paretian Liberal, 78 *Journal of Political Economy* 152 (1970), and is elaborated in Amartya Sen, Liberty, Unanimity, and Rights, 43 *Economica* 217 (1976).

3. Though Dworkin criticizes law and economics in liberal terms, the economic approach is itself strongly rooted in the liberal tradition, in that the central problem liberalism addresses is how individuals with widely divergent wants and beliefs can live together in a pluralistic society. As Adam Smith observed more than two centuries ago in his paean to the invisible hand of the marketplace, economic exchange provides a paradigm for such cooperation. Indeed, Dworkin's critique was provoked by Richard Posner's attempt to justify the efficiency criterion on liberal grounds in terms of an ideal social contract. Do you think that the Kaldor-Hicks principle can be justified in this way? For Posner's argument that it can be, and his response to Dworkin, see *The Economics of Justice* (Cambridge, Mass.: Harvard University Press, 1985), chap. 4 ("The Ethical and Political Basis of the Efficiency Norm in Common Law Adjudication"). See also Jules Coleman, *Risks and Wrongs* (Cambridge: Cambridge University Press, 1992) [arguing that the market is best defended as a rational way of fostering stable relationships among diverse members of society].

4. In order for paternalism to form a basis for public policy, policymakers must have some special access to the beneficiary's objective wants—access that the beneficiary herself lacks. For some categories of individuals, such as children and the intoxicated, this access is plausible, but as Kennedy observes, for categories such as race, gender, and socioeconomic class it is problematic. There is always the danger that the paternalist's judgment is itself clouded by lack of information, misguided sympathy, or conscious or unconscious self-interest. Consider in this regard Kennedy's discussion of *Steelworkers v. United States Steel,* in which he argues that collective bargaining agreements should be interpreted to contain a nonwaivable term granting unions a conditional property right in any individual plant that the employer wishes to close. Are you persuaded by his claim that such an implied term is justified on paternalist grounds? Could such a term, or analogous restrictions on plant closings, be justified on efficiency, distributional, or other grounds?

5. Kennedy concludes that the paternalist's ultimate goal should be to share power and knowledge with the intended beneficiaries, mobilizing them to the point where they can protect their own interests. If this is so, do paternalist motives ultimately reduce to distributive or efficiency ones? Is there reason to think that encouraging disadvantaged or ill-informed persons to act through the legal or political process, which Kennedy appears to advocate, is more effective than assisting them to act through market institutions?

6. Does the existence of a distinction between offer and asking price make cost-benefit analysis impossible, as Mark Kelman suggests, or otherwise undermine the contributions of the economic approach to law? Does the value of the economic approach really stand or fall on its ability to solve the larger jurisprudential and political problems of legal indeterminacy and value conflicts among citizens? Why is it important to the radical critique to claim that it does so depend? See A. Mitchell Polinsky, Economic Analysis as a Potentially Defective Product: A Buyer's Guide to Posner's "Economic Analysis of Law", 87 *Harvard Law Review* 1655 (1974) [discussing limitations of efficiency criterion and resulting implications for the economic approach to law]; Herbert Hovenkamp, Positivism in Economics and Law, 78 *California Law Review* 815 (1990) [suggesting that lawyers can use economics without adopting or endorsing the efficiency criterion]; Avery Katz, Positivism and the Separation of Law and Economics, 94 *Michigan Law Review* 2229 (1996) [arguing for use of economics on pragmatic grounds, notwithstanding claims that value is socially determined].

7. A major reason why the majority of economists have had difficulty studying changes in preferences is the absence of any agreed-upon normative framework to deal with the problem. If economic policies or institutions can change preferences, it is unclear whether the results should be judged by the ex ante or ex post viewpoint. To privilege either set of preferences over the other based on priority in time seems arbitrary; to decide based on which set is "objectively" better seems equally arbitrary. Some have suggested that the choice can be made based on *metapreferences,* or preferences over preferences [see, for example, Amartya Sen, Rational Fools: A Critique of the Behavioral Foundations of Economic Theory, 6 *Philosophy and Public Affairs* 317 (1977)], but such second-order preferences are difficult to measure and may themselves change over time. Accordingly, the mainstream of the profession has remained agnostic on the issue, and has been content to leave problems of preference formation to sociologists and psychologists. An influential minority, however, continues to raise the issue. See, for example, John Kenneth Galbraith, *The Affluent Society* (Boston: Houghton Mifflin, 1984); E. F. Schumacher, *Small is Beautiful: Economics as If People Mattered* (New York: Harper & Row, 1973), especially chap. 4 ("Buddhist Economics"); Tibor Scitovsky, *The Joyless Economy* (New York: Oxford University Press, 1976).

8. In recent years, a number of commentators in law and economics have been more willing to discuss explicitly the ethical and policy issues raised by preference formation. For such discussions, see Cass Sunstein, Public Regulation of Private Preferences, 53 *University of Chicago Law Review* 1129 (1986); Symposium on Law, Economics, and Norms, 144 *University of Pennsylvania Law Review* 1643 (1996). For specific applications to legal doctrine, see John Donohue, Prohibiting Sex

Discrimination in the Workplace: An Economic Perspective, 56 *University of Chicago Law Review* 1337 (1989); and Kenneth Dau-Schmidt, An Economic Analysis of the Criminal Law as a Preference-Shaping Policy, 1990 *Duke Law Journal* 1 (1990). Does the fact that individual preferences are socially influenced justify state intervention to change them? Compare Sunstein supra [arguing that government preference-shaping policies are often legitimate], with Friedrich Hayek, The Non Sequitur of the "Dependence Effect," 27 *Southern Economic Journal* 346 (1961) and Milton Friedman, *Capitalism and Freedom* (Chicago: University of Chicago Press, 1982), chap. 7 ("Capitalism and Discrimination") [arguing on libertarian grounds that such policies threaten both political and economic freedom].

9. Does the behavior of Ellickson's Shasta County correspondents suggest that their motivations were not primarily economic ones, or does it simply mean that the operative constraints within which they maximized their welfare were societal rather than legal? How might societal norms of reciprocity and fair dealing help reinforce economic efficiency? See Robert M. Axelrod, *The Evolution of Cooperation* (New York: Basic, 1984); Robert Ellickson, A Hypothesis of Wealth-Maximizing Norms: Evidence from the Whaling Industry, 5 *Journal of Law, Economics, & Organization* 83 (1989) ; Robert Cooter, Structural Adjudication and the New Law Merchant: A Model of Decentralized Law, 14 *International Review of Law and Economics* 215 (1994).

10. Both the sociological and communitarian critiques suggest a possible tension between economic, political, and sociological methods of social control. In economic terms we might describe this tension as a tradeoff—because people's attitudes toward society are not fixed, a greater reliance on economic incentives might encourage self-interest over public-spiritedness, working to undercut the political and sociological ties that are necessary for basic cooperation among citizens. In this regard, there is some empirical evidence that persons specially trained in economics behave in a more self-interested fashion than the general population—for example, the experiments reported by Gerald Marwell and Ruth Ames, Economists Free Ride: Does Anyone Else? Experiments on the Provision of Public Goods, IV, 15 *Journal of Public Economics* 295 (1981). What might account for such phenomena; and how might an economist such as Becker or Schelling respond to this general claim? See Robert H. Frank et al., Does Studying Economics Inhibit Cooperation?, 7 *Journal of Economic Perspectives* 159 (1993). For a classic discussion of political, economic, and sociological modes of interaction and the tradeoffs among them, see Albert O. Hirschman, *Exit, Voice, and Loyalty* (Cambridge, Mass.: Harvard University Press, 1970).

11. In the 1980s and 1990s, the Environmental Protection Agency, under Congress's direction, substantially expanded its use of market-based approaches to environmental regulation, ranging from "bubble" policies (under which a polluting enterprise can allocate emissions among the individual sources of pollution under its control, subject to the restriction that total emissions from the enterprise as a whole do not exceed 80% of the sum of source-specific emission limits) to "netting" policies (under which polluters can avoid the tighter restrictions governing new sources of pollution by reducing emissions from existing sources below regulatory requirements) to "banking" policies (under which polluters who reduce pollution by more than the law requires in a given year can bank the excess reduction as a credit against

compliance efforts in future years) to "offset" policies (under which new and modified sources of pollution must offset their emissions by explicitly buying reduction credits from existing polluters.) For descriptions and evaluations of these policies, see Robert W. Hahn and Roger G. Noll, Barriers to Implementing Tradable Air Pollution Permits, 1 *Yale Journal on Regulation* 63 (1983); Robert W. Hahn and Gordon L. Hester, Where Did all the Markets Go? An Analysis of EPA's Emissions Trading Program, 6 *Yale Journal on Regulation* 109 (1989); Bernard S. Black and Richard J. Pierce Jr., The Choice Between Markets and Central Planning in Regulating the U.S. Electricity Industry, 93 *Columbia Law Review* 1341 (1993). How would Steven Kelman evaluate these various policies? Which of them would be most objectionable, and which least objectionable, from the viewpoint of a communitarian environmentalist? Which is most desirable on efficiency and distributional grounds?

12. Part of the communitarian critique of law and economics is a cultural critique. Economics, as a distinct discipline, has a particular culture and rhetoric, features of which include a strong commitment to instrumental reasoning, a liberal attitude toward the content of individual tastes, a preference for abstract and general explanations rather than particularized ones, a preference for reductionist theories rather than holistic ones, and a commitment to philosophical positivism that leads economists sharply to distinguish between fact and value and to disparage arguments based on interpretation compared to direct observation. Legal culture shares only some of these features, however, so some lawyers may find themselves resisting economic methods out of a desire to resist economic culture. Is this reaction appropriate, or can the culture and method of economics be distinguished from one another? Can you think of other important differences between legal and economic culture? See James Boyd White, Economics and Law: Two Cultures in Tension, 54 *Tennessee Law Review* 161 (1987); Donald McCloskey, The Rhetoric of Law and Economics, 86 *Michigan Law Review* 752 (1988); Avery Katz, Positivism and the Separation of Law and Economics, 94 *Michigan Law Review* 2229 (1996).

13. Leff claims that the assumption of utility maximization, together with economics' definition of value in terms of individual willingness to pay, combine to make the efficiency criterion a circular one. Is this criticism correct? How can it be squared with the model of market failure? Consider in particular his argument that, because willingness to pay can be measured in units of force just as easily as in units of money, robbery can be viewed as efficient (supra at page 363). Can it so be viewed? How would Demsetz, Calabresi and Melamed, or Cooter respond to this argument?

14. Leff's main concern about of the economic analysis of law—that it is excessively reductionist and slights important noneconomic values—has been frequently reiterated in the critical literature. One important variant of this criticism is that the model of rational choice that underlies economic analysis, while ostensibly neutral in theory, in practice favors those values that are easily expressible in economic or pecuniary terms. Laurence Tribe has coined the term "dwarfing of soft variables" to describe this phenomenon. In his view, such "dwarfed" values tend systematically to be distributional, procedural, dignitary, and aesthetic ones. See Laurence Tribe, Trial by Mathematics: Precision and Ritual in the Legal Process, 84 *Harvard Law Review* 1329 (1971). Is this criticism fair? Does it provide a reason to avoid using the economic approach in some or all situations? How could economic practice be improved

to better take account of such values, and what institutional arrangements would best assure that lawyers who use economics in their work do so wisely? See, for example, Richard Pildes and Cass Sunstein, Reinventing the Regulatory State, 62 *University of Chicago Law Review* 1 (1995) [suggesting guidelines for the use of cost-benefit analysis in administrative proceedings].

9

An Application on the Frontier: Family Law

This concluding chapter presents three readings that discuss various aspects of the economics of family law. The discussion is intended to serve three main purposes: first, to provide readers with an occasion to review and to integrate the ideas of the book as a whole; second, to show, in more depth and with a new set of applications, how taking an economic perspective can reorient one's views of a particular doctrinal or policy area; and third, to explore the scope and the limits of the economic approach.

Focusing on family law helps underscores the fundamental lesson with which this reader began: that economic analysis can shed light on any sphere of human interaction in which individuals pursue their goals subject to constraint. From this perspective, family relations make up a vital part of economic life. They determine the workings of what might be called the household sector of the economy, in which market commodities are transformed through human time and effort into consumable form, and in which such basic goods as education and child care are produced and consumed. Interactions among and within families raise all of the economic issues presented in earlier chapters of this book: externalities imposed by individual families on the rest of society and by individual family members on others in the household; incentives to invest in the family's material assets and in the human assets of its individual members; strategic behavior arising from family members' efforts to influence each others' conduct; insurance against the financial and emotional risks of disability, unemployment, and household dissolution; and the effects of limited information and bounded rationality on such crucial personal decisions as family formation and career choice.

Notwithstanding the economic importance of families and of family relations, however, the economic analysis of law so far has had relatively less influence in family law than in other doctrinal fields. In part this is because in our culture the market and the family have traditionally been taken to constitute distinct social realms; and such customary habits of thought are slow to change. But an equally important reason is that the considerations outlined in the previous chapter have particular force within the domain of family life. The inevitable conflicts between the family unit and the interests of its individual members raise fundamental issues of liberalism. The recurring need to make decisions on behalf of family members incapable of protecting their own interests, such as children and the infirm, makes the family the archetypical paternalist institution. The radical, sociological, and communitarian worldviews, all of which stress the social origin of individual preferences, partake of our common understanding that the family is the crucible of value. And the Legal Realist claim that reality is too complex to be reduced to formal economic models may seem specially persuasive in a setting where the insights of competing disciplines such as psychology and biology are of plain relevance. For all these reasons, family law can fairly be regarded as falling on the frontier of law and economics; it offers, perhaps more than any other application, an opportunity though which to compare the strengths and weaknesses of the economic approach.

The readings in this chapter address these themes from a variety of economic perspectives. Gary Becker and Kevin Murphy present some basic ideas about the economics of the family, and discuss the consequences of these ideas for government policies regulating family decisions about education, retirement savings, bequests, and divorce. In Becker and Murphy's view, such policies are best analyzed through the paradigm of an intergenerational contract. Because the incapacity of children and the mortality of elders interfere with efficient private arrangements between parents and children, state regulation in this area is justified on a theory of market failure. The authors present their theory as a positive one, conjecturing that the state intervenes in family relations when, and only when, doing so increases the combined welfare of parents and children.

Elisabeth Landes and Richard Posner, focusing on state regulation of adoption, reach a rather different conclusion. They argue that the adoption policies in force in the U.S. at the date of their writing are substantially inefficient. In their view, existing law allows state adoption agencies to act as monopolists, with predictable economic consequences. Specifically, by interfering with a free market in adoption, state regulations have created both a shortage of adoptable babies and a surplus of other children who, for various reasons—including age, health, and race—are viewed by potential adoptive parents as less desirable. The authors discuss the advantages and disadvantages of deregulation, concluding that movement in the direction of a free market would improve the welfare of both parents and children.

The third reading, by Michael Trebilcock and Rosemin Keshvani, also considers the merits of freedom of contract in the domain of familial rights and obligations. It concentrates not on adoption, however, but on separation agreements and divorce settlements entered into by couples whose marriages have broken down. Trebilcock and Keshvani argue that the economic and practical advantages of private ordering make such settlements generally desirable, but a variety of contractual and market failures such as externalities, imperfect information, and strategic behavior justify public oversight of the terms of the bargains that are thereby reached. Their analysis, combined with that of Landes and Posner's, nicely recapitulates the tension between the paradigms of market failure and cooperation that has served as an organizing theme for this book.

The Family and the State

GARY BECKER AND KEVIN MURPHY

Children are incapable of caring for themselves during many years of physical and mental maturation. Since their mental development is not sufficient to trust any contractual arrangements they may reach with caretakers, laws and social norms regulate the production and rearing of children. Laws punish child abuse, the sale of children, and unauthorized abortions. They provide compulsory schooling, welfare payments to families with dependent children, stringent rules about divorce when young children are involved, and minimum ages of marriage.

Trades and contracts are efficient if no deviation from the terms would raise the welfare of all participants. An alternative criterion for efficiency is that the monetary gains to those benefiting from a deviation do not exceed the monetary loss to those harmed. Unfortunately, the immaturity of children sometimes precludes efficient arrangements between children and parents or others responsible for child care.

This difficulty in establishing efficient relations within families provides the point of departure for our interpretation of the heavy state involvement in the family. We believe that a surprising number of state interventions mimic the agreements that would occur if children were capable of arranging for their care. Stated differently, our belief is that many regulations of the family improve the efficiency of family activities. To be sure, these regulations raise the welfare of children, but they also raise the welfare of parents, or at least they raise the combined welfare of parents and children.

The efficiency perspective implies that the state is concerned with justice for children, if "justice" is identified with the well-being of children, for their well-being is the prime factor in our analysis. The efficiency perspective does not imply, however, that the effect on children alone determines whether the state intervenes. The effect on parents is considered too. The state tends to intervene when both gain or when the gain to children exceeds the loss to their parents. . . .

Gary Becker and Kevin Murphy, The Family and the State, 31 *Journal of Law & Economics* (1988), pp. 1–18. Copyright © 1988 The University of Chicago Law School. Reprinted with permission.

We cannot prove that efficiency guides state involvement in the family. We will show, however, that state interventions in the market for schooling, the provision of old-age pensions, and access to divorce are consistent on the whole with the efficiency perspective. . . .

Investments in the Human Capital of Children

Since parents must reduce their own consumption (including leisure) to raise the time and resources they spend on child care and children's education, training, and health, even altruistic parents have to consider the trade-off between their consumption and the human capital of children. But altruistic parents who plan to leave bequests can avoid this trade-off by using [reductions in those] bequests to help finance their investments in children. In effect, they can force even selfish children to repay them for expenditures on the children's human capital. These parents would want to invest efficiently in children because that raises children's utility without costing them anything.

To make this clear, assume a 4 percent rate of return on assets accumulated over the life cycle to provide either old-age consumption or gifts and bequests. If the marginal rate of return on investments in children exceeds 4 percent, parents who give gifts and bequests could invest more in children without lowering their own consumption by accumulating fewer assets. For example, if the marginal rate on human capital is 7 percent, an additional $1,000 invested in children raises their adult earnings by about $70 per year. If parents finance this investment through reduced savings of $1,000 and by reducing annual gifts by $40, their consumption at all ages would be unaffected by greater investment, while their children's income increases by $30 per year. . . .

Some altruistic parents do not leave bequests because they get less marginal utility from consumption by their adult children than from their own consumption when elderly. They would like to raise their own consumption at the expense of their children's, but they cannot do this if unable to leave debts to children. Although children have been responsible for parents' debts in some societies, that is uncommon nowadays. Selfish and weakly altruistic parents would like to impose a large debt burden on their children. Social pressures can discourage this in closely knit societies where elderly parents live with and depend on the care of children, but these pressures are not effective in mobile modern countries where the elderly do not live with children.

Parents who cannot leave debt can substitute their own consumption for their children's by investing less in the children's human capital and instead saving more for old age. Therefore, in families without bequests, the equilibrium marginal rate of return on investments in children must exceed the rate on assets saved for old age; otherwise, parents would reallocate some resources from children to savings. These parents underinvest in the human capital of children.

When the rate of return on savings is less than the marginal rate on human capital, both children and parents could be better off with a "contract" that calls for parents to raise investments to the efficient level in return for a commitment by children to repay their elderly parents. Unfortunately, young children cannot be a party to such contracts. Without government intervention, social norms, or "guilt" by parents and children, families without bequests would underinvest in children's human capital. . . .

State intervention in the provision of education and other human capital could raise investments in children to the efficient levels. Since poor parents are least likely to make efficient investments, such intervention would also reduce the inequality in the opportunities between children from richer and poorer families. The compulsory schooling laws in the United States that began in the 1880s and spread rapidly during the subsequent thirty years tended to have this effect. A state usually set minimum requirements at a level that was already exceeded by all but the poorest families in that state. These laws raised the schooling of poor children but did not tend to affect the schooling of other children.

Subsidies to public elementary schools in the United States also began to grow in the latter half of the nineteenth century, and subsidies to public high schools expanded rapidly during the twentieth century. These subsidies appear to have raised the schooling of poorer families relative to richer ones, for the effect of parental wealth and education on the education of children declined over time as public expenditures on schooling grew.

Strong altruism of parents contributes to efficient investments in children by raising the likelihood that parents give gifts or bequests to adult children. Strong altruism may reduce efficiency in other ways, however, if children recognize that they will be rescued by parents when they get into trouble. For example, children who do not receive gifts now but expect gifts in the future from altruistic parents will save less and borrow more to increase their current consumption and reduce their future resources since altruistic parents tend to increase their gifts when children are poorer. Similarly, children may have fun in school and neglect their studies if they expect greater future support from their parents when their earnings are lower. Or children who receive gifts from altruistic parents may take big risks because they expect large gifts if they fail and yet can keep most of their gains if they succeed since gifts cannot be negative.

Parents will not give children such perverse incentives if they can precommit the amount of future gifts and bequests. With precommitment, children cannot rely on parents to bail them out of bad gambles or other difficulties. Precommitment is unnecessary if parental altruism declines enough when they believe that children caused their own difficulties by gambling excessively, neglecting their studies, and so on. . . . Parents may choose not to precommit, however, . . . because they want to help children who get into difficulties through no fault of their own.

When precommitment is either not feasible or not desirable, parents may take other actions to give children better incentives in the future. They would overinvest in education and other training if children cannot run down human capital as readily as marketable wealth. They would also invest more in other illiquid assets of children. such as their housing.

Public policies can also discourage children from inefficient actions. Many countries require parental approval when children want to marry early, drop out of school, get an abortion, or purchase alcoholic beverages. Presumably, one reason is to prevent children who do not anticipate delayed consequences from taking actions that will make them worse off in the future. Another reason, however, is that children may anticipate all too well the future help they will receive from parents if they get into trouble. The state then tries to reproduce the effects on children's behavior of an optimal degree of commitment by parents.

Social Security and Other Old-Age Support

Throughout history, children have been a major help to elderly parents. The elderly frequently have lived with children who care for them when ill and provide food and other support. In the United States a mere thirty years ago, only about 25 percent of persons over age sixty-five lived alone.

Richer families who leave bequests rely less on children because they are insulated from many risks of old age. For example, parents who live longer than expected can reduce bequests to finance consumption in the additional years. The opportunity to draw on bequests provides an annuity-like protection against an unusually long life and other risks of old age. If bequests are not a large part of children's assets, elderly parents get excellent protection against various hazards through the opportunity to reduce bequests, and yet this does not have much influence on children's welfare. In effect, children would help support their parents in old age, although their support is not fully voluntary.

Children in poorer and many middle-level families would be willing to help support parents who agree to invest the efficient amount in the children's human capital. Few societies have contracts or other explicit agreements between parents and children, but many societies have social "norms" that pressure children to support elderly parents. Although little is known about how norms emerge, it is plausible that norms are weaker in modern societies with anonymous cities and mobile populations. Public expenditures on the elderly together with public expenditures on children's education and other human capital can fill the void left by the breakdown in norms.

Expenditures on the elderly in Western countries have grown rapidly in recent decades. United States governments now spend more than $8,000 on each person aged sixty-five or over, largely in the form of medical and pension payments. Is the rapid growth in expenditures on the elderly mainly due to the political power of a growing elderly population? The media contains much discussion of generations fighting for a limited public purse. Some economists support a balanced budget amendment to prevent present generations from heavy taxation of children and other future generations. . . .

We would like to suggest the alternative interpretation that expenditures on the elderly are part of a "social compact" between generations. Taxes on adults help finance efficient investments in children. In return, adults receive public pensions and medical payments when old. This compact tries to achieve for poorer and middle-level families what richer families tend to achieve without government help; namely, efficient levels of investments in children and support to elderly parents. . . .

Divorce

Practically all societies forbid marriage prior to specified ages; many countries have banned marriages between men and women of different races, religions, and social classes; and Christian countries have not allowed polygamy. Regulation of divorce is equally common. The United States and other Western countries essentially did not allow divorce until the mid-nineteenth century. There were fewer than two (!) divorces per year in England from 1800 to 1850. Gradually, divorce laws in the West liberalized toward allowing divorce when one party committed adultery, abandoned his or her spouse, or otherwise was seriously "at fault." Divorce by mutual consent also began to be possible, especially when there were no young children. About twenty years ago, the United States and other countries started to allow either spouse to divorce without proving fault or getting consent.

Although some divorces badly sear the children involved, little is known about the usual effects of divorce on children. Among other things, the available evidence cannot distinguish the effect of a divorce from the effect of having parents who do not get along. All altruistic parents consider the interests of children and are less likely to divorce when their children would be hurt badly. Nevertheless, even if we ignore the conflict between divorced parents in determining how much time and money each spends on their children, altruistic parents might still divorce when their children are harmed. Parents who do not leave bequests might divorce even when the money value of the cost to children exceeds the money value of the gain to parents. The reason is that children do not have a credible way to "bribe" their parents to stay if they cannot commit to old-age support or other future transfers to parents contingent on the parents not getting a divorce. . . .

On the other hand, children may suffer from a divorce even by parents who give bequests if the divorce reduces the nontradable goods consumed by children. For example, children may be unhappy after a divorce because they seldom see their fathers. Parents cannot directly compensate children for the effect of a divorce on their happiness or other consumption. Indeed, if the effect on nontradables lowers the marginal utility to children of tradable resources, altruistic parents who divorce would reduce their gifts of tradables to children and thereby make children still worse off. . . .

We claimed earlier that the degree of altruism is not fixed but often responds to the frequency and intensity of contacts with beneficiaries. In particular, over time a divorced father might become less altruistic toward his children as his contact with them declines. This would explain why many divorced fathers are delinquent in child-support payments, and it strengthens our conclusion that a divorce may make children worse off even when their parents are quite altruistic prior to a divorce and even if they continue to give bequests after a divorce.

A divorce may greatly harm a wife who has many children and cannot earn much in the labor force or when her ex-husband fails to meet his financial and other obligations to the children. This is true even when divorce requires mutual consent because in many societies husbands could intimidate wives into agreeing to a divorce under unfavorable terms for them.

It does not seem farfetched to suggest that the state often regulates divorce to mimic the terms of contracts between husbands and wives and parents and children that are not feasible. Such contracts, for example, might greatly reduce the incidence of divorce when families have many children since the aggregate loss to children (and mothers) from divorce would rise with the number of children. Many countries did prohibit divorce when the typical family was large. Moreover, even when a divorce could not be easily obtained, marriages without children often could dissolve—could be "annulled." Divorce laws eased as birth rates began to decline in the nineteenth century. In recent decades, low birth rates and the much higher labor force participation of women stimulated a further easing toward no-fault divorce. . . .

Optimal Population

With a heroic amount of additional imagination, we can consider not only the relation between parents and actual children but also contracts between parents and *potential* children. Such a thought experiment provides a new way of determining optimal family size and optimal population. The literature on optimal population has lacked an attractive guiding principle.

Suppose that a potential child could commit to compensating his parents eventually if he is born. This "contract" would be Pareto improving (we assume that third parties are not hurt by births) if the child would still prefer to be born after compensation to parents that makes them better off. Since such contracts are impossible, some children may not get born even when both parents and children could be better off. Both fertility and population growth are too low when compensation from unborn children to their parents would be Pareto improving. . . .

The seemingly bizarre thought experiment with unborn children has a very concrete implication. We have shown that poorer families are less likely than richer ones to leave bequests. If commitments for compensation from unborn children are not feasible, fertility in poorer families is too low, and fertility in richer families (who give bequests) is optimal. Therefore, our approach implies—with any third-party effects ignored—that the aggregate private-fertility rate is below the Pareto-efficient rate.

A conclusion that poorer families may have too few children will shock some readers because poorer families already have larger families than richer ones. But other factors raise fertility by poorer families, including welfare programs, subsidies to education, and limited birth control knowledge. . . .

Political Competition between Generations

Since public policy results from competition among interest groups, how does competition for political favors lead to efficiency-raising state interventions in the family? In this section we sketch out a possible answer when parental altruism is important.

Political competition between adults and children is hardly a contest since children cannot vote and do not have the means and maturity to organize an effective political coalition. If adults use their political power to issue bonds and other obligations, they can help support themselves when old by selling these obligations to the next generation of younger adults. Some economists support balanced government budgets and limits on debt issue to control such exploitation of the political weakness of children and later generations. Of course, this is not a problem if each generation can repudiate debt issues by previous generations. Since the issues involved in debt repudiation are beyond the scope of this article, we will just assume that debt is not repudiated.

Although present generations may be able to exploit future generations, altruism limits their desire to do so. Indeed, if all parents are altruistic and leave bequests, present generations have no desire to exploit future generations. After all, if they want to, they may take resources from future generations by leaving smaller bequests. Although families who do not leave bequests favor debt and other exploitation of the political weakness of future generations, their degree of altruism may greatly affect how they use their political power against future generations.

We showed [above] that families who do not leave bequests underinvest in the human capital of their children. They can increase the wealth of the children's generation by using their political power to raise education and other training through state schools and subsidies to other investments in children. Then the present generation may, if it wishes, issue obligations to future generations that extract this increase in children's wealth.

Although selfish parents try to extract as much as they can from children, altruistic parents may prefer to share some of the increased wealth with children. This means that future generations may also benefit from the political power of present generations. Therefore, even if the altruism of many parents is not strong enough to lead to positive bequests and efficient investments in human capital, it could be strong enough to ensure that future generations also gain when the present generation uses its political power to issue debt and other obligations to future generations.

This overly simplified analysis of political power and political incentives may help explain why public expenditures in the United States on children are not small compared to public expenditures on the elderly. The discussion [above] indicates that the next generation gains enough from public expenditures on children by the current generation to pay social security and other help to the elderly of the current generation, and yet the next generation still has some profit left over from the public investment in their human capital.

Summary

We have tried to understand the widespread intervention by governments in families. We conclude that many public actions achieve more efficient arrangements between parents and children. Clearly, parents and children cannot always make efficient arrangements because children are unable to commit to compensation of parents in the future.

Families who leave bequests can "force" children to repay parents for investments in human capital by reducing bequests. Therefore, these families do not underinvest in children's human capital. By contrast, families who do not leave bequests, often poorer families, do underinvest in children. The state may subsidize schools and other training facilities to raise investments in children by poorer families to efficient levels. . . .

The Economics of the Baby Shortage

ELISABETH LANDES AND RICHARD POSNER

Although economists have studied extensively the efforts of government to regulate the economy, public regulation of social and personal life has largely escaped economic attention. With the rapid development of the economic analysis of nonmarket behavior, the conceptual tools necessary for the economic study of social (as distinct from narrowly economic) regulation are now at hand. Nor is there any basis for a presumption that government does a good job of regulating nonmarket behavior; if anything, the negative presumption created by numerous studies of economic regulations should carry over to the nonmarket sphere. An example of nonmarket regulation that may be no less perverse than the widely criticized governmental efforts to regulate imports, transportation, new drugs, bank entry, and other market activities is the regulation of child adoptions—the subject of this paper. . . .

Ordinarily, potential gains from trade are realized by a process of voluntary transacting—by a sale, in other words. Adoptions could in principle be handled through the market and in practice, as we shall see, there is a considerable amount of baby selling. But because public policy is opposed to the sale of babies, such sales as do occur constitute a black market. . . .

[T]his paper develops a model of the supply and demand for babies for adoption under the existing pattern of regulation and shows (1) how that regulation has created a baby shortage (and, as a result, a black market) by preventing a free market from equilibrating the demand for and supply of babies for adoptions and (2) how it has contributed to a glut of unadopted children maintained in foster homes at public expense. . . .

Disequilibrium in the Adoption Market

The Baby Shortage and the Baby Glut

Students of adoption agree on two things. The first is that there is a shortage of white babies for adoption; the second is that there is a glut of black babies, and of children who are no longer babies (particularly if they are physically or mentally handicapped), for adoption. . . .

Contrary to popular impression, [federal government statistics show] that the increased availability of contraception and abortion has not perceptibly diminished the number of illegitimate births. A partial explanation may be that the availability of contraception and abortion, by reducing the risk of producing an unwanted child (but not to zero), has reduced the expected cost and hence increased the incidence of sexual

Elisabeth Landes and Richard Posner, The Economics of the Baby Shortage, 7 *Journal of Legal Studies* (1978), pp. 323–348. Copyright © 1978 The University of Chicago Law School. Reprinted with permission.

intercourse outside of marriage. However, while the illegitimate birth rate remains high the availability of babies for adoption has declined, apparently because a larger proportion of parents of illegitimate children are keeping them. This trend may be due to inexplicable (on economic grounds) changes in moral standards; or it may be due to the fact that the increased opportunities for women in the job market have made them less dependent on the presence of a male in raising a child. An additional feature is that, given the increased availability of contraception and abortion, an illegitimate baby is more likely than formerly to be a desired baby.

Students of adoption cite factors such as the declining proportion of illegitimate children being put up for adoption as the "causes" of the baby shortage. But such factors do not create a shortage, any more than the scarcity of truffles creates a shortage; they merely affect the number of children available for adoption at any price. At a higher price for babies, the incidence of abortion, the reluctance to part with an illegitimate child, and even the incentive to use contraceptives would diminish because the costs of unwanted pregnancy would be lower while the (opportunity) costs to the natural mother of retaining her illegitimate child would rise.

The principal suppliers of babies for adoption are adoption agencies. Restrictive regulations governing nonagency adoption have given agencies a monopoly (though not a complete one) of the supply of children for adoption. However, while agencies charge fees for adoption, usually based on the income of the adoptive parents, they do not charge a market-clearing (let alone a monopoly-profit-maximizing) price. This is shown by the fact that prospective adoptive parents applying to an agency face waiting periods of three to seven years. And the (visible) queue understates the shortage, since by tightening their criteria of eligibility to adopt a child the agencies can shorten the apparent queue without increasing the supply of babies. Thus some demanders in this market must wait for years to obtain a baby, others never obtain one, and still others are discouraged by knowledge of the queue from even trying. Obtaining a second or third baby is increasingly difficult.

The picture is complicated, however, by the availability of independent adoptions. An independent adoption is one that does not go through an agency. Most independent adoptions are by a relative, for example a stepfather, but some involve placement with strangers and here, it would seem, is an opportunity for a true baby market to develop. However, the operation of this market is severely curtailed by a network of restrictions, varying from state to state (a few states forbid independent adoption by a nonrelative) but never so loose as to permit outright sale of a baby for adoption.

Just as a buyer's queue is a symptom of a shortage, a seller's queue is a symptom of a glut. The thousands of children in foster care . . . are comparable to an unsold inventory stored in a warehouse. Child welfare specialists attribute this "oversupply" to such factors as the growing incidence of child abuse, which forces the state to remove children from the custody of their natural parents, and the relatively small number of prospective adoptive parents willing to adopt children of another race, children who are no longer infants, or children who have a physical or mental handicap. No doubt these factors are important. However, some children are placed in foster care as infants and remain there until they are no longer appealing to prospective adoptive parents. We believe that the large number of children in foster care is, in part, a manifestation of a regulatory pattern that (1) combines restrictions on the sale of babies with the effective monopolization of the adoption market by adoptive agencies, and (2) fails to provide effectively for the termination of the natural parents' rights.

A Model of the Adoption Market

. . . Whereas in 1957 only 53 percent of all nonrelative adoptions went through adoption agencies, in 1971 the proportion was almost 80 percent. This would be a matter of limited significance from the economic standpoint if adoption agencies were both numerous and free from significant restrictions on their ability to operate as efficient profit-maximizing firms. The first condition is more or less satisfied but not the second. While agencies are generally not limited in the fees they may charge prospective adoptive parents, they are constrained to other inefficient restrictions. For example, they are constrained to operate as "nonprofit" organizations which presumably retards, perhaps severely, their ability to attract capital, and may have other inefficient effects as well. The most significant restriction is the regulation of the price at which the agencies may transact with the natural parents. Adoption agencies that are also general child-welfare agencies must accept all children offered to them at a regulated price (but may place them in foster care rather than for adoption); and they may offer no additional compensation to suppliers (the natural parents) in order to increase the supply of babies. The regulated price is generally limited to the direct medical costs of pregnant women plus (some) maintenance expenses during the latter part of the pregnancy. To be sure, agencies have some flexibility in the kinds of services they may offer the natural parents, such as job counseling, but they cannot thereby transfer to the natural parents anything approaching the free-market value of the child.

There are rough counterparts to such regulation in many explicit markets. . . . Similar regulatory patterns are found in industries as otherwise diverse as taxi service and television broadcasting. Nevertheless the regulation of adoption has several peculiar characteristics reflected in our model: collusion among agencies, including market division (often along religious lines), is permitted; there exists a very close substitute for the good supplied by the agencies—independent adoption; and the agency has, as mentioned, no power to refuse to take the children tendered to it. . . .

The Effects of the Baby Shortage

The baby shortage generates social costs in excess of the traditional welfare loss of monopoly. The counterpart to that loss would be the lost consumer surplus from sales not made at all because of the artificial unavailability of the product. . . . [But there is a further welfare loss if] the nonprice rationing methods used by agencies to allocate children are random with respect to willingness to pay. . . . To be sure, some of this loss is offset by the availability of children in the independent market, but the search costs in the independent market also represent a substantial social cost. . . .

In these circumstances, the economist expects a black market to emerge. Some fraction—we do not know what—of the 17,000 independent adoptions are indeed black-market adoptions in the sense that the compensation paid either the natural parents or the middlemen, or both, exceeds the lawful limits. However, the potent criminal and professional sanctions for the individuals involved in baby selling not only drive up the costs and hence the price of babies (and so reduce demand) but necessarily imply a clandestine mode of operation. This imposes significant information costs on both buyers and sellers in the market, which further raise the (real) price of black-market babies to buyers and reduce the net price to sellers. . . .

A further consideration is that there will be more fraud in a black market for babies than in a lawful market, so fear of being defrauded will further deter potential demanders. In lawful markets the incidence of fraud is limited not only by the existence of legal remedies against the seller but also by his desire to build a reputation for fair dealing. Both the clandestine mode of operation of current baby sellers and the lack of a continuing business relationship between seller and buyer reduce the seller's market incentives to behave reputably. To summarize, we cannot, simply by observing the black market, estimate the market-clearing prices and quantities of babies in a lawful baby market.

The constraints on the baby market may also be responsible in part for the glut of children in foster care—and this quite apart from the possible incentives of adoption agencies to place children in foster care rather than for adoption. Since the natural parents have no financial incentive to place a child for adoption, often they will decide to place it in foster care instead. This is proper so long as they seriously intend to reacquire custody of the child at some later date. But when they do not the consequence of their decision to place the child in foster care may be to render the child unadoptable, for by the time the parents relinquish their parental rights the child may be too old to be placed for adoption. This would happen less often if parents had a financial incentive to relinquish their rights at a time when the child was still young enough to be adoptable.

The total effect of the baby-market constraints on the number of foster children is, to be sure, a complicated question. In particular, the limited supply of desirable babies for adoption may lead some prospective adoptive parents to substitute children who would otherwise be placed in foster care. We suspect that this substitution effect is small, but in any event it is partly controlled by the agencies; they can manipulate the relative "prices" of infants and children residing in foster care by modifying the criteria for eligibility that must be satisfied by prospective adoptive parents.

Objections to a Free Baby Market

The foregoing analysis suggests that the baby shortage and black market are the result of legal restrictions that prevent the market from operating freely in the sale of babies as of other goods. This suggests as a possible reform simply eliminating these restrictions. However, many people believe that a free market in babies would be undesirable. . . . The objections to baby selling must be considered carefully before any conclusion with regard to the desirability of changing the law can be reached.

Criticisms Properly Limited to the Black Market

We begin with a set of criticisms that in reality are applicable not to the market as such, but only, we believe, to the black market. The first such criticism is of the high price of babies and the bad effects that are alleged to flow from a high price, such as favoring the wealthy. This criticism of the use of the price system is based on the current prices in the black market. There is no reason to believe that prices would be so high were the sale of babies legalized. On the contrary, prices for children of equivalent quality would be much lower.

The current black-market price is swollen by expected punishment costs which would not be a feature of a legalized baby market. In a legal and competitive baby market, price would be equated to the marginal costs of producing and selling for adoption babies of a given quality. These marginal costs include certain well-known items, such as the natural mother's medical expenses and maintenance during pregnancy and the attorney's fee for handling the legal details of the adoption proceeding, that are unlikely to exceed $3,000 in the aggregate. The question marks are the additional fees that would be necessary (1) to compensate a woman either for becoming pregnant or, if she was pregnant already, for inducing her to put the baby up for adoption rather than abort or retain it, and (2) to cover the search costs necessary to match baby and adoptive parents.

With regard to the first item (the natural mother's opportunity costs of adoption), the most important point to be noted is that these costs may be no greater than the cost savings to the adoptive mother of not undergoing pregnancy and childbirth herself. Adoption is a process by which the adoptive mother in effect contracts out one of the steps in the process of child production and rearing, namely the actual pregnancy and childbirth. The anxieties and inconveniences of pregnancy are a cost to the biological mother but a cost saving to the adoptive mother. Equally, all or most of the out-of-pocket expenses of the natural mother, including the obstetrician's fee, represent a cost saving to the adoptive mother. Therefore, at least as a first approximation, the only net cost of purchasing a baby in a free and competitive market should be the cost of the search, which would presumably be low.

Also, because the adoption agencies give substantial emphasis to the employment and financial situation of adoptive parents, a baby market might actually provide more opportunities for the poor to adopt than nonprice rationing does. If we are correct that the (acquisition) costs of babies in a lawful and competitive market would often be small, perhaps no more than the cost of an automobile, low-income families who would normally be considered financially ineligible by adoption agencies would be able in a free market to obtain a child.

Another prevalent criticism of the market, and again one that pertains primarily to the operations of the black market, is that fraud and related forms of dishonesty and overreaching pervade the market method of providing children for adoption. It is contended, for example, that the health of the child or of the child's mother is regularly misrepresented and that frequently after the sale is completed the seller will attempt to blackmail the adoptive parents. Such abuses are probably largely the result of the fact that the market is an illegal one. Sellers cannot give legally enforceable guarantees of genealogy, health, or anything else to the prospective parents, and even the seller's adherence to the negotiated price is uncertain given the buyer's inability to enforce the contract of sale by the usual legal procedures. Any market involving a complex and durable good (i.e., one that yields services over a substantial period of time) would probably operate suboptimally in the absence of legally enforceable contracts or, at a minimum, regular, repetitive business relations between (the same) sellers and (the same) buyers. Both conditions are absent from the illegal baby market and this is the likeliest explanation for the number of complaints about the honesty of the sellers in that market. . . .

Criticisms of a Legal Market

We now consider criticisms of baby selling that are applicable to a legal market rather than just to the present illegal market. The first is that the rationing of the supply of babies to would-be adoptive parents by price is not calculated to promote the best interests of the children, the objective of the adoption process. This criticism cannot be dismissed as foolish. The ordinary presumption of free-enterprise economics is no stronger than that free exchange will maximize the satisfaction of the people trading, who in this case are the natural and adoptive parents. There is no presumption that the satisfactions of the thing traded, in most instances a meaningless concept, are also maximized. If we treat the child as a member of the community whose aggregate welfare we are interested in maximizing, there is no justification for ignoring how the child's satisfactions may be affected by alternative methods of adoption.

Very simply, the question is whether the price system would do as good a job as, or a better job than, adoption agencies in finding homes for children that would maximize their satisfactions in life. While there is no direct evidence on this point, some weak indirect evidence is provided in a follow-up study of independent adoptions which suggest that children adopted privately do as well as natural children. . . . It is true that some, perhaps most, independent adoptions do not involve price rationing, but the most important thing is that independent adoption involves a minimum of the sort of screening of prospective parents that the adoption agencies do. If children adopted without the screening seem nevertheless to do about as well as natural children, then one is entitled to be skeptical of the need for or value of the screening.

This conclusion is reinforced by the way in which adoption agencies screen. Agencies attempt to allocate children only to "fit" or caring parents. But after determining the pool of fit, or eligible-to-adopt, couples, they allocate available children among them on a first-come, first-served basis. The "fittest" parents are not placed at the head of the queue.

Further, and perhaps most important, agencies have no real information on the needs of a particular child they place for adoption beyond its need for love, warmth, food, and shelter. One cannot read from the face of a newborn whether he or she will be of above or below normal intelligence, or be naturally athletic, musical, or artistic. Hence agencies cannot be presumed to match these very real, if inaccessible, qualities of infants with the qualities of the adoptive parents any more effectively than a market would.

One valuable function agencies may perform is screening out people whose interest in having children is improper in an uncontroversial sense—people who wish to have children in order to abuse or make slaves of them. The criminal statutes punishing child abuse and neglect would remain applicable to babies adopted in a free market, but the extreme difficulty of detecting such crimes makes it unlikely, at least given current levels of punishment, that the criminal statutes alone are adequate. This may make some prescreening a more effective method of prevention than after-the-fact punishment. But the logical approach, then, is to require every prospective baby buyer to undergo some minimal background investigation. This approach would be analogous to licensing automobile drivers and seems as superior to the agency monopoly as licensing is to allocating automobiles on a nonprice basis.

Moreover, concern with child abuse should not be allowed to obscure the fact that abuse is not the normal motive for adopting a child. And once we put abuse aside, willingness to pay money for a baby would seem on the whole a reassuring factor from the standpoint of child welfare. Few people buy a car or a television set in order to smash it. In general, the more costly a purchase, the more care the purchaser will lavish on it. Recent studies suggest that the more costly it is for parents to obtain a child, the greater will be their investment in the child's quality attributes, such as health and education.

A further point is that today some fetuses are probably aborted because the cost to the mother of carrying them to term and placing them for adoption exceeds the permissible return. In a free adoption market, some of the 900,000 fetuses aborted in 1974 would have been born and placed for adoption. If the welfare of these (potential) children is included in the calculation of the welfare of adopted children, both actual and potential, the heavy costs imposed on the market by adoption regulation may actually decrease child welfare.

Another objection to the market for babies is the alleged vulnerability of both natural and adoptive parents to overreaching by middlemen. Parenthood is thought to be so emotional a phenomenon that people cannot reason about it in the same way they reason about the goods and services normally traded in the market. But many of those goods and services, such as medical care, also involve a strong emotional component, yet it has rarely been thought appropriate to exclude such goods from market exchange. And studies of marriage and procreation have shown that people in fact calculate in family matters, whether implicitly or explicitly, in the same way they do when purchasing ordinary goods and services.

Other objections to legalizing the market in babies are more symbolic than pragmatic. For example, to accord a property right in the newborn child to the natural parents seems to some observers to smack of slavery. But allowing a market in adoptions does not entail giving property rights to natural parents for all purposes. Laws forbidding child abuse and neglect would continue to be fully applicable to adoptive parents even if baby sales were permitted. Further, we are speaking only of sales of newborn infants, and do not suggest that parents should have a right to sell older children. The creation of such a right would require identification of the point at which the child is sufficiently mature to be entitled to a voice in his placement. However, the question is largely academic given the lack of any significant market for adopting older children.

Moreover, it is incorrect to equate the possession of property rights with the abuse of the property, even if the property is a human being. For example, a serious problem with foster care is the foster parents' lack of any property rights in the foster child. The better the job the foster parents do in raising the child, the more likely are the natural parents to reclaim the child and thereby prevent the foster parents from reaping the full fruits of their (emotional as well as financial) investment. This possibility in turn reduces the incentive of foster parents to invest in foster children, to the detriment of those children's welfare.

The antipathy to an explicit market in babies may be part of a broader wish to disguise facts that might be acutely uncomfortable if widely known. Were baby prices quoted as prices of soybean futures are quoted, a racial ranking of these prices would be evident, with white baby prices higher than nonwhite baby prices. One is reminded of Professor Tribe's objection to instructing the jury on the numerical probability implicit in the concept of proof beyond a reasonable doubt. He argues that while the system of criminal justice would be unworkable if subjective certainty of guilt were required, to acknowledge explicitly that people are convicted on less than such certainty might tear the social fabric. Similarly, anyone who thinks about the question will realize that prices for babies are racially stratified as a result of different supply and demand conditions in the different racial groups, but perhaps bringing this fact out into the open would exacerbate racial tensions in our society.

Some people are also upset by the implications for the eugenic alteration of the human race that are presented by baby selling. Baby selling may seem logically and inevitably to lead to baby breeding, for any market will generate incentives to improve the product as well as to optimize the price and quantity of the current quality level of the product. In a regime of free baby production and sale there might be efforts to breed children having desirable characteristics and, more broadly, to breed children with a known set of characteristics that could be matched up with those desired by prospective adoptive parents. Indeed, one can imagine, though with some difficulty, a growing separation between the production and rearing of children. No longer would a woman who wanted a child but who had a genetic trait that might jeopardize the child's health have to take her chances on a natural birth. She could find a very close genetic match—up to her and her husband's (healthy) genetic endowment in the baby market. However, so long as the market for eugenically bred babies did not extend beyond infertile couples and those with serious genetic disorders, the impact of a free baby market on the genetic composition and distribution of the human race at large would be small.

The emphasis placed by critics on the social costs of a free market in babies blurs what would probably be the greatest long-run effect of legalizing the baby market: inducing women who have unintentionally become pregnant to put up the child for adoption rather than raise it themselves or have an abortion. Some of the moral outrage directed against the idea of "trafficking" in babies bespeaks a failure to consider the implications of contemporary moral standards. At a time when illegitimacy was heavily stigmatized and abortion was illegal, to permit the sale of babies would have opened a breach in an otherwise solid wall of social disapproval of procreative activity outside of marriage. At the same time, the stigma of illegitimacy, coupled with the illegality of abortion, assured a reasonable flow of babies to the adoption market. Now that the stigma has diminished and abortion has become a constitutional right, not only has the flow of babies to the (lawful) adoption market contracted but the practical alternatives to selling an unwanted baby have increasingly become either to retain it and raise it as an illegitimate child, ordinarily with no father present, or to have an abortion. What social purposes are served by encouraging these alternatives to baby sale?

The symbolic objections to baby sale must also be compared with the substantial costs that the present system imposes on childless couples, aborted fetuses (if they can be said to incur costs), and children who end up in foster care. In particular, many childless couples undergo extensive, costly, and often futile methods of fertility treatment in order to increase their chances of bearing a child. Some people produce unhealthy offspring (due to various genetic disorders) because of their strong desire to have children. And no doubt many people settle for childlessness because of the difficulties of obtaining an adopted child.

The Sources of Opposition to Baby Selling

Even though the benefits of free baby selling might well outweigh the costs, still it will come as no surprise to students of government regulation to find that there are well-organized interests opposed to an improvement in social welfare. The most vocal and organized opponents of the baby market are the adoption agencies. This is logical: . . . both the supply of babies to agencies and agency revenues from adoption would be greater if the private market were regulated out of existence. Assuming that agencies would have no cost or efficiency advantage over private firms in an unregulated market, they would be reduced to operating at the competitive margin if such a market were permitted. They might even be competed out of the market.

To be sure, adoption agencies are generally not specialized in adoptions but engage in a variety of child welfare services—the primary one being foster care. Children placed in foster care are maintained at agency expense, although some fraction of the maintenance expenditures may be offset by government reimbursement. Today some 350,000 children are in foster care at an annual expense to the U.S. government alone of some $700 million. Clearly, healthy infants and older, perhaps less healthy, children are substitutes in adoption, albeit imperfect substitutes. By obtaining exclusive control over the supply of both "first-quality" adoptive children and "second-quality" children residing in foster care but available for adoption, agencies are able to internalize the substitution possibilities between them. Agencies can charge a higher price for the children they place for adoption, thus increasing not only their revenues from adoption but also the demand for children who would otherwise be placed or remain in foster care at the agency's expense. Conversely, if agency revenues derive primarily from foster care, the agencies can manipulate the relative price of adopting "first-quality" children over "second-quality" children to reduce the net flow of children out of foster care.

The group that has the largest stake in the adoption agencies' net revenues is their professional personnel. If the principal effect of eliminating the agency monopoly in adoptions was to force agencies to operate at the competitive margin, it would surely reduce any rents now being received by agency personnel. Nor can it be argued that if baby selling were legalized the agency personnel would simply become the middlemen of the legal market; if the Securities and Exchange Commission were abolished, few of its personnel would become stockbrokers. One is not surprised that professional social workers' organizations have been strong proponents of governmental restrictions on nonagency adoptions.

Potentially allied to the agencies and the social welfare professionals who staff them in opposition to baby selling are those prospective adoptive parents who by virtue of their contacts and general sophistication are able to jump to the head of the queue or procure a baby easily in the (lawful) independent market, either way paying less than they would have to pay in a free market. The analogy is to the effect of usury laws in reducing the interest rate paid by the most credit-worthy borrowers.

The potential supporters of baby selling are difficult to organize in an effective political coalition. They consist of unborn babies, children in foster care, taxpayers (each only trivially burdened by the costs of foster care), and people who have only a low probability of ever wanting to adopt a baby, as well as couples currently wanting to adopt one. The members of this last group have the most concentrated interest in a free baby market, but they are relatively few and widely scattered at any given time.

Interim Steps toward a Full-fledged Baby Market

We close by speculating briefly on the possibility of taking some tentative and reversible steps toward a free baby market in order to determine experimentally the social costs and benefits of using the market in this area. Important characteristics of a market could be simulated if one or more adoption agencies, which typically already vary their fees for adoption according to the income of the prospective parents, would simply use the surplus income generated by the higher fees to make side payments to pregnant women contemplating abortion to induce them instead to have the child and put it up for adoption.

This experiment would yield evidence with respect to both the demand and supply conditions in the adoption market and would provide information both on the value that prospective adoptive parents attach to being able to obtain a baby and on the price necessary to induce pregnant women to substitute birth for abortion. Follow-up studies of the adopted children, comparing them with children that had been adopted by parents paying lower fees, would help answer the question whether the payment of a stiff fee has adverse consequences on the welfare of the child.

Some states appear not to limit the fees that adoption agencies pay to natural parents. The experiment we propose could be implemented in such states without new legislation.

The Role of Private Ordering in Family Law:
A Law and Economics Perspective
MICHAEL TREBILCOCK AND ROSEMIN KESHVANI

. . . [O]ver the last several decades family law has evolved quite dramatically in the direction of permitting spouses, on marriage breakup, to settle between themselves, by agreement, most aspects of the termination arrangements, and to treat such agreements as legally binding and enforceable and not to be lightly upset subsequently by the courts. Broader trends towards no-fault divorce, a diminished role for the courts in resolving marital disputes, and an enhanced role for mediation and conciliation all recognize an enlarged role for private ordering in domestic relations.

With respect to separation agreements, it is not difficult to identify the major advantages of private ordering in this context. First, given the infinitely varied nature of marital relations and individual preferences regarding such relations, private ordering holds out the potential for parties on marriage breakdown to fashion an agreement as to its consequences that best suits their particular preferences and life plans. . . . Second, while the separation agreement marks the end of a marital relationship, it often does not mark the end of the rights and obligations that flow between the parties with respect to matters such as support payments and custody and visitation arrangements. Because these arrangements require some minimum ongoing cooperation between the parties, it seems more likely that durable arrangements will emerge from an agreed resolution of the issues than one imposed by a third party (a court), where one party will often perceive himself or herself as the loser who is coerced into accepting obligations to the advantage of the other party, which the first party might not otherwise have found acceptable. . . . Third, negotiated separation agreements conserve on various kinds of costs that may be loosely grouped under the rubric of "transaction costs." These include the risks, uncertainties, delays, financial costs, invasive public scrutiny, and emotional trauma of litigation. Fourth, separation agreements, if binding and enforceable, provide a degree of finality to the resolution of the consequences of marriage breakdown that encourages the parties to look to the future, to build new relationships, and to undertake new commitments with some confidence in the security of prior arrangements.

With such powerful advantages favouring private ordering in the context of marital breakup, what is to be said on the other side of the case? Obviously, it would be indefensible to adopt the position that all separation agreements should be enforced willy-nilly. The law of contracts at large recognizes a rich array of defences and excuses with respect to contract enforcement. Moreover, even economists who strongly favour the contractual paradigm recognize that issues such as coercion, imperfect information, and externalities may generate contracting failures, in the sense of not yielding Pareto-superior bargains. Thus, from a law and economics perspective, the challenge is to identify with some precision those features of bargaining over the consequences of marital breakdown that might rebut a presumption of enforceability. We believe that it is useful to think of these potential sources of contracting failure as falling into three broad categories: (1) inadequacy of background legal entitlements (or default rules); (2) transaction-specific failures (including externalities); and (3) post-agreement contingencies. We will address each in turn.

Inadequacy of Background Legal Entitlements

. . .

Two sources of inadequacy in background entitlements in the matrimonial context can readily be identified. First, to the extent that background entitlements are uncertain (custody and visitation rights or support obligations, for example), this uncertainty may have either of two undesirable effects. Either it may generate widely divergent expectations of likely legal outcomes on the part of the affected parties in the event of litigation, which may make it difficult for them to negotiate a mutually acceptable settlement, thus leading to litigation, with all the transaction costs entailed that have been previously noted. Or, alternatively, if one of the affected parties is significantly more risk averse than the other, the uncertainty associated with the possible legal outcomes may induce that party to enter into a settlement that is disadvantageous to him or her relative to a neutral assessment of the likely legal outcomes. A second and much more substantial concern about the inadequacy of background entitlements is that while the rules may be clear and certain, they do not represent the entitlements that rational parties would have chosen ex ante (that is, at the time of marriage) had they negotiated a fully specified contingent claims contract, so that ex post separation negotiations that occur in the shadow of these inappropriate entitlements may well yield sub-optimal outcomes. This issue is sufficiently important to warrant a somewhat extended comment.

While family law reforms over the last two decades have moved strongly in the direction of equal division of family assets, that is, the value of all assets acquired during the marriage and any interim increase in the value of premarital assets, the principal focus of these provisions is on tangible assets, such as savings, homes, farms, businesses, now even pension entitlements and sometimes educational degrees. However, current laws much less satisfactorily address a more general class of family assets: human capital (of which educational degrees are a small component). As our economy becomes more service-oriented, increasingly major family assets comprise the earning capacity of one or other spouse embodied in the human capital that this spouse has accumulated through formal training and on-the-job experience. . . . While courts in many jurisdictions have begun to grapple with the problem of valuing professional degrees acquired by one spouse as a result of financial and other support by the other spouse, there is a high degree of inconsistency and uncertainty associated with current judicial policy on valuation and division of these forms of intangible assets. . . .

We believe that the impact of marriage on the value of human capital is one of the major unresolved issues on the family law reform agenda. Moreover, it is an issue where law and economics scholars and feminist scholars can make common cause. . . .

Feminist concerns generally in this area reflect a good deal of ambivalence, indeed moral anguish, over the "dilemma of difference." On the one hand, the concern is that recognizing the state of economic dependency in which wives in more traditional marriage relationships find themselves, through more generous property division or support rules, risks legitimizing and reinforcing the subordinate roles of women in our society. On the other hand, failure to recognize the differences that actually exist in the current roles of men and women in many relationships risks discriminating against women who find themselves in this state of dependency. . . .

A law and economics perspective on the merits of economically dependent relationships would be normatively somewhat more agnostic than at least some feminist perspectives. The premise would be that individuals have a myriad of different utility functions and preference sets in the real world, and if a particular couple find a particular kind of relationship mutually congenial and beneficial, that choice should be respected. Moreover, to the extent that there is concern that preferences in many of these kinds of relationships have been distorted by socialization processes excessively dominated or influenced by men, law and economics scholars might be inclined to question whether correcting these biases of the socialization process is a goal that can be meaningfully advanced by the choice of legal regime to govern the consequences of marriage breakdown. In other words, serious responses to this problem may lead us into quite different policy fields (such as public education policy).

However, that said, we believe that a law and economics perspective has useful contributions to offer in the marital breakdown context in the valuation of non-market contributions to a spousal relationship. One approach to the valuation issue, espoused by Duclos, is to focus on the opportunity costs incurred by a woman in an economically dependent relationship in forgone career opportunities that translate into reduced earning capacity on marriage breakdown relative to the earning capacity the woman would probably have enjoyed had her career not been interrupted by marriage, and attempt to compensate for this reduced earning capacity in the future. This focus on the opportunity costs of marriage is consistent with the approach advocated by law and economics scholars, for example, in valuing the loss of a housewife's time in fatal accident litigation. . . .

To restate the opportunity-cost approach slightly more formally. . . , if the wife is to be fully compensated for her investment in household production, she must be compensated, upon divorce, an amount that equals the following: the present value of the reduction in earnings she has experienced up to the date of the divorce because of her investments in household production (both because she has not worked or has accepted jobs with below opportunity-cost levels of earnings or because she has already suffered a loss in her full-time earnings because of her reduced investment in her own human capital) plus the present value of her future loss of earnings resulting from her reduced investments in human capital minus any gain in consumption benefits that she has obtained, up to the date of the divorce, from the incremental income earned by her husband (because she has freed his time to concentrate on his career).

But even this measure of compensation for investment in household production may be less than fully compensatory. In many economically dependent relationships, the wife, by investing in household production in the early periods of the relationship, may reasonably expect to realize returns on these investments in later periods of the relationship when the husband's income will provide economic security for herself and the family. . . .

In designing a default rule that would apply unless the parties have validly contracted away from it by prenuptial or separation agreement . . . , it may be useful conceptually to construct a hypothetical contract at the time that marriage is entered into, when the parties decide on a specialization of functions involving one devoting his time to market production and the other to household production. If we assume that the wife is more risk averse than the husband, given the differential impact of divorce on the economic welfare of men and women reflected in the data cited above, the wife may rationally demand an "insurance policy" from the husband to the effect that if there is no marriage dissolution the wife will share in the economic returns to the husband over her life, but if there is marriage dissolution she should be no worse off than had she not married, which would seem to require a focus on the opportunity costs to her of the marriage. . . . One vice of the opportunity-cost approach is, of course, that it will reflect enduring differentials, in opportunities in the labour market for men and women. For example, a waitress who marries at the age of nineteen may conceivably have suffered no opportunity costs during marriage or after divorce, but to the extent that her husband enjoys prospects of enhanced earning capacity in the future, she will be denied the benefits of this. Here a difficult issue must be confronted. In constructing the hypothetical contract at the time of marriage, would the two parties agree that the wife should share in the husband's economic returns both where there is no marriage dissolution and where there is? Pursuing the insurance analogy, no insurer would write such a policy because the wife may well be rendered largely indifferent to sustaining or terminating the marriage, given the assumption that her entitlements on divorce are not contingent on proof of absence of fault on her part for the marriage dissolution. This is the standard problem of moral hazard in insurance markets, or in the present context reciprocal opportunism. Such a policy would be analogous to a fire insurance policy on a widget factory that provided coverage not only for the replacement costs of the factory in the event of fire, but also the present value of all expected future widget sales. With such coverage, incentives on the part of the insured to preserve and protect the insured asset are highly attenuated. . . .

We realize that our approach risks resurrecting the "partnership for life" conception of marriage, which the recent thrust of family law reform measures has sought to exorcise and which most feminists, recognizing one horn of the dilemma of difference, also generally reject. The alternative and more powerful argument, however, is that by making economically dependent relationships much more expensive for men in the event of their termination (and more reflective of actual investments), one may discourage many such relationships where these are driven by considerations of exploitation or subordination rather than by genuine considerations of mutual advantage. We also do not seek to minimize the methodological problems entailed in developing a rigorous and operational valuation methodology for compensating for investments in non-market time, even if this approach has merit. However, these difficult valuation problems confront the courts daily in other contexts, such as personal injury and fatal accident litigation, and we are confident that if the legal system ordained a requirement of compensation, reasonably robust valuation techniques would in due course emerge. . . .

Transaction-specific Failures

Assuming that the problem of appropriate background entitlements has been satisfactorily resolved, the question then arises as to whether further deficiencies might be observed in a private ordering regime. Clearly, as with any other form of contracting, contracting failures can occur. Mnookin identifies a range of such failures in the separation agreement context. First, there may be problems of transactional incapacity. At the time of marriage breakup, either or both spouses may be emotionally destabilized and incapable of rational assessment of the implications of the proposed separation agreement for their future welfare. Second, problems of strategic behaviour may arise, where one or [the] other party lies to the other about either his or her future intentions, or about the current status of financial assets, or, alternatively, suppresses other highly material information. In addition, one party may trade on the other's risk aversion to litigation, or the latter's inability to absorb the financial costs and emotional trauma associated with litigation, to extract an unfair bargain. Third, there is the problem of externalities. Specifically, in a matrimonial context, this is likely to involve the future support and care of children. Obviously, there is the potential in a separation agreement for the immediate parties thereto to weight inappropriately the welfare of these unrepresented interests. However, Mnookin persuasively argues that custody issues are in general an appropriate subject for mutual bargaining, and are more likely to be satisfactorily resolved by the parties than the courts, given high degrees of indeterminacy in the social science literature on what the best interests of the child require, subject to a judicial override to ensure that private agreements not entail custody, support, or visitation arrangements that violate the established minimum standards for protecting children from neglect and abuse that the state imposes on all families (an unfitness test).

Unlike the problem of the inadequacy of background entitlements, most of these transaction-specific failures seem reasonably well able to be accommodated under a variety of contract doctrines pertaining to defences and excuses, such as fraud, misrepresentation, material nondisclosure, inequality of bargaining power, and unconscionability. . . . In addition to ex post judicial review of the terms of these transactions, one might also impose some ex ante constraints on the potential for unfairness, by, for example, requiring that both spouses obtain independent legal advice before signing an enforceable agreement and perhaps by providing for some limited, say, sixty-day, cooling-off period after the signing of the agreement, during which period either party can withdraw from it without legal consequences.

Post-agreement Contingencies

Here we are assuming that background entitlements have been appropriately specified, and that the separation agreement in question was fairly negotiated, so that issues of unfairness, either systemic or transaction-specific, do not arise. Still, a problem that may arise is that some contingency that was not addressed in the separation agreement and perhaps was entirely unforeseen at that time, has now arisen and its occurrence substantially diminishes the welfare of one of the parties. This is the kind of problem that in general contracting contexts has traditionally been addressed under the rubric of the doctrine of frustration. However, it should be noted that in the standard frustration cases arising, for example, in a commercial contract setting, typically the alleged frustrating event occurs during the performance of the contract, rendering further performance of the contract by one or other of the parties either impossible or substantially more costly. In the separation agreement context, . . . often these post-agreement contingencies may arise well after the agreement has been fully performed on both sides and the relationship between the parties long since terminated.

The law relating to the doctrine of frustration generally in contract law is confused and difficult, and it seems likely that similar difficulties of application will arise in the family law context, especially in cases where the contingency has occurred long after marriage breakdown. The law and economics approach to frustration issues would tend to emphasize as a criterion for deciding whether to reopen a contract and reallocate a risk that has now materialized from one party to the other considerations of who is the most efficient risk bearer with respect to that risk, on the assumption that the risk was not foreseen at the time that the contract was entered into and therefore not explicitly allocated by the contract. Adopting this approach, it is perhaps useful to proceed by way of example. First, suppose that a separation agreement provides for support payments to the children of the marriage. Suppose that during the course of the agreement one of the children is severely injured in an automobile accident and requires extremely expensive specialized medical and therapeutic care. A court would no doubt under current law vary the terms of the agreement, and for good reason. Had the parties to the agreement addressed their minds to this contingency when the agreement was entered into, given ongoing mutually accepted obligations of support to the children, it seems obvious that both would have agreed that some sharing of the additional costs would be appropriate.

Second, suppose that two lawyers after a short marriage involving no children separate, terminate the marriage, and enter into a separation agreement pertaining to division of family assets. Ten years later one of the former spouses, who has successfully pursued her legal career in the meantime, is unfortunately struck by a bus and rendered a paraplegic. Assuming no tort claims against the transit operator for negligence and no substantial disability insurance coverage, the spouse now finds herself in impoverished circumstances. To what extent should the separation agreement be subject to reopening so as to allocate to the former husband some of the burdens of this risk that has now materialized to his former spouse? This seems to us to be an easy case where reopening is not justified. Had the parties addressed themselves to this contingency at the time of entering into the agreement, the two of them would surely have agreed that each should bear separate responsibility for insuring adequately against future contingencies of this kind. Neither would wish to be an insurer of the other for life with respect to such a contingency given that one of the virtues of a binding separation agreement is to provide some finality and certainty as to the termination of the prior relationship so that each former spouse can go on to build new relationships and take on new commitments, without these new relationships and commitments being placed at the risk of one spouse suddenly being required to assume responsibility for adverse contingencies that the other former spouse might suffer. Here the law and economics perspective would take the view that either the former spouse suffering the contingency, as between the two former spouses, is the superior riskbearer (in the sense of being able to buy whatever insurance she considers appropriate to her circumstances), or alternatively the state, through medical and social welfare programmes, is the superior riskbearer.

A third, and more problematic, example may involve a case where the conduct of one of the spouses towards the other during the marriage was responsible for causing an incipient form of physical or emotional vulnerability in the other spouse that manifests itself sometime after the marriage breakdown. . . . Here, this seems to us much less appropriately the domain of frustration-type doctrines than the domain of tort law. For example, if one former spouse alleges that the other former spouse was guilty of cruelty or intentionally inflicted mental suffering towards the other spouse during marriage, causing a nervous breakdown sometime after the termination of the marriage, thus preventing the latter spouse from engaging in remunerative employment, one would have thought that various tort doctrines could be invoked to secure compensation for this form of injury. However, we acknowledge that this runs the risk of resurrecting the concept of the matrimonial offence, and the attendant concerns that this raises of encouraging all sorts of allegations and counter-allegations requiring courts to perform distant post-mortems on long-since failed relationships, although one would expect that the available tort doctrines would normally require a reasonably demanding burden of proof and therefore are unlikely to be satisfied beyond a narrow range of cases. . . .

NOTES AND QUESTIONS

1. There is by now a substantial body of literature on the economics of the family, much of which has relevance to family law. Surveys of the economic literature can be found in Theodore W. Schultz, ed., *Economics of the Family: Marriage, Children, and Human Capital* (Chicago: University of Chicago Press, 1974); Gary Becker, *A Treatise on the Family* (Cambridge, Mass.: Harvard University Press, 1981); and Robert A. Pollak, A Transaction Cost Approach to Families and Households, 23 *Journal of Economic Literature* 581 (1981). For an early and influential application of this literature to the legal setting, see Robert Mnookin and Lewis Kornhauser, Bargaining in the Shadow of the Law: The Case of Divorce, 88 *Yale Law Journal* 950 (1979).

2. The readings in this chapter offer substantially different views of the efficiency of government regulation of the family. Both Becker and Murphy and Trebilcock and Keshvani present family interactions as subject to market failure, and state regulation as a possible Pareto improvement, whereas Landes and Posner see private individuals as generally acting in the social interest and state regulators as monopolists sub-

ject to the influence of special interests. Which of these views do you find more persuasive? Which of the various types of market failure we have discussed—externalities, public goods, imperfect information, strategic behavior—are most important in the family setting? Can the presence of love and altruism among family members help to mitigate market failure?

3. Are Becker and Murphy correct in their suggestion that parents—even altruistic ones—lack sufficient incentives to invest in their children's human capital? Are there any countervailing factors that might encourage excessive investment in children? Compare Lynn Stout, Some Thoughts on Poverty and Failure in the Market for Children's Human Capital, 81 *Georgetown Law Journal* 1945 (1993) [arguing that laissez-faire approach to child education and training leads to inadequate investment] with Robert H. Frank, *Choosing the Right Pond: Human Behavior and the Quest for Status* (New York: Oxford University Press, 1985), and Robert H. Frank and Philip J. Cook, *The Winner-Take-All Society* (New York: Free Press, 1995) [both arguing that private rewards flowing from relative status and achievement encourages excess expenditure on so-called "positional" goods such as education]; A. Michael Spence, Job Market Signaling, 87 *Quarterly Journal of Economics* 355 (1973) [suggesting that it may be profitable to invest in otherwise useless educational credentials in order to signal one's preexisting skills or motivation]. Would such effects be mitigated if children were made legally responsible for their parents' debts, as the authors hint? What if parents were made responsible for the debts of their adult children?

4. Evaluate Becker and Murphy's claim that social security and other public expenditures on the elderly can be understood as part of an implicit intergenerational compact designed to encourage efficient investments in children. How can such a compact be maintained, given that no individual parent is obligated by virtue of accepting social security payments to invest in or leave bequests to his or her children? What prevents the younger generation from abrogating the compact when it is their turn to perform?

5. What restrictions on divorce would be consistent with an optimal contract among husbands, wives, parents, and children? Are the incentives provided by modern no-fault divorce laws better or worse in this regard than the fault-based regimes they replaced? If no-fault divorce makes it easier to get divorced, is this efficient or inefficient? Compare H. Elizabeth Peters, Marriage and Divorce: Informational Constraints and Private Contracting, 76 *American Economic Review* 437 (1986) [suggesting that costless bargaining between spouses could reduce incidence of divorce without reliance on fault]; Martin Zelder, Inefficient Dissolutions as a Consequence of Public Goods: The Case of No-Fault Divorce, 22 *Journal of Legal Studies* 503 (1993) [arguing that the destruction of public goods in divorce interferes with efficient Coasian bargaining between spouses]; and Elizabeth S. Scott, Rational Decisionmaking about Marriage and Divorce, 76 *Virginia Law Review* 9 (1990) [applying models from economics and cognitive psychology to question of whether social norms supporting lasting marriage should be legally enforced].

6. What rules for allocating marital assets upon divorce or separation give the best incentives for investing in such assets ex ante? Does Harold Demsetz's theory of property rights, outlined in chapter 3, or Oliver Hart's theory of business organization, set out in chapter 4, suggest an answer? Does the issue pose a tradeoff between

efficiency and equity, or is a Pareto superior outcome possible? See Severin Borenstein and Paul Courant, How to Carve a Medical Degree: Human Capital Assets in Divorce Settlements, 79 *American Economic Review* 992 (1989).

7. In a portion of their article not reproduced here, Trebilcock and Keshvani discuss whether the justifications for state regulation of marital separation agreements are stronger or weaker when such agreements are negotiated prior to entering the marriage. Are there special reasons, economic or otherwise, to be wary of prenuptial agreements? Conversely, given the substantial increase in the frequency of marital dissolutions in recent years and the private and social costs of divorce, should prospective spouses be encouraged to consider and provide for such contingencies in advance? See, for example, Jeffrey Stake, Mandatory Planning for Divorce, 45 *Vanderbilt Law Review* 397 (1992) [proposing, inter alia, that marital partners be required to choose at the time they enter into marriage whether divorce will be available to them on a fault or no-fault basis].

8. The economic approach to family law has drawn substantial criticism from liberal, radical, and communitarian quarters. A common theme in the critical literature has been that economic analysis tends to commodify the family; that is, by recharacterizing family relations as exchange, economic rhetoric works to undercut the love and altruism that are essential to the healthy upbringing of children and to the maintenance of cooperation and good citizenship in society at large. This criticism has been especially well articulated by feminist critics; see, for example, Frances Olson, The Family and the Market: A Study of Ideology and Legal Reform, 96 *Harvard Law Review* 1497 (1983); and Margaret Jane Radin, Market Inalienability, 100 *Harvard Law Review* 1849 (1987). Is the criticism fair? Is economic analysis inconsistent with altruism, or can cooperative as well as competitive goals be incorporated into economic models? For an account that tries to incorporate feminist concerns into the economic analysis of family law, see Margaret Brinig, Comment on Jana Singer's "Alimony and Efficiency," 82 *Georgetown Law Journal* 2461 (1994).

9. Landes and Posner's discussion of adoption as a market has proved a particular flashpoint for critics of the economic approach. See, for example, Daniel Farber's parody, An Economic Analysis of Abortion, 3 *Constitutional Commentary* 1 (1986) [satirizing economic approach as advocating freedom of contract between fetuses and their potential parents, with the former represented by guardians for purposes of negotiation, as solution to abortion controversy]. An excellent survey of objections to a free market in adoption can be found in J. Robert Prichard, A Market for Babies, 34 *University of Toronto Law Journal* 341 (1984). Prichard divides the objections into categories based on market failure, distribution, commodification, oppression of children and mothers, and what he calls, drawing on Calabresi and Bobbitt's work on "tragic choices," the "cost of costing." Posner's response to the critics can be found in his article The Regulation of the Market for Adoptions, 67 *Boston University Law Review* 59 (1987), and his book *Sex and Reason* (Cambridge, Mass.: Harvard University Press, 1992), chap. 15 ("Separating Reproduction from Sex"). Which of these various objections, if any, do you find most persuasive? If a free market in adoption is ruled out on one or more of these grounds, does Landes and Posner's supply-and-demand model suggest alternative acceptable reforms? Conversely, do the ob-

jections to buying and selling parental or custodial rights also apply to market provision of child care?

10. The objections to open adoption also raise the more general issue of inalienability, which Calabresi and Melamed discuss in chapter 3 and Steven Kelman alludes to in chapter 8. As we have seen, mainstream economics has a difficult time justifying the existence of inalienable rights, since any voluntary trade made in a free market is presumptively Pareto superior. While some limited support for inalienability can be found in the concept of market failure (specifically, market failures such as externality, imperfect information, and monopoly may make particular trades inefficient), economists' usual policy preference in such situations is to address the specific market failure directly, not to ban trading. In modern Western political culture, however, the concept of inalienable rights has longstanding appeal. One possible explanation, in addition to those already identified, may lie in what James Tobin has called *specific egalitarianism.* In this view, despite (or perhaps because of) the fact that our society does not guarantee equal income, wealth, or opportunity to its members, we are committed to everyone having an equal share of certain particular goods. Making entitlements to these specific goods inalienable, even when this has the effect of decreasing the welfare of the poor by preventing them from engaging in mutually beneficial exchange, is in this account a way to preserve their equal shares in such goods. See James Tobin, On Limiting the Domain of Inequality, 13 *Journal of Law and Economics* 263 (1970); for additional discussions of the problems of inalienability, see generally Susan Rose-Ackerman, Inalienability and the Theory of Property Rights, 85 *Columbia Law Review* 931 (1985), and Michael Trebilcock, *The Limits of Freedom of Contract* (Cambridge, Mass.: Harvard University Press, 1993). Does Tobin's explanation help account for resistance to markets in adoption and child custody? Does it explain resistance to using markets in the environmental area?

11. Which of the goods of society are fit and which are unfit for market allocation? Consider the examples of human blood [see Richard Titmuss, *The Gift Relationship: From Human Blood to Social Policy* (New York: Pantheon, 1981)], human organs [see Henry Hansmann, The Economics and Ethics of Markets for Human Organs, 14 *Journal of Health Politics, Policy, and the Law* 57 (1989)], obligations of military service [see *The Military Draft: Selected Readings on Conscription,* ed. Martin Anderson (Stanford, Calif.: Hoover Institution Press, 1982)], voting [see Tobin, supra, note 10], and the right to sue in tort for physical injury [see Mark Shukaitis, A Market in Personal Injury Tort Claims, 16 *Journal of Legal Studies* 329 (1987)].